Springer Series in Materials Science

Volume 297

The Springer Series in Materials Science covers the complete spectrum of materials research and technology, including fundamental principles, physical properties, materials theory and design. Recognizing the increasing importance of materials science in future device technologies, the book titles in this series reflect the state-of-the-art in understanding and controlling the structure and properties of all important classes of materials.

More information about this series at http://www.springer.com/series/856

Ziqi Sun · Ting Liao
Editors

Responsive Nanomaterials for Sustainable Applications

Springer

Editors
Ziqi Sun
School of Chemistry and Physics
Queensland University of Technology
Brisbane, QLD, Australia

Ting Liao
School of Mechanical, Medical
and Process Engineering
Queensland University of Technology
Brisbane, QLD, Australia

ISSN 0933-033X ISSN 2196-2812 (electronic)
Springer Series in Materials Science
ISBN 978-3-030-39993-1 ISBN 978-3-030-39994-8 (eBook)
https://doi.org/10.1007/978-3-030-39994-8

This Springer imprint is published by the registered company Springer Nature Switzerland AG
The registered company address is: Gewerbestrasse 11, 6330 Cham, Switzerland

Preface

The rapid progress of science and technology achieved over the past decades has pushed the sophisticated artificial intelligent or smart machines into our routine life, such as intelligent household, self-piloted cars, and self-learning robots. These intelligent systems share one common characteristic that the non-biological systems would achieve mankind-mimicking functions through their information collecting system, information processing system, and movement executive system. In the information collection and movement executive systems, responsive functional materials, or smart materials, play the key roles in constructing the most important components of sensors, transducers, actuators, and so on. The responsive functional materials are a class of materials which have one or more properties that can be responsible for an external stimulus, such as stress, strain, temperature, light, and electric or magnetic fields. The responsibility endorses the materials have the capability to sense disturbances, evoke reactions, and execute movements. In contrast to the widely studied polymer-based responsive materials, inorganic responsive materials have been rarely summarized and reviewed, even though they have been innovated for emerging sustainable technologies.

It is thus of great interest that, beyond the application of responsive functional materials in smart systems, smart materials have also been used in some sustainable energy and environmental technologies, such as next generation rechargeable batteries, nanoelectronics, smart buildings, sustainable energy generation and conversion, and so on. This monograph entitled *Responsive Nanomaterials for Sustainable Applications* thus intends to give an overview on the application of responsive functional materials, particularly inorganic responsive nanomaterials, in sustainable technologies. Nine chapters are included in this monograph, which cover different materials with targeting functions for the defined applications.

Chapter 1 introduces "green" and earth abundant photonic responsive semiconductor materials which can capture photons to generate electrons and holes for the application in third-generation solar cells, such as dye-sensitized solar cells (DSSCs), quantum dot sensitized solar cells (QDSSCs), and perovskite solar cells (PSCs), together with the discussion of the key features of the third-generation solar cells and the overview of the physical, chemical, and optoelectrical properties of the photonic responsive semiconductor nanomaterials.

Chapter 2 focuses on the latest progress in microwave responsive nanomaterials, which integrate the catalyst function and the microwave absorption property into one single material, for the application of heterogeneous catalysis. It first introduces the principles of microwave response and the mechanism of microwave in catalytic reactions, and then the principles of designing microwave responsive catalysts are discussed. Followed with these principles, the state of the art of the microwave responsive catalysts in liquid-phase, gas-phase, and solid-solid reactions was presented to address the validity of this novel responsive materials and the associated challenges and possible solutions.

Chapter 3 gives us an idea of designing and constructing different types of flexible/stretchable supercapacitors based on responsive nanomaterials including flexible/stretchable substrates, electrode materials, and electrolyte. Meanwhile, the application of these flexible/stretchable devices in the emerging field of integrated systems is also simply illustrated, such as the integration of micro-planar supercapacitors with pressure sensors, linear supercapacitors with photoelectric detectors and supercapacitors with gas sensors, and the corresponding in situ signal acquisition and display system. Furthermore, the challenges and perspective of this class of self-responding devices have also discussed and outlined.

Chapter 4 introduces magnetic responsive nanomaterials by taking MnO_2 nanomaterials as a typical example. In this chapter, the origins and the structural/phase effect on the magnetism of MnO_2 nanomaterials with different phases have been discussed in-depth. It is also reported that the electronic structure, morphologies, and the introduction of hetero-ions have great impact on the magnetic properties of the MnO_2. The applications of the magnetic responsive MnO_2 nanomaterials in magnetic resonance imaging have also demonstrated in this chapter.

Chapter 5 describes stimuli-responsive hydrogels for colorimetric chemical sensors. Combining the stimuli-responsive hydrogels, a class of three-dimensional hydrophilic polymeric networks with a fascinating property that can undergo obvious and reversible color or shape variations in response to external environmental stimuli, with photonic crystals, the responses to external stimuli can be converted into color changes, thus creating a kind of colorimetric sensors. This chapter gives detailed introduction on the basic concepts, synthesis methods, and sensing mechanisms and some typical applications, including detecting pH value, ionic species, solvents, humidity, and biomolecules. This is a very attractive sensing technology due to its simple operation and visualized readout.

Chapter 6 deals with functional interfaces which are considered as "materials within materials" and active at nanoscale and responsive to applied external stimuli. In this chapter, nano-electro-mechanical properties of the morphotropic phase boundaries in bismuth ferrite thin films are reviewed, and particularly, the factors governing nano-mechanical and electro-mechanical properties are identified, together with the potential applications of the responsive functional interfaces in oxide nanoelectronics.

Chapter 7 introduces chromogenic responsive nanomaterials for the application of smart windows. Smart windows are one of the key components of smart home system for saving energy by dynamically managing the light transmittance and solar heat exchange through chromogenic materials employed in the window glass to respond for temperature, light, or electricity. In this chapter, facile and low-cost preparation methods have been introduced to promote the widespread adoption of smart windows. Moreover, nanostructure engineering on improving the thermochromic and electrochromic properties of the chromogenic responsive nanomaterials has been outlined. Finally, the authors have also summarized the current challenges in this field and proposed potential solutions to address those challenges to achieve greater market penetration and commercial success.

Chapter 8 brings our attention on proton conductive nanomaterials for the application of solid oxide fuel cells (SOFCs), which convert hydrogen directly into electricity by catalyzing H_2 into protons and then transporting them to the other side to combine with oxygen to form H_2O. Compared with the SOFCs with oxygen-ion conductors, the SOFCs based on proton responsive nanomaterials can work at a much lower temperature and thus have extended lifetime as well as quick start-up. In this chapter, the authors gave detailed introduction on the proton responsive electrolytes, anode materials, and cathode materials for high-performance low-temperature protonic SOFCs, together with the strategies on how to improve the overall performance through rational materials design and nanostructure engineering.

Chapter 9 performs the concept of construction of thermoelectric generators based on thermo-responsive nanomaterials. The authors first introduced the recent advances in thermoelectric materials and the facing challenges, such as low the dimensionless of merit (zT), high cost, and environmentally unfriendly; then they proposed some general methods to enhance the performance of thermoelectric materials through proper electronic structure design and optimization; and finally, they gave examples of prototype modules to validate these strategies. Furthermore, the potential applications of thermoelectric generator in powering the implantable and wearable devices have also been presented.

This book is organized to deliver the recent progress in responsive functional materials for targeted applications to the readers in wide fields of materials science, chemistry, nanotechnology, engineering, and sustainable technologies. We hope this book will be a good reference for not only undergraduate students and research students but also researchers and engineers by providing readers with valuable insights into responsive functional materials or smart materials from materials synthesis and design, structure characterizations, property evaluations, to engineering applications.

Brisbane, Australia

Ziqi Sun
Ziqi.sun@qut.edu.au
Ting Liao

Contents

Contributors

Lei Bi Institute of Materials for Energy and Environment, College of Materials Science and Engineering, Qingdao University, Qingdao, China

Zhi-Gang Chen Centre for Future Materials, University of Southern Queensland, Springfield Central, QLD, Australia

Tuo Ji Department of Chemical and Biomolecular Engineering, The University of Akron, Akron, OH, USA

Wenxian Li Institute of Materials & Institute for Sustainable Energy, Shanghai University, Shanghai, China

Yue Li Key Laboratory of Materials Physics and Anhui Key Laboratory of Nanomaterials and Nanotechnology, Institute of Solid State Physics, Hefei Institutes of Physical Science, Chinese Academy of Sciences (CAS), Hefei, Anhui, People's Republic of China

Ji Liang Institute for Superconducting and Electronic Materials, University of Wollongong, Wollongong, NSW, Australia;
Key Laboratory for Advanced Ceramics and Machining Technology of Ministry of Education, School of Materials Science and Engineering, Tianjin University, Tianjin, China

Daolan Liu Institute for Superconducting and Electronic Materials, University of Wollongong, Wollongong, NSW, Australia

Wei-Di Liu Materials Engineering, The University of Queensland, Brisbane, QLD, Australia

Dandan Men Jiangxi Key Laboratory of Surface Engineering, Jiangxi Science and Technology Normal University, Nanchang, People's Republic of China

Jiadong Qin Centre for Clean Environment and Energy, School of Environment and Science, Griffith University, Gold Coast, QLD, Australia

Jan Seidel School of Materials Science and Engineering, The University of New South Wales, Sydney, NSW, Australia

Pankaj Sharma School of Materials Science and Engineering, The University of New South Wales, Sydney, NSW, Australia;
ARC Centre of Excellence in Future Low-Energy Electronics Technologies, The University of New South Wales, Sydney, NSW, Australia

Vincent Tiing Tiong School of Chemistry, Physics and Mechanical Engineering, Queensland University of Technology, Brisbane, QLD, Australia

Yueyu Tong Institute for Superconducting and Electronic Materials, University of Wollongong, Wollongong, NSW, Australia

Hongxia Wang School of Chemistry, Physics and Mechanical Engineering, Queensland University of Technology, Brisbane, QLD, Australia

Lei Wen Shenyang National Laboratory for Materials Science, Institute of Metal Research, Chinese Academy of Sciences, Shenyang, Liaoning, China

Xi Xu Institute of Materials for Energy and Environment, College of Materials Science and Engineering, Qingdao University, Qingdao, China

Honghua Zhang Jiangxi Key Laboratory of Surface Engineering, Jiangxi Science and Technology Normal University, Nanchang, People's Republic of China

Yu Lin Zhong Centre for Clean Environment and Energy, School of Environment and Science, Griffith University, Gold Coast, QLD, Australia

Jiahua Zhu Department of Chemical and Biomolecular Engineering, The University of Akron, Akron, OH, USA

Jin Zou Materials Engineering, The University of Queensland, Brisbane, QLD, Australia;
Centre for Microscopy and Microanalysis, The University of Queensland, Brisbane, QLD, Australia

Chapter 1
Photon-Responsive Nanomaterials for Solar Cells

Vincent Tiing Tiong and Hongxia Wang

Abstract The global issue of the utmost exhaustion of fossil fuels on earth has driven research towards the development of alternative energy resources to meet the increasing demand for energy required in modern society. Among the different types of renewable sources, solar energy is the largest energy source which is unlimited and clean. Currently solar cells or photovoltaic (PV) technologies that generate electricity by harnessing sunlight is one of the fastest growing power generation sources in the energy sector. In this chapter we review the application of nanomaterials in some types of solar cells including dye-sensitized solar cells, quantum dots solar cells and perovskite solar cells. Semiconductor materials such as TiO2, ZnOx, SnOx, NiOx etc have been widely used as electron or hole transport materials in these type of solar cells. The morphology, shape, size, crystal structure of particles of these materials can significantly influence the device performance. The outlook of the future research direction is provided at the end of the review.

1.1 Introduction

The global issue of the utmost exhaustion of fossil fuels on earth has driven research towards the development of alternative energy resources to meet the increasing demand for energy required in modern society. Well, the solution is undoubtedly clean renewable energy sources. Among the different types of renewable sources, solar energy is the largest energy source which is unlimited and clean. Based on this consideration, it is not surprising that solar cells or photovoltaic (PV) technologies that generate electricity by harnessing sunlight is one of the fastest growing power generation sources in the energy sector. In fact, in 2017 alone saw more new capacity of solar energy deployed than all fossil fuels and nuclear combined together, which is nearly twice as much capacity as its renewable peer, wind power [1].

V. T. Tiong · H. Wang (✉)
School of Chemistry, Physics and Mechanical Engineering, Queensland University of Technology, Gardens Point, Brisbane, QLD 4000, Australia
e-mail: Hx.wang@qut.edu.au

Z. Sun and T. Liao (eds.), *Responsive Nanomaterials for Sustainable Applications*, Springer Series in Materials Science 297,
https://doi.org/10.1007/978-3-030-39994-8_1

A solar cell may take different structures. Nevertheless, photo-responsive light absorbing material is always the heart of the device. According to the nature of the light absorbing material, the PV technology can be classified into three categories or generation. The first-generation solar cells are based on crystalline silicon which was invented in 1954 by Bell Laboratory. Over the six decades of development, monocrystalline silicon solar cells have achieved a certified power conversion efficiency (PCE) of more than 25% and have dominated the commercial PV market with more than 90% market share [2]. Nevertheless, the technology of crystalline silicon solar cells still has drawbacks. For example, the tedious process and cost associated with the production of silicon cells have motivated research in search of alternative cheaper and more efficient solar cells to make solar electricity economically competitive to fuels. This effort of development of cheaper solar electricity has led to the invention of subsequent second-generation based on thin-film and third-generation solar cells that involves nanomaterials such as extremely small inorganic semiconductor nanoparticles (also known as quantum dots (QDs)). In addition, semiconductor materials with electron (n-type) and hole (p-type) transport characteristics have also been extensively investigated as charge extractive materials in third-generation solar cells. The nanomaterials exhibit unique, fascinating properties compared with the corresponding bulk materials, thus allowing higher theoretical energy conversion efficiency [3].

In the past decades, the majority of research in the field of third-generation PV technology has been focused to realise high-efficient, low-cost PV cells through the development of new nanomaterials and photon management to enhance device performance. The chemical, physical, and optoelectronics properties of the nanomaterials are critical to the device performance. Through delicate experimental design and material synthesis, the energetic properties (band gap, band energy position), morphology, particle size, and chemical composition of the semiconductor material can be tailored to achieve efficient light absorption, charge separation, and collection. These eye-catching properties of nanostructured semiconductor materials have created revolutionarily new research field that has the potential to deliver cost-effective solar electricity by using low-cost chemical solution-based production process.

This chapter will introduce the recent advances in third-generation solar cells, in particular dye-sensitised solar cells (DSCs), quantum dot solar cells (QDSCs), and perovskite solar cells (PSCs) where nanotechnology plays important roles through influencing the photonic-responsive light absorbing materials, the electron transporting materials (ETMs), and hole transport materials. A general overview of the physical, chemical, and optoelectrical properties of nanomaterials will be given. The key features of the semiconductor nanomaterials used in the solar cells will be discussed. Research direction of the development of third-generation solar cells in the future will be shown at the end of the chapter.

1.2 Dye-Sensitised Solar Cells

The studies of dye-sensitised solar cells (DSSCs) can be traced back to 1968 where zinc oxide was used as the photoanode and perylene as the light sensitisers [4]. Nevertheless, the development of DSSCs was at slow pace due to the insufficient surface area available to adsorb large amount of light sensitisers (also called dye in the literature). A breakthrough was made in 1991 when O'Regan and Grätzel reported a DSSC [5] which demonstrated power conversion efficiency of 7.1–7.9% by using large surface area, nanostructured mesoporous titanium dioxide (TiO_2) film coated with a monolayer of ruthenium dye. The mesoporous TiO_2 (m-TiO_2) architecture was proven to be critical to the device performance which possessed a surface area that is more than a thousand time higher than a flat film, thus offering efficient dye loading for efficient light absorption and energy conversion efficiency of DSSC. Since then, DSSC has been regarded as a promising PV technology delivering cost-effective solar electricity in the future. In the past 30 years, extensive research has been carried out to improve the cell's efficiency and stability. Currently, a power conversion efficiency (PCE) over 14% has been achieved for liquid electrolyte-based DSSC with TiO_2 nanostructure [6].

1.2.1 DSSC Device Structure and Working Principle

A typical DSSC consists of five key material components (Fig. 1.1): (1) a transparent conductive substrate, generally made up of fluorine-doped tin oxide (FTO) or indium tin oxide (ITO)-coated glass; (2) a semiconductor film, normally based on metal-oxide nanomaterials, such as TiO_2 or ZnO; (3) light sensitisers or dyes which are adsorbed onto the surface of the nanostructured metal-oxide film; (4) an electrolyte (liquid or solid state); and (5) a counter-electrode, such as platinum and carbon [5]. The working principle of DSSCs is closely related to natural photosynthesis which involves four basic steps: light absorption, electron injection, charges carrier transportation, and current collection. The constituent components and fundamental processes of DSSCs are schematically illustrated in Fig. 1.1.

The key fundamental processes occurring in a DSSC can be described in the following [8–10]:

(1) photoexcitation of sensitisers to produce excited dye;
(2) injection of electron into the conduction band of metal oxide, leading to oxidised dye;
(3) transportation of electron to the anode and reach the counter-electrode through external circuit;
(4) regeneration of the oxidised dye by electron donation from the reduced redox couple;

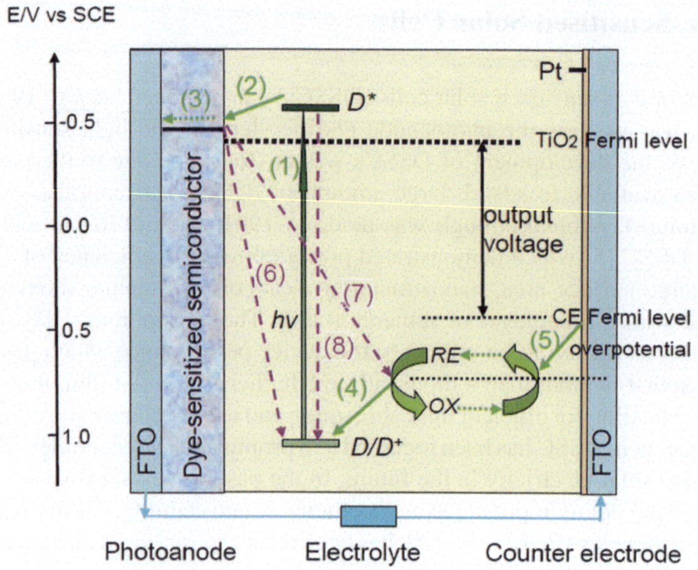

Fig. 1.1 Fundamental processes and constituent components of a dye-sensitised solar cells (DSSC). Reprinted with permission from [7]

(5) regeneration of oxidised redox couple at cathode by accepting electrons from anode;
(6) recombination of electron by donating electrons to oxidised dye;
(7) recombination of electron by donating electrons to the oxidised redox couple;
(8) relaxation of the excited dye to its ground state by a non-radiative decay process.

Processes (1)–(5) contribute to the conversion of light to electricity in the DSSC. The parasitic processes (6) and (7) lead to charge carrier recombination either with oxidised dye or with the redox couple. In addition, process (8) is the direct recombination of the excited photoelectron, reflected by the excited state lifetime. The aforementioned recombination process is detrimental to the efficiency of electron transfer and thus the device performance.

In the quest to improve the photovoltaic performance of DSSCs, most researches have been emphasised on tailoring the physicochemical properties of the four main material components of the cell, namely the metal-oxide semiconductor, the dye, the redox couple in the electrolyte, and the counter-electrode, in particular the metal-oxide semiconductor which determines the dye loading, thus light-harvesting efficiency and the open-circuit photovoltage of the DSSC devices. Theoretically, the metal oxides can be n-type (electron transporting material) or p-type (hole transport) as long as they have energetic properties that match the adjacent light absorbing material. Due to the challenge of finding p-type semiconductor material that have energy level matches the state-of-the-art ruthenium-based dye, most of DSSCs use n-type large band gap semiconductor such as TiO_2 and ZnO in the photoanode. Rich research on tuning the properties of these metal oxides through material engineering has resulted in many interesting discoveries.

1.2.2 Electron Transporting Materials in DSSCs

Photoanode also known as working electrode in DSSC acts as a scaffold for dye adsorption while functions as charge transport medium to extract electrons from dye molecules followed by transporting them to the current collected based on transparent conductive oxide (TCO) substrate. In order to achieve optimum dye adsorption and smooth transportation of electrons without undergoing recombination, the photoanode should possess the following properties: (1) a suitable conduction band gap as it will determine the effectiveness of electron extraction process and the V_{oc} of the cell device; (2) large surface area to enable sufficient dye loading for efficient light harvesting; (3) high electron mobility to allow smooth and fast extraction of electron from the absorbers and collection by the conductive substrate; (4) chemical stability, low toxicity, low cost; (5) high transparency to reduce the loss of incident photon; (6) contains hydroxyl group or defects for the attachment of dye molecules onto its surface [10–12]. Furthermore, the photoanode architecture should be carefully engineered to minimise the loss of incident light when passing through the photoanode before being absorbed by the dye. Metal oxide with large band gap such as TiO_2, ZnO, SnO_2, Nb_2O_5, ZrO_2, $SrTiO_3$, and Zn_2SnO_4 has been used as electron transporting materials (ETMs) in DSSCs due to their good chemical stability, nontoxicity, and simple yet mature preparation. Among them, TiO_2 and ZnO are the most popular choice due to their large band gap and high electron mobility with some of their representative DSSC parameters and performances which were shown in Table 1.1.

1.2.3 TiO_2-Based Photoanodes

TiO_2-based mesoporous films (m-TiO_2) built from TiO_2 nanoparticles (TiO_2NP) are the most popular ETMs in DSSCs due to their superior characteristics such as chemical stability, low toxicity, and low cost [26]. Additionally, attractive features including the large surface area, suitable conduction band (CB) edge, and electron affinity of TiO_2 make it the most studied ETMs in DSSCs compared to other metal-oxide materials.

As depicted in Fig. 1.2, there are three main polymorphs of crystalline TiO_2, namely anatase (tetragonal), rutile (tetragonal), and brookite (orthorhombic) [27–29]. Anatase is an indirect band gap semiconductor while rutile and brookite are direct band gap semiconductors [27]. Owing to the indirect band gap nature of anatase, the transition of photoexcited electrons from conduction band (CB) to valance band (VB) is slower, leading to a longer lifetime of the photoexcited electrons compared to rutile and brookite. Moreover, the average effective mass of photoexcited electrons

Table 1.1 Representative photovoltaic parameters and materials used in DSSCs based on different ETMs

ETM	QDs	Electrolyte	CE	V_{oc} (V)	J_{sc} (mA/cm^2)	FF	PCE (%)	References
TiO_2NP	ADEKA-1 + LEG4	Polyiodide	Au/Graphene	1.013	18.36	0.77	14.3	[6]
TiO_2NT	N719	Polyiodide	Pt	0.770	18.5	0.64	9.1	[13]
TiO_2NR	N719	Polyiodide	Pt	0.71	20.49	0.5454	7.91	[14]
TiO_2NW	N719	Polyiodide	Pt	0.787	18.25	0.65	9.4	[15]
TiO_2NS	N719	Polyiodide	Pt	0.840	18.53	0.679	10.57	[16]
TiO_2B	C106	Polyiodide	Pt	0.7459	18.38	0.769	10.7	[17]
TiO_2HS	N719	Polyiodide	Pt	0.77	16.52	0.73	9.4	[18]
ZnONP	N719	Polyiodide	Pt	0.621	18.11	0.585	6.58	[19]
ZnOHA	N719	Polyiodide	Pt	0.64	19.8	0.59	7.5	[20]
ZnONT	N719	Polyiodide	Pt	0.65	5.5	0.61	2.2	[21]
ZnONR	N719	Polyiodide	Pt	0.71	10.7	0.62	4.7	[22]
ZnONW	N719	Polyiodide	Pt	–	–	–	7.0	[23]
ZnOHS	N719	Polyiodide	Pt	0.63	14.2	0.63	5.6	[24]
ZnONS	D149	Polyiodide	Pt	0.53	18.01	0.634	6.06	[25]

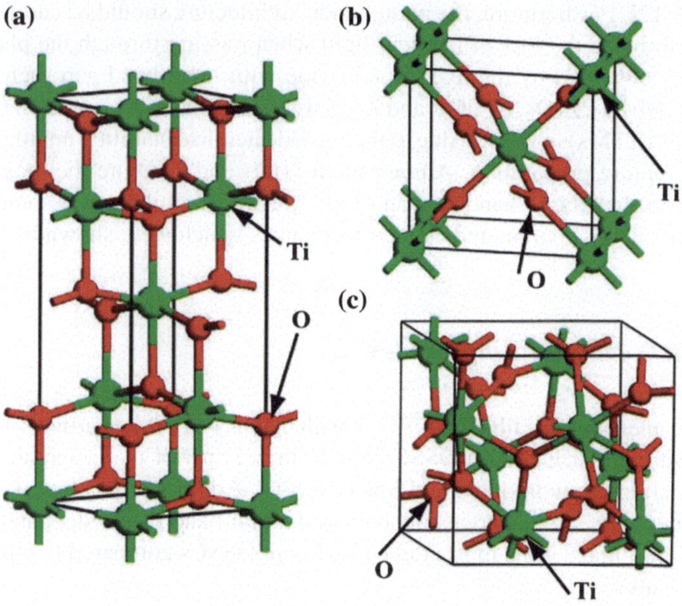

Fig. 1.2 Schematic conventional cells for **a** anatase, **b** rutile, and **c** brookite TiO_2. The green spheres represent Ti atoms, and the red spheres represent O atoms [27]

of anatase is also the lightest among the three polymorphs, facilitating faster transportation of photoexcited electrons and thus lower recombination rate in contrast with rutile and brookite. As a result, anatase TiO_2 is the most popular choice in the application of DSSC even though rutile is the most thermodynamically stable form with higher refractive index among them.

1.2.4 *Effect of Morphology of TiO_2 on DSSC Performance*

TiO_2 nanoparticles (TiO_2NPs) are the most widely used nano-architectures of TiO_2 in DSSC due to the large surface area for high dye loading, which is favourable for achieving high performance. Different particle sizes of TiO_2NPs have been reported to have significant influences on photovoltaic performance of DSSCs. Park et al. revealed that larger TiO_2NPs show low electron transport resistance and high recombination resistance, which benefit the charge collection efficiency [30]. Similar finding was revealed by Tahay et al. where charge injection was found to be accelerated for larger TiO_2NPs, thus enhancing the overall performance of the DSSCs [31]. Nevertheless, there is an optimum TiO_2NPs size whereby further increment of TiO_2NPs size over the limit is detrimental to the overall cell performance, owing to the reduction of available surface area for dye loading, and thus the light-harvesting efficiency. Therefore, it is very crucial to determine a suitable TiO_2NPs size so that the charge collection and light harvesting could be balanced to acquire superior cell performance.

In addition, recent studies have found that the anatase (001) facets have higher surface energy and stronger ability to absorb the COOH group of the dye molecules over the dominating (101) facets, leading to high dye adsorption and thus improved DSSC performance [32]. Moreover, the highly exposed (001) facets were also reported to effectively retard the charge recombination process in DSSCs besides improving the amount of dye loading [33]. Nevertheless, the numerous grain boundaries, defects, and traps induced in nanoparticle-based films have the downside of inefficient charge transportation, which promotes electron recombination at the grain boundaries, resulting in lower DSSC efficiency. Owing to these disadvantages, one-dimensional (1D) structures, such as nanotubes, nanowires, nanorods, and nanofibers, which usually possess higher diffusion coefficients than disordered nanoparticles have attracted significant attention in the research community of DSSCs.

The 1D array structures can act as electron transport highway, allowing fast transport of electron to the conductive substrate, reducing charge recombination and hence improve the overall performance of DSSCs. Researchers have proven this concept through the production of DSSC with PCE over 9% using TiO_2 nanotubes (TiO_2NTs) grown Ti foil as photoanode (Fig. 1.3a, b) [13], which is higher than those reported TiO_2NPs-based DSSC performance under the same fabrication conditions, confirming the effect of prolonging electron lifetime and minimising interfacial resistance by nanotube structure [35, 36]. Furthermore, through controlling the TiO_2NTs length (from 15 to 57 μm), the efficiency of DSSCs increased from

Fig. 1.3 SEM images of **a** top view and **b** cross section of TiO_2NT arrays after immersion in 33 wt% H_2O_2 for 20 s [13]; **c** TiO_2 porous beads [17]; **d** cross section of pine tree-like TiO_2NT arrays [34]; **e** cross section and **f** top view of etched porous TiO_2NRs [14]; **g** TEM and **h** high-resolution TEM images of multi-shell TiO_2HSs (inset: d-spacing of the outer shell of TiO_2HSs) [18]

4.5 to 6.1% due to the improved surface area and better light-scattering effect of the extended TiO_2NTs [35]. The performance of the TiO_2NTs-based DSSCs was further improved to 9.86% through filling the TiO_2NT arrays with TiO_2NPs, owning to the increment of surface area, thus leading to the enhancement of dye loading and cell's performance [37]. In contrast to nanotubes, TiO_2 nanowires (TiO_2NW) and nanorods (TiO_2NR) only offer their outer walls for dye adsorption, which result in lower surface area for dye loading, thus limiting their application in DSSC [38]. For example, Chen et al. confirmed that TiO_2NR film exhibits longer electron life-time, better light scattering, and fast electron diffusion efficiency as compared to the TiO_2NPs [39]. However, the efficiency of TiO_2NR-based DSSCs (6.07%) was lower than TiO_2NP-based DSSCs (7.79%) due to poor dye loading. In order to improve the dye loading, researchers adopted similar strategy as TiO_2NT-based DSSCs by increasing the length of TiO_2NRs and TiO_2NWs. Lv et al. developed DSSCs using ~30 µm thick TiO_2NR arrays with large internal surface area and produced a cell efficiency of 7.91% (Fig. 1.3e, f) [14]. Similarly, Wu et al. investigated the effect of various length of TiO_2NWs (15–55 µm) on the performance of DSSCs and revealed that DSSC fabricated from long TiO_2NWs (~50 µm) achieved high PCE of 9.4% which is not only superior to that of short TiO_2NWs-based DSSCs but also higher than TiO_2NPs-based counterparts (7.23%) [15]. The outstanding performance of longer TiO_2NWs-based DSSCs is due to the higher dye loading, prominent light-scattering capacity, and suppressed charge recombination. On the other hand, high porosity hierarchical TiO_2 spheres/beads (TiO_2Ss/TiO_2Bs) were synthesised and used as the ETMs in DSSCs. The high surface area TiO_2NSs were believed to be responsible for the extended electron lifetime, along with the increment of electron diffusion lengths, thus delivering DSSC PCE over 10% (Fig. 1.3c) [16, 17]. Another interesting hierarchical TiO_2 hollow spheres (TiO_2HSs) with single or multiple shell structures have also been employed as alternatives to nanoparticles as photoanodes in DSSCs. These TiO_2HSs were discovered to exhibit extremely high surface area which facilitate electron transfer while inducing light-scattering effect owning to the light reflective effect introduced by the shells (Fig. 1.3g, h) [18]. Huang et al. revealed that PCE as high as 9.4% can be achieved from multi-shell TiO_2HS-based DSSC due to enhanced multi-reflectance effect. Other interesting structures such as pine tree-like TiO_2NTs were reported to improve the surface area, electron transport of the ETM while at the same time, reduce the electrode/electrolyte interfacial resistance to yield a solid-state DSSC with PCE of 8% (Fig. 1.3d) [34].

1.2.5 TiO_2-Based Light-Scattering Materials

Employing large particles is a popular approach to enhance the light absorption of the photoanode and thus the performance of DSSCs through light scattering. The large particles are capable of promoting light reflection and/or scattering when incident light passes through the photoanode film, thus prolonging light path and allowing enhanced light harvesting of the dye. Generally, two structures, namely multilayer structure and mixture structure composed of same or different materials with different sizes and/or morphology, have been applied to introduce light-scattering effect in photoanodes as demonstrated in Fig. 1.4.

In the bilayer or multilayered structure, usually extra film(s) composed of larger particles is placed on top of the dye adsorbing layer of the DSSCs. For example, Park et al. fabricated a bilayered photoanode-based DSSCs with a light-scattering layer consist of three different sizes of TiO_2NPs (587, 757, and 1554 nm in diameter) deposited on a transparent nanocrystalline TiO_2 film [41]. The authors discovered that the light-scattering layer not only promotes the light absorption, but also enhances the dye loading, resulting in an extraordinary improvement of cells' PCE of 9.37% versus 6.80% for the reference cell without the light-scattering layer. Mathew et al. further

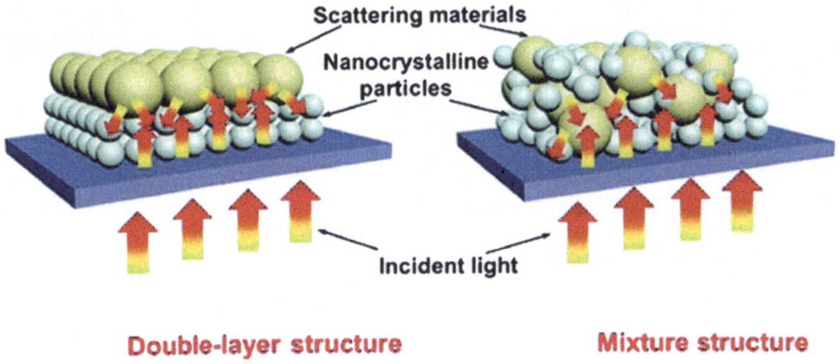

Fig. 1.4 Typical structure of photoanodes in DSSCs containing light-scattering materials [40]

optimised the bilayer structural photoanode-based DSSC to an impressive PCE of 13% using a 3.5 μm scattering layer composed of 400-nm-sized TiO_2NPs on top of a 3.5 μm transparent layer with 32-nm-sized TiO_2NPs [42]. Apart from different-sized nanomaterials, Yu et al. showed that the light-scattering effect produced by TiO_2HSs is larger compared to TiO_2NPs, thus enhancing the light-harvesting efficiency of DSSCs when used as an overlayer on the TiO_2NP film (Fig. 1.5g, h) [43]. Similarly, light-scattering material based on 1D materials such as TiO_2NTs [44], TiO_2NRs (Fig. 1.5e, f) [45], TiO_2NFs [46], and TiO_2NWs [47] were also applied in DSSCs and displayed similar phenomenon in improving cells' performance. Meanwhile, an interesting three-dimensional-structured architecture, consisting of urchin-like, hyper-branched TiO_2HSs (top layer), rambutan-like TiO_2HSs (intermediate layer) and tree-like, hyper-branched TiO_2NWs (underlayer), was fabricated as a photoanode for DSSCs (Fig. 1.5a–d) [48]. A high PCE of 11.01% was attained with the DSSCs, which was attributed to the favourable charge collection, superior light-harvesting efficiency, and extended electron lifetime. Besides different morphologies, different materials such as SnO_2 and CeO_2 were also utilised as a scattering layer in the photoanodes of DSSCs [49, 50]. Song et al. demonstrated that a PCE of 9.86% can be achieved through a bilayer structured photoanode consisting of an overlayer of porous flowerlike CeO_2 microspheres on TiO_2NP film, owning to enhanced light scattering as well as the reduced electron recombination [49].

In the mixture structure, large particles are embedded into the electron transporting layer (ETL) instead of depositing them as an overlayer to improve light absorption. The embedded particles act as light scatterers to generate random degrees of light scattering, aiming to prolong the light travelling distance within the photoanode film, thus increasing the probability of photons to be absorbed by the dye [40]. Similar to multilayer structure, nanomaterials with different sizes and morphology were assorted to optimise the light scattering while maintaining high dye adsorption

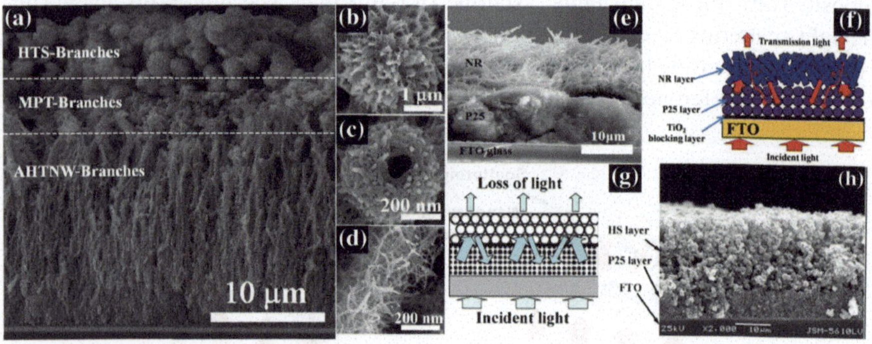

Fig. 1.5 SEM images of **a** cross section of three-layer stacking of TiO_2 photoanodes consist of **b** hierarchical TiO_2 microspheres, **c** TiO_2HS, and **d** hierarchical TiO_2NWs [48]; **e** cross section and **f** schematic illustration of light-scattering of bilayered composite (TiO_2NRs top and TiO_2NPs underlayer) [45]; **g** schematic illustration of light-scattering and **h** cross-sectional of double-layered photoanode (TiO_2HSs top and TiO_2NPs underlayer) [43]

onto photoanode. Miranda-Muñoz et al. mixed large-sized light scatterers, TiO_2 spheres with m-TiO_2 to form photoanode for DSSCs, attaining 25% and 35% PCE enhancement under front and rear illumination [51]. On the other hand, Tan et al. fabricated DSSC with the composite of TiO_2NPs and TiO_2NWs as photoanode and discovered that the TiO_2NWs offer light-scattering effect and highways for swift electron transport, thus delivering an improved PCE of 8.6% from 6.7% (without TiO_2NW) [52]. Similar finding was also reported by Wang et al. when DSSC based on TiO_2NP/TiO_2NS gradient film photoanode was prepared [53].

Although both multilayer and mixture structures can improve the DSSC performance through similar light-scattering effect, both structures have yet to yield high-efficient DSSC compared to the record efficiency DSSC without light scatterers. This is due to the increment of internal resistance which hindering charge transport within solar cells introduced by the large-sized scatterers in multilayer structure [54]. Meanwhile, the large-sized scatterers in mixed structure can also affect the internal surface area of the photoanode, limiting the dye loading.

1.2.6 ZnO-Based Photoanodes

Zinc oxide (ZnO) is another metal-oxide material that has been broadly investigated and applied as an ETM in DSSCs. ZnO is a wide band gap semiconductor which has similar energy band structure and physical properties as TiO_2 [55]. However, ZnO has very high electron mobility, which is more than an order of magnitude higher than TiO_2. This means that a faster electron transport in the DSSCs is expected. These extraordinary properties make ZnO an ideal candidate to replace TiO_2. Moreover, ZnO is relatively easy to crystallise due to its anisotropic growth, allowing easy tuning of ZnO morphologies through different experimental conditions. Nevertheless, the application of ZnO in DSSCs is limit because of its lower chemical stability than TiO_2. The amphoteric nature of ZnO means that it can dissolve in acidic environment which is often the case in DSSCs [56]. This chemical instability is unfavourable for long-term lifetime of DSSC devices.

1.2.7 Effect of ZnO Morphology on DSSC Performance

Even though ZnO is chemically unstable in acidic electrolyte, several ZnO photoanode-based DSSCs with promising performance have been reported. Adopting the concept from TiO_2-based DSSC, ZnONP is considered to have great surface area, allowing extensive dye loading and capable of harnessing the visible light to the maximum extent. Similar to TiO_2, different particle sizes of ZnO were found to have significant influence on the performance of DSSC. Sakai et al. showed that DSSCs with 40 nm ZnO particle yielded best PCE among the particle size range from 20 nm to 2 μm [57]. Nevertheless, Saito et al. demonstrated a high DSSC PCE of 6.58%

could be attained through using 20–30 nm ZnONP [19]. The PCE of the DSSC was further boosted to an unprecedented value of 7.5% through using a photoanode consisting of polydispersed ZnO hierarchical aggregates (ZnOHA) of nanocrystallites over a compact ZnO buffer layer [20].

Atomic surface termination of semiconductors is highly relevant to their chemical reactivity and electronic behaviour. It is found that certain facet with high surface energy has strong affinity to some dye molecules and, therefore, helps increase light-harvesting efficiency. Chang et al. investigated the influence of different facets of ZnO through synthesising ZnO nanocones (Fig. 1.6b) and nanorods (ZnONRs) (Fig. 1.6a) with [1011] and [1010] facets, respectively. They discovered that ZnO with [1011] facets has stronger binding with dye molecules compared to ZnO with predominated [1010] facets, due to the highly reactive exposed O-terminated [1011] ZnO facets [58]. As a result, PCE of 4.36% was produced with the DSSC based on the ZnO with [1011] facet which is three times more efficient than the ZnONR counterpart.

1D nanostructures demonstrated promising properties which favour charge transport and electron collection while reducing recombination, due to the exceptional boundary-free morphology. Nevertheless, 1D nanostructured ZnO, such as ZnO nanotubes (ZnONTs) (Fig. 1.6c) [21, 61, 62], nanorods (ZnONRs) [22, 58, 63], and nanowires (ZnONWs) [64–66]-based DSSCs, normally shows lower performance due to the inadequate available surface area for high dye loading, leading to insufficient light harvesting. To solve this problem, Xu et al. fabricated multilayer ZnONW arrays (with thickness up to 40 μm) which possess high internal surface area, allowing higher dye loading. The PCE of the corresponding DSSC was increased to 7 from 2.1% produced by single-layered ZnONW-based DSSCs (Fig. 1.6g) [23]. Furthermore, two-dimensional (2D) and three-dimensional (3D) nanostructured ZnOs were synthesised and applied as photoanode in DSSCs, aiming to improve the dye loading while maintaining fast electron transport and collection. Among the 2D nanostructured ZnO including ZnO nanosheet (ZnONSs) (Fig. 1.6e) [25, 59, 67] and

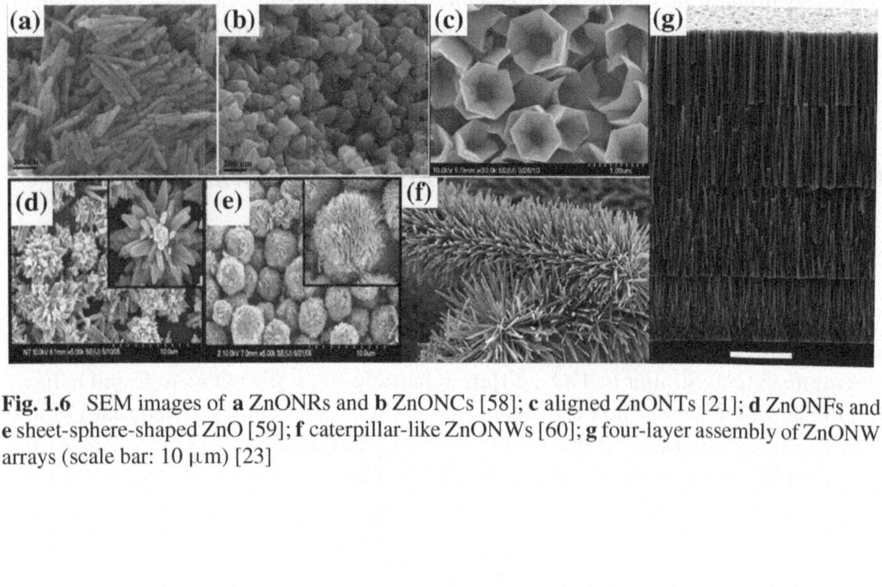

Fig. 1.6 SEM images of **a** ZnONRs and **b** ZnONCs [58]; **c** aligned ZnONTs [21]; **d** ZnONFs and **e** sheet-sphere-shaped ZnO [59]; **f** caterpillar-like ZnONWs [60]; **g** four-layer assembly of ZnONW arrays (scale bar: 10 μm) [23]

nanoplates [59], the ZnONS-based DSSCs displayed promising result, yielding PCE up to 6.06%, which is more than double the PCE of ZnONP-based DSSCs (2.92%) [25]. The superior PCE produced by ZnONS-based DSSCs was attributed to the higher effective electron diffusion coefficient and dye loading of ZnONSs compared to its counterpart, ZnONPs. Meanwhile, the 3D nanostructured ZnO, for instance nanoflowers (ZnONFs) (Fig. 1.6d) [59, 68], ZnO nanoburgers [69], caterpillar-like nanowires (Fig. 1.6f) [60], hierarchical panel [70], nanourchin [71], and 3D sponge [72], were created to take the advantages of both 1D and 2D nanostructure, which resulted in great improvement in cell's performance. In particular, the 3D sponge-like ZnO and hierarchical ZnO nanourchins (ZnONUs) presented relatively high specific surface area value, providing a high number of adsorption sites for dye molecules, thus making them ideal candidates as photoanodes for DSSCs. PCE as high as 6.40 and 6.67% was reported with the ZnONUs and 3D sponge-like ZnO-based DSSC, respectively, owing to their superior light-harvesting efficiency as well as higher electron lifetime and longer charge diffusion length [71, 72].

1.2.8 Effect of Light Scattering of ZnO

Addition of light-scattering large particles is one of the commonly used strategies to improve the light-harvesting efficiency by redirecting and prolonging the incident light when it passes through the photoanode. Similar to TiO_2-based DSSCs, bilayer and mixture structure were also applied in ZnO-based DSSCs. Application of hierarchical ZnO nanocrystallites aggregates is one of the most widely used methods to provide the photoanode film with both large surface area and the capability of generating light scattering. As a matter of fact, DSSCs fabricated based on ZnOHA photoanode generate the most efficient DSSC with PCE of 7.5%, due to its high light scattering of the ZnO film, extending the light path inside the sensitised film and improving light absorption in addition to other merits such as inhibition of back electron transfer between the conducting layer at the anode and the electrolyte [20]. A separate research was conducted by Zhang et al. to evaluate the performance of ZnOHA with various sizes and size distributions in DSSC (Fig. 1.7b, c) [73]. The authors discovered that the aggregates with sizes closer to the wavelengths of incident light could result in more intensive light scattering and hence produce higher cell's PCE. The photoanode containing well-packed different-sized aggregates was also recommended by the authors to further boost the cell performance. Zheng et al. reported a unique ZnONUs nanoarchitecture which has the combined advantages of high surface area, direct electron transport networks as well as strong light scattering, leading to an outstanding PCE of 6.4%, which is 31.1% improvement in comparison with ZnOHA-based DSSC [71]. Similar observation was reported by Wu et al. when fabricating DSSC based on ZnOHA assembled by orderly aligned nanorods [74].

ZnO photoanode with bilayer structure which normally involves deposition of a layer of larger particles as scattering layer on top of the dye adsorbing layer is another popular method to enhance light absorption in DSSCs. For example, bilayer ZnO

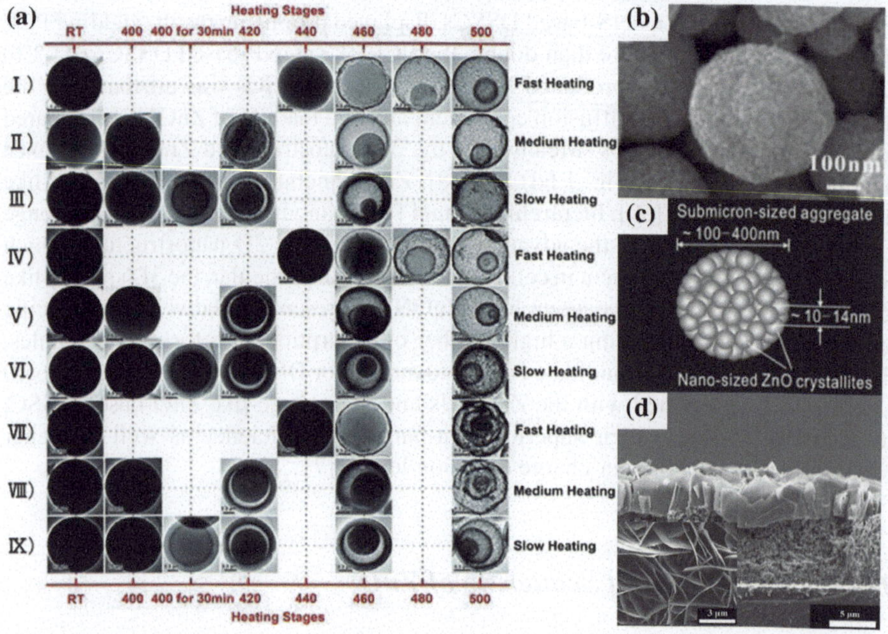

Fig. 1.7 **a** Evolution process of the family of multi-shell ZnO hollow microspheres [24]; **b** SEM image; and **c** schematic illustration of ZnO aggregates structure formed by closely packed nanocrystallites [73]; **d** cross-sectional SEM image of ZnONSs on top of ZnONPs (inset: top-view SEM images) [75]

photoanode consists of ZnO micro-flowers as scattering layer and ZnONP as underlayer exhibited excellent light-scattering effect, leading to better light absorption, and thus improved PCE of the bilayered ZnO-based DSSC [76]. Similar outcome was reported by Kung et al. when ZnONSs were applied as scatterers, deposited on top ZnOPs, delivering a PCE of 4.65% (Fig. 1.7d) [75]. In the meantime, hollow core/shell structure is considered as a special type of bilayer structure which allows internal nanostructured scattering effects while preserving large specific surface. However, the balance between dye adsorption and light penetration is critical to achieve high device performance as the excess adsorbed dye will hinder light scattering. Dong et al. synthesised a series of multishelled ZnO hollow spheres (ZnOHS) with controlled shell number and inter-shell spacing, which enable sufficient dye loading and multi-light reflection for enhancing light harvesting and thus realising high PCE (Fig. 1.7a) [24]. As a result, quadruple-shelled ZnOHS-based DSSC yielded PCE of 5.6%.

1.2.9 Other Metal Oxides Used as Photoanode in DSSCs

Besides the most popular TiO_2 and ZnO, other metal oxides such as SnO_2 [77, 78], Nb_2O_5 [79, 80], Fe_2O_3 [81], $SrTiO_3$ [82], and Zn_2SnO_4 [83, 84] have also been utilised as a photoanode in DSSCs. However, DSSCs fabricated from these ETMs normally produce low cell performance (below 4%) due to the inefficient electron mobility and/or unfavourable band edge position of the ETMs. Thus far, only DSSCs based on SnO_2 and Zn_2SnO_4 photoanode have successfully produced cell PCE over 6% through morphology engineering [77, 84].

1.3 Semiconductor-Sensitised Solar Cells

The concept and structure of semiconductor quantum dot-sensitised solar cells (QDSSCs) are derived from dye-sensitised solar cells (DSCs) by substitution of dye molecules with semiconductor nanocrystallite or quantum dots (QDs). The attractive optoelectronic features of nanostructured semiconductor sensitisers include excellent stability in the presence of light, thermal and moisture, tunable light absorption range, high absorption coefficient, multiple exciton generation, and solution processability as well as facile and low-cost fabrication [85]. In the past decade, we have witnessed a dramatic boost in the PCE of QDSCs from 3% to nearly 17%, which is comparable to other kinds of emerging solar cells [86]. The exploration of new QD light-harvesting materials has contributed to this tremendously fast improvement.

Typically, QD-based solar cells can be classified into four categories: Schottky junction solar cells, p-n junction solar cells (both homojunction and heterojunction), hybrid QD-polymer solar cells, and quantum dot-sensitised solar cells (QDSCs) [87]. Among all, QDSCs possess the advantages of low-cost device fabrication process due to the simple device structure derived from their analogous DSSC counterparts.

1.4 Photo-Response Nanomaterials Used Quantum Dot-Sensitised Solar Cells (QDSCs)

1.4.1 Electron Transporting Materials

Metal-oxide-based electron transporting materials (ETMs) play a crucial role in QDSCs. ETMs are normally deposited on a TCO substrate, which act as the matrix for anchoring the QD sensitisers in the photoanode. In addition, ETM has the role of extracting photogenerated electrons from the QDs followed by transporting them to the current collector, such as FTO or ITO [88]. Therefore, ETM should possess the following properties: (1) a suitable energy level that matches the conduction band (CB) and valence band (VB) of the QD light absorber; (2) large surface area to facilitate sufficient QD loading for efficient light harvesting; (3) high electron mobility to allow fast electron extraction collection; (4) good chemical stability, low toxicity, low cost. With these properties in mind, our passionate researchers have been putting unremitting efforts in selecting best ETMs for QDSCs and made promising progress as displayed in Table 1.2.

1.4.2 TiO_2-Based Photoanodes for QDSCs

1.4.2.1 The Effect of Morphology

Metal-oxide-based nanomaterials are the most widely studied and employed ETM in QDSCs due to their good chemical stability, nontoxicity, and simple preparation [100]. In general, mesoporous TiO_2 film (m-TiO_2) fabricated from TiO_2 nanoparticle (TiO_2NP) is the most commonly used material in QDSCs due to their large surface area, which allow high QD loading and thus high light-harvesting capability [101]. In fact, QDSCs fabricated from TiO_2 nanoparticle mesoporous films yield some of the highest performance QDSCs with power conversion efficiency reaching more than 12% [89, 102–104]. Although crystalline TiO_2 exists in three polymorphs, namely anatase (tetragonal), rutile (tetragonal), and brookite (orthorhombic), anatase TiO_2 is the most popular crystal phase used in QDSCs due to its higher electron mobility and less electron–hole recombination [28, 29]. Recent reports have also revealed that different crystal planes of anatase TiO_2 would also affect the overall performance of QDSCs [105]. QDSCs fabricated with (001) faceted anatase TiO_2 produced a power conversion efficiency of 4.7% due to the higher conduction band and surface energy over conventional anatase TiO_2 with (101) crystal plane.

Compared to anatase TiO_2, rutile TiO_2 possesses the advantages of thermodynamically stable with higher refractive index and superior light-scattering feature which is capable of enhancing the light harvesting [28]. Although anatase TiO_2 normally produced better performance in QDSCs, a recent publication reveals that rutile TiO_2 used in CdS/CdSe-based QDSC produced higher short-circuit photocurrent density and open-circuit photovoltage, leading to a 30% higher power conversion efficiency than the QDSC made from anatase TiO_2 [106]. The studies discovered that rutile is more favourable for the higher loading of CdS/CdSe QDs, thus leading to better light harvesting. In addition, electron injection from QDs to TiO_2 was found to have improved using the rutile phase with reduced charge recombination at the interface. This is attributed to its lower surface/interface defects in the photoanode film and larger conduction band offset between the QD and rutile TiO_2.

Even though mesoporous nanoparticle photoanode achieved higher reported power conversion efficiency in QDSCs, it has a major downside of inefficient electron transport due to defects such as grain boundaries and surface traps, thus increasing the charge recombination probability [92]. Therefore, one-dimensional (1D)-structured TiO_2 materials, such as nanotubes (TiO_2NT) [108–114], nanorods (TiO_2NR) [115–120], and nanowires (TiO_2NW) [92, 121–123], have drawn a lot of attention in

Table 1.2 Representative QDSCs based on different ETMs and their photovoltaic parameters

ETM	QDs	Electrolyte	CE	V_{oc} (V)	J_{sc} (mA/cm^2)	FF	PCE (%)	References
TiO_2NP	ZCISe-CdSe	Polysulfide	Carbon	0.752	27.39	0.619	12.75	[89]
TiO_2NT	CdS/CdSe	Polysulfide	Cu_2S	0.689	10.81	0.62	4.61	[90]
TiO_2NR	CdSe/ZnS	Polysulfide	$Cu_2ZnSnS(Se)_4$	0.563	17.5	0.605	5.96	[91]
TiO_2NW	CdS/CdSe	Polysulfide	Cu_2S	0.465	17.98	0.502	4.20	[92]
TiO_2NS	CdSe	Polysulfide	Cu_2S	0.591	16.95	0.50	5.01	[93]
ZnONP	CdS/CdSe	Polysulfide	Cu_2S	0.683	10.48	0.623	4.46	[94]
ZnONR	CdS/CdSe	Polysulfide	Cu_2S	0.642	13.28	0.567	4.83	[95]
ZnONW	CdS/CdSe	Polysulfide	Au	0.627	17.3	0.383	4.15	[96]
ZnONC	$Zn_xCd_{1-x}Se$	Polysulfide	Cu_2S	0.65	18.05	0.4	4.74	[97]
ZnOND	CdS/CdSe	Polysulfide	Pt	0.62	16.0	0.49	4.86	[98]
ZnOTP	ZnSe/CdSe/ZnSe	Polysulfide	Cu_2S-GO	0.761	17.3	0.471	6.2	[99]

recent years and have been explored as alternative ETMs in QDSC to realise a faster electron transfer and thus reduce the charge recombination loss (Fig. 1.8) [107]. Baker et al. have shown an improvement of QDSCs performance when TiO_2NT is employed as an ETM over TiO_2NP due to the better charge separation and electron transport capacity [124]. On the other hand, Tao et al. produced a QDSC efficiency of 4.61% with 3 μm long, $TiCl_4$-treated TiO_2NT film by increasing the QDs loading in the TiO_2NT array [111]. Zhang et al. fabricated QDSCs with a PCE of 5.96% using TiO_2NR as building blocks of photoanodes and sensitised it with CdSe QDs [117]. They found that the TiO_2NR possesses the beneficial properties of higher QDs loading and fluent electrolyte diffusion due to wider pore size distribution, higher electron diffusion coefficient, and longer electron lifetime thanks to the 1D network with limited grain boundaries. Rao et al. found that branched hierarchical structured TiO_2NW is capable of enhancing the J_{sc} value due to the efficient light-harvesting and charge collection, delivering a PCE of 4.2% (Fig. 1.9) [92]. By using a photoanode composed of three-dimensional (3D) hierarchically branched TiO_2NW-coated hollow spheres structure, the PCE of QDSCs was further boosted to 6.01%. This is ascribed to the unique properties of the film structure including high specific surface area for dye loading, roomy space, and ample porosity for efficient electrolyte infiltration [122]. Besides 1D-structured TiO_2 materials, two-dimensional (2D)-structured TiO_2 materials such as nanosheets (TiO_2NSs) have also been utilised as ETMs for QDSCs. Both Etgar et al. and Zhou et al. discovered that higher PCE of the QDSC could be obtained through employing anatase TiO_2NSs with (001) facets as the ETM compared to anatase (101) TiO_2NPs, because of the higher conduction band and surface energy which increases QD loading onto TiO_2NSs and thus enhances the short-circuit photocurrent [93, 105]. Moreover, TiO_2NSs were found to have less defects, efficient electron transfer, and prolong electron lifetime due to their good crystallisation, large particle size, and low surface area.

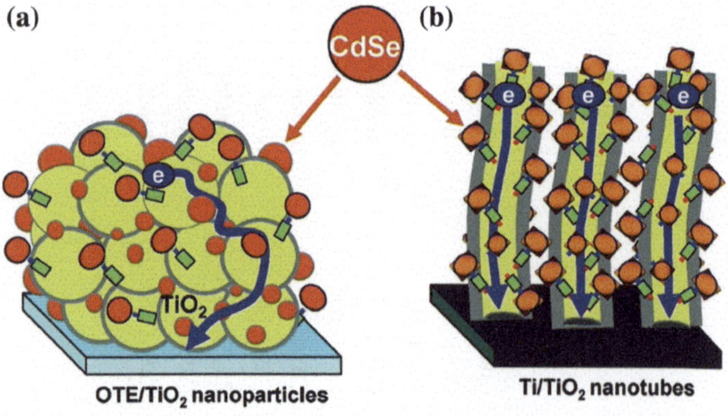

Fig. 1.8 Random versus directed electron transport through support architectures, **a** TiO_2-particle and **b** TiO_2-nanotube films modified with CdSe quantum dots [107]

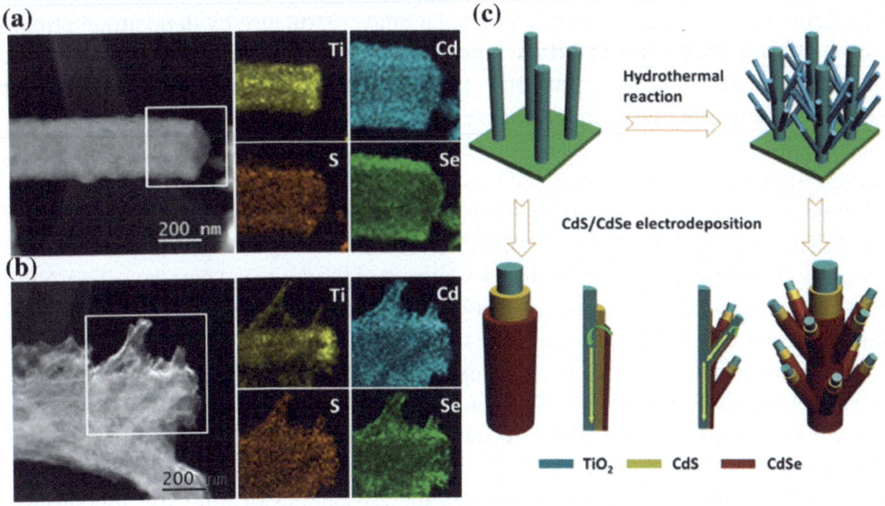

Fig. 1.9 High-angle annular dark-field scanning transmission electron microscopy (HAADF-STEM) and energy-dispersive X-ray spectroscopy (EDX) mappings of a TiO_2/CdS/CdSe photoanode based on **a** smooth and **b** hierarchical nanowires. **c** Schematic diagram showing the preparation process and structure of CdS/CdSe-co-sensitised smooth and hierarchical TiO_2 nanowires as well as the path of electron injection (green arrow) and electron transport (yellow arrow) [92]

1.4.2.2 Effect of Light Scattering

Light scattering is another factor which affects the light-harvesting efficiency of QDSCs. In QDSCs, hierarchical 3D TiO_2 architectures were often used to construct the photoanode by taking advantage of their light-scattering ability, in addition to their large surface area and smooth electron transport features. As an example, hierarchically branched TiO_2NWs when applied as photoanode were found to improve the J_{sc} of the QDSCs profits from superior light-scattering ability, thus obtaining 30% higher PCE than that of smooth NW devices (Fig. 1.9) [92]. Similar phenomenon was discovered by Xu et al. when QDSCs based on multi-dimensional hierarchically branched HS-NW hybrid TiO_2 photoanode were fabricated, showing an improvement of PCE as high as 6.01% [122]. Meanwhile, Feng et al. created a special 3D hyper-branched TiO_2/ZnO heterostructured arrays as photoanode in QDSCs. The hyper-branched array films show great improvement of light-scattering abilities, which is beneficial to trap the incident light, thus enhancing the J_{sc} and overall device performance [125].

Construction of a bilayer structure is another popular way to increase light absorption in the photoanode of QDSCs through depositing a layer of larger particles, also known as scattering layer, on top of QD-sensitised layer. The integrated scatterers are capable of prolonging the incident light travel distance, either through refractions or reflections, thus increasing the light absorption by the QD sensitisers. As a matter of fact, most high-efficient QDSCs adopted a bilayer structure by depositing a light-scattering layer of larger TiO_2NPs with size in the range of hundred nanometres on top of the dye absorbing layer [89, 102–104]. This demonstrates the importance and necessity of scattering layer in producing high-efficient QDSCs. Zhang et al. investigated the different configuration of bilayered structure of TiO_2NP-based photoanode and revealed that the large-sized TiO_2 scatterers are responsible for the increment of light-harvesting efficiency in the long-wavelength region [128]. Nevertheless, the authors stressed the importance of balancing between the light scattering and surface area as it can significantly restrain the dark current of the device. Apart from TiO_2NPs, other nanostructured TiO_2s, such as TiO_2HSs, have also been applied as scatterers and shown some promising results (Fig. 1.10a, b) [126, 129–131]. Zhou et al. studied different configurations of m-TiO_2 beads combined with TiO_2NPs, namely double layer (TiO_2 beads on top of TiO_2NPs) and mixture structure (Fig. 1.10c, d) [127]. As opposed to the finding in TiO_2NPs, the authors discovered that the mixture structure delivered a slight better PCE of 4.65% in comparison to double-layered structure (4.33%), due to the m-TiO_2 beads possess the advantages of high specific surface

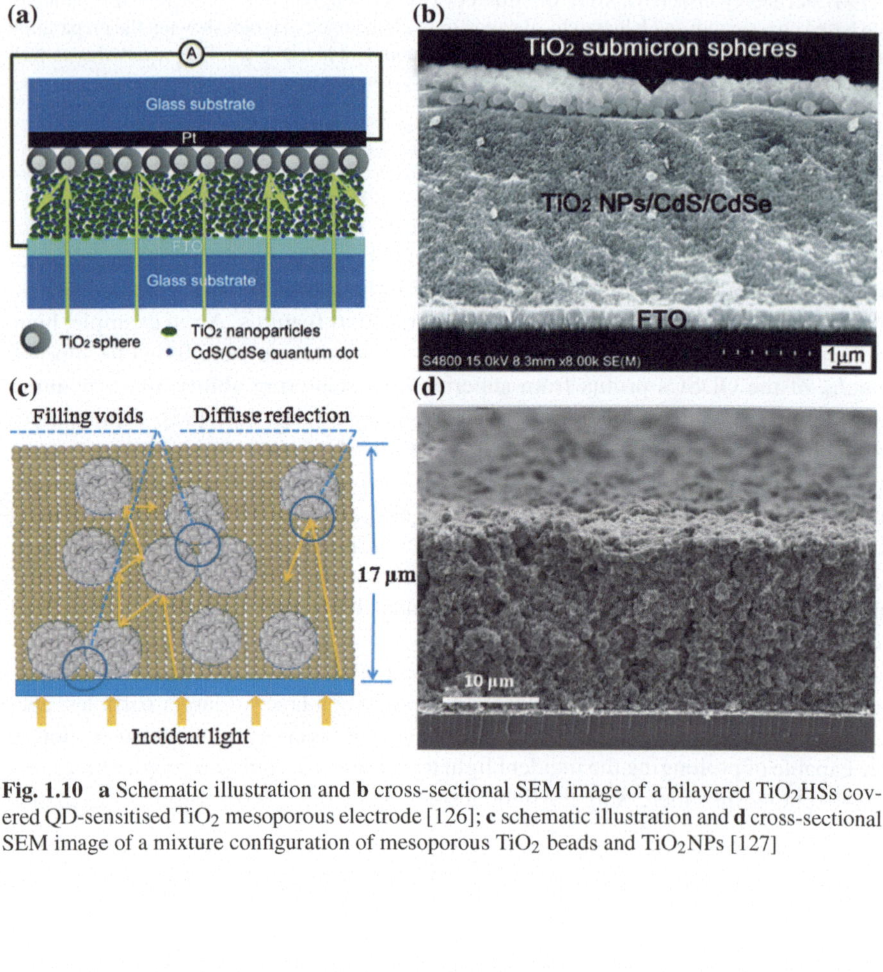

Fig. 1.10 **a** Schematic illustration and **b** cross-sectional SEM image of a bilayered TiO_2HSs covered QD-sensitised TiO_2 mesoporous electrode [126]; **c** schematic illustration and **d** cross-sectional SEM image of a mixture configuration of mesoporous TiO_2 beads and TiO_2NPs [127]

area for high QD loading and effective light scattering as opposed to other large-sized scatterers that have low surface area.

1.4.3 *ZnO-Based Photoanodes in QDSCs*

1.4.3.1 Effect of Morphology

ZnO is another metal-oxide material that has been broadly investigated as an ETM in QDSCs due to its higher electron mobility (130–200 cm^2/VS) compared to TiO_2 [132]. Additionally, ZnO morphologies can be easily modified through different experimental conditions due to the anisotropic growth of ZnO. These exceptional properties make ZnO a more suitable ETM than TiO_2. However, ZnO amphoteric nature which causes it to dissolve in both acidic and basic media has limited its application in QDSCs. This chemical instability can affect the long-term stability of QDSC devices, thus resulting in low report efficiency shown by ZnO photoanode-based QDSCs in comparison to TiO_2-based QDSC counterpart.

Although ZnO is chemically unstable, there are a number of reports on high efficiency ZnO photoanode-based QDSCs. Similar to TiO_2, ZnO nanoparticles (ZnONPs) are popular choice of material morphology in photoanodes of QDSCs [94, 133–135]. Karageorgopoulos et al. prepared a thin transparent ZnONP films by utilising an amino double-edged polypropylene oligomer and zinc precursor [135]. A solid-state QDSCs using the co-sensitised ZnONP films with CdS/CdSe were then constructed, producing a PCE of 4.5%. Similar to TiO_2, other material morphologies of ZnO such as nanorods (ZnONRs) [95, 136], nanowires (ZnONWs) [96, 137–141], nanocables (ZnONCs) [97, 142], nanodiscs (ZnONDs) [98], and nanotetrapods (ZnOTP) [99] have also been employed as ETMs in QDSCs. Wu et al. prepared ZnONRs which consist of oriented aligned elongated nanoparticles along the [0001] direction through conventional solvothermal method [95]. The large pore and rough surface ZnONR films were discovered to enhance the QDs loading, light-scattering effect as well as suppressing recombination losses, leading to a CdS/CdSe-sensitised cell's PCE of 4.83%. Similarly, well-aligned ZnONW array structure was also found to be able to improve the light absorption and charge carrier transfer efficiency while reducing charge recombination, thus generating PCE of 4.15% with CdS/CdSe-sensitised device [96]. A simple ion-exchange approach was employed to fabricate $ZnO/Zn_xCd_{1-x}Se$ core/shell nanocable-structured arrays, where the band gap of the nanocables could be tuned by the composition in the shell layer without changing the nanocable size [97]. A PCE of 4.74% was obtained when the nanocable arrays were applied as the photoanode. Photoanode based on hexagonal shaped mesoporous ZnONDs with exposed ±[0001] polar facets was discovered to be capable of producing PCE of 4.86%, which is around 35% increment in PCE compared to ZnONP-based QDSC (3.14%) [98]. The enhancement of cell's performance is attributed to an effective charge separation and collection, owning to the exposed ±[0001] facets, better light-scattering effects, and high surface area

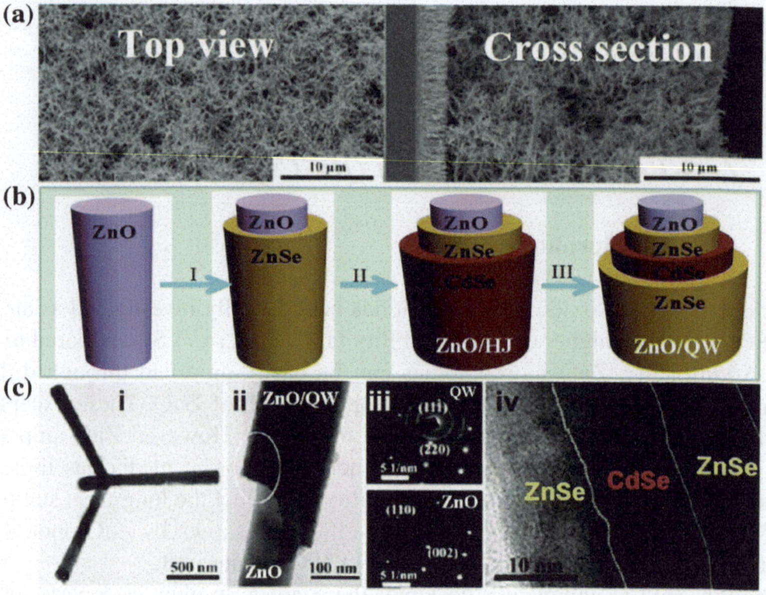

Fig. 1.11 **a** Top-view and cross-sectional SEM images of a ZnO nanotetrapods photoanode for SSSCs, **b** schematic diagram of layer-by-layer formation of a ZnSe/CdSe/ZnSe quantum well (QW) sensitiser, **c** TEM images of a QW-sensitised tetrapod, the arm part from the tetrapod, and the electron diffractions corresponding to different sections with and without the QW shell and structure [99]

for QD loading. Yan et al. created a novel photoanode using a quasi-quantum well (QW)-structured ZnSe/CdSe/ZnSe as the sensitiser and quasi-epitaxially deposited on ZnOTP to produce a QDSC PCE of 6.20% (Fig. 1.11) [99]. The QW structure was revealed to yield longer electron diffusion length due to high electron transport and charge recombination resistance, resulting in higher V_{oc}, J_{sc}, and FF.

1.4.3.2 Effect of Light Scattering in ZnO-Based QDSCs

Similar to TiO_2-based film architecture, ZnO-based photoanode consisting of a double layer, namely a light-scattering layer of large particles on top of an underlayer of small particles, is widely used in order to combine the advantages of the two layers. Bilayer-structured photoanode such as ZnONPs/ZnOMSs, [144], ZnONWs/ZnOHSs [143], and ZnONRs/ZnOTPs [145] has demonstrated their capability in improving the light scattering. For instance, bilayer structure-based photoanode consists of mesoporous ZnOHSs and ZnONW arrays which were discovered to have enhanced the photogenerated current of QDSCs, owing to the high surface area, efficient light-scattering effects, and fast electron transport ability (Fig. 1.12) [143]. Meanwhile, Yu et al. studied the effect of different-sized ZnOHS on the QDSCs performance and

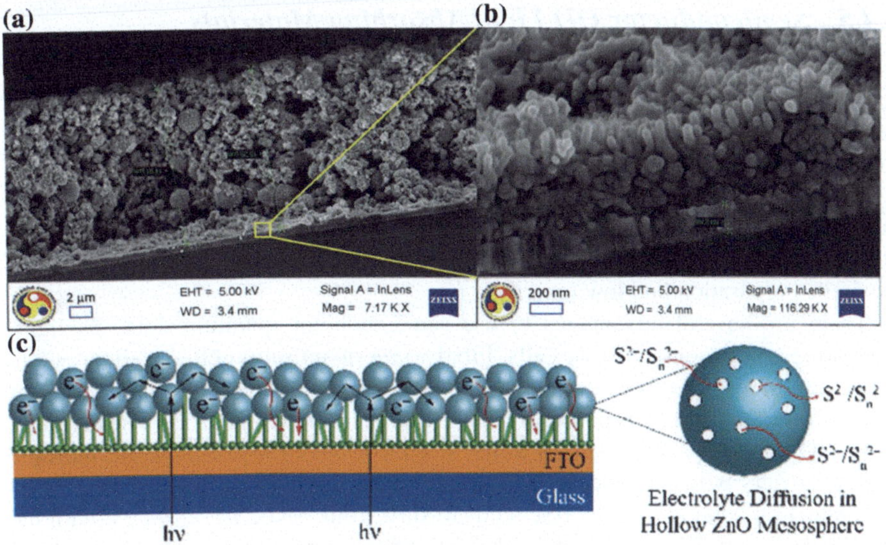

Fig. 1.12 **a** Cross-sectional SEM and magnified images of the ZnOHS layer over ZnONW network **b** and **c** schematic illustration of charge migration from 3D ZnOHSs to 1D ZnONWs injected by CdS QDs, light scattering, and diffusion of electrolyte [143]

revealed that ZnOHS with size around 500 nm displayed best light-scattering effect compared to other sized HSs [146]. Moreover, the PCE of QDSCs can be further improved by using a mixture of hollow spheres with different sizes, ranging from 300 to 800 nm, due to the enhanced light scattering in different light wavelengths induced by the various sized HSs.

1.4.4 Other Types of ETMs Used in QDSCs

Apart from TiO_2 and ZnO, a variety of metal oxides, such as SnO_2 [147], ZrO_2 [148], $BaTiO_3$ [149], and Zn_2SnO_4 [150], have been exploited as photoanodes in QDSCs. Nonetheless, the cell performance of these ETMs is normally poor, which is due to the low electron mobility and/or unfavourable band edge position of the ETMs. Among these, SnO_2-based QDSCs provide the best cell performance with PCE of 4.37% by using highly ordered SnO_2 inverse opal films in the photoanode to reduce charge recombination and increase V_{oc} and FF and thus the overall cell performance [147].

1.4.5 *Semiconductor QD Light Absorbing Materials in QDSCs*

Semiconductor QDs are the core of QDSCs that harvest sunlight and generate electron–hole pairs. Therefore, ideal QD sensitisers should possess the following characteristics: (1) an appropriate band gap to absorb sunlight over a broad range of the solar spectrum and a high absorption coefficient to maximise the light-harvesting efficiency of the incident light; (2) decent stability towards light, heat, and electrolyte; (3) facile preparation and low toxicity.

There are several advantages to use semiconductor quantum dots as light-harvesting components in solar cells. Firstly, size quantisation effects allow one to tune the light response and vary the band offsets to achieve vectorial charge transfer across different-sized QDs particles (Fig. 1.13a, b, d) [107, 151, 153]. Furthermore, these QDs open up new ways to utilise hot electrons [155] or generate multiple charge carriers with a single photon [156]. Tvrdy et al. showed that energy level of conduction band (E_{CB}) of CdSe tends to shift from −3.65 to −3.94 eV with the increasing size of the CdS nanocrystals from 2.8 to 4.2 nm, whereas the energy level of valence band (E_{VB}) merely shifts from −6.16 to −6.13 eV [157]. Since charge transfer between QDs and metal oxides is mainly driven by the difference in E_{CB} of the QDs and metal oxides, the size of QDs is one of the most critical factors to influence the charge transfer rate. Both Kongkanand et al. and Yun et al. studied the size effect of QDs on the performance of QDSCs and revealed that larger CdSe

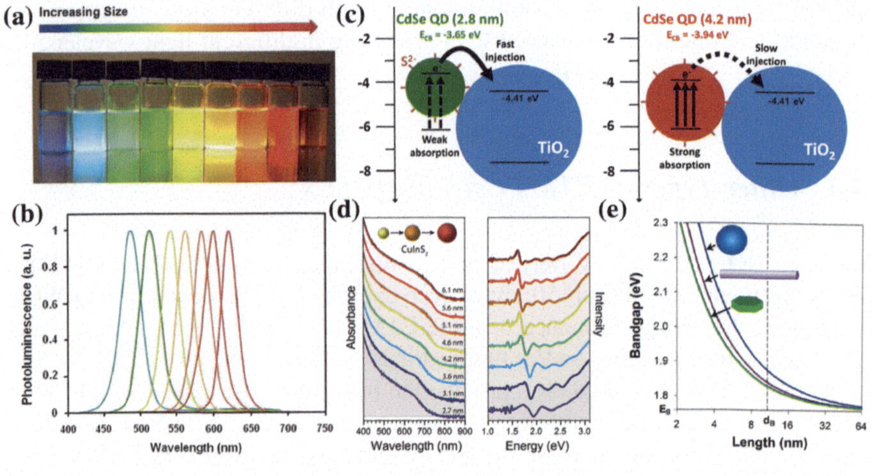

Fig. 1.13 **a** Suspensions of colloidal CdSe/ZnS core/shell heterostructured nanocrystals of different sizes under UV excitation and **b** relative size-dependent emission spectra [151]; **c** schematic illustration of charge transfer between TiO_2 and different-sized CdSe QDs [152]; **d** absorption spectra and corresponding second derivative spectra of wurtzite CIS QDs with different sizes [153]; **e** band gaps of CdSe quantum wells, wires, and dots versus the length of confined dimension with the bulk band gap and exciton diameter noted on the axes [154]

has better absorption under the irradiation of long wavelength; nonetheless, smaller QDs exhibit faster electron injection kinetics from CdSe QDs to TiO_2 owing to the high E_{CB} of CdSe (Fig. 1.13c) [107, 152]. Similar findings were published by Peng et al. and Yang et al. when CIS and CISe were utilised as the QDs in their devices [158, 159]. Nevertheless, Jara et al. discovered that the photocurrent and PCE of CIS devices were mainly affected by the internal and surface defect sites dictating the excited state dynamics of CIS QDs, rather than the participation of quantised conduction band electrons in photocurrent generation of CIS [160]. On the other hand, the band gap may also be tuned to a precise energy level depending on the dimensionality and degree of confinement through confining the exciton of a semiconductor [161, 162]. Figure 1.13e depicts the shift of the band gap of CdSe nanocrystals confined in three dimensions (QDs), two dimensions (quantum wires or rods), or one dimension (quantum wells or discs) [154]. The increase in the number of confined dimensions yields a stronger degree of electronic confinement and thus a wider range of tunability in the band gap. These discoveries suggested that size and shape of QDs have to be carefully tuning in order to optimise the performance of QDSCs.

Throughout decades of studies and investigation, researchers have successfully fabricated QDSCs with numerous different types of absorbers and some of the representative QDSCs are included in Table 1.3.

1.4.6 Binary QD-Based Light Absorber

Binary QDs are most commonly used as sensitisers in the early stage of QDSCs due to their established synthesis recipe and well-known photoelectronic properties. Moreover, binary QDs can be directly grown on a metal-oxide substrate from an ionic precursor solution, such as chemical bath deposition (CBD) or successive ionic layer adsorption and reaction (SILAR) at low temperature. Alternatively, ex situ approach including direct adsorption, electrophoretic deposition, or linker-assisted deposition methods can also be used to deposit colloidal QDs that are pre-synthesised onto the metal-oxide substrate. These features make them feasible and attractive as a light absorbers in QDSCs. Various binary QDs, mostly metal chalcogenide-based QDs, such as Ag_2S [163, 190], CdS [124, 164, 191], CdSe [165, 173, 192, 193], CdTe [166, 194], In_2S_3 [167, 195], PbS [169, 175, 196–198], PbSe [169, 199, 200], Sb_2S_3 [170, 201, 202], and Sb_2Se_3 [203, 204], have been applied as sensitisers in QDSCs. Among these, CdS and CdSe QDs are the most widely investigated and employed due to their simple preparation methods and high stability in QDSCs. Jiang et al. demonstrated PCE as high as 7.54% using pure CdSe QD in their devices through modifying the addition of poly(vinyl pyrrolidone) (PVP) in the polysulfide electrolyte, which reduces the charge recombination losses at TiO_2/QDs/electrolyte interfaces [165]. Although CdS and CdSe are suitable for charge separation due to their higher conduction band edge, their band gap is relatively wide which limits their light-harvesting range. Therefore, PbS and PbSe QDs that have the abilities to extend their light absorption into infrared (IR) region are favourite choices of

Table 1.3 Photovoltaic parameters and materials of QDSCs based on different absorbers

QDs	ETMs	Electrolyte/HTM	CE	V_{oc} (V)	J_{sc} (mA/cm^2)	FF	PCE (%)	References
Ag_2S	ZnONP	Polysulfide	Pt	0.509	9.28	0.52	2.41	[163]
CdS	TiO_2NP	Polysulfide	SiO_2 Gel–d C-fabric/WO_{3-x}	0.899	11.045	0.63	6.28	[164]
CdSe	TiO_2NP	PVP/Polysulfide	Cu_2S	0.693	16.01	0.677	7.54	[165]
CdTe	TiO_2NP	Polysulfide	Cu_2S	0.609	11.83	0.676	4.87	[166]
In_2S_3	TiO_2NP	Polysulfide	Pt	0.57	4.810	0.474	1.30	[167]
InP	TiO_2NP	Polysulfide	Cu_2S	0.59	10.58	0.567	3.54	[168]
PbS	ZnONP	1,2-ethanedithiols	Au	0.62	22.6	0.63	8.8	[169]
PbSe	ZnONP	1,2-ethanedithiols	Au	0.49	25.3	0.63	7.8	[169]
Sb_2S_3	TiO_2NP	PEDOT/PCPDTBT	Au	0.711	16.1	0.65	7.5	[170]
Ag_2S/ZnS	TiO_2NP	Polyiodide	Pt	0.27	28.5	0.238	1.76	[171]
CdS/CdSe	TiO_2NP	Polysulfide	Cu_2S	0.61	21.49	0.55	7.24	[172]
CdSe/ZnS	TiO_2NP	Polysulfide	Cu_2S	0.561	16.96	0.566	5.42	[173]
CdTe/CdS	TiO_2NP	Polysulfide	Cu_2S	0.642	13.88	0.669	5.96	[166]
CdTe/CdSe	TiO_2NP	Polysulfide	Cu_2S	0.606	19.59	0.569	6.76	[174]
CdTe/CdSeS	TiO_2NP	Polysulfide	Cu_2S	0.629	16.58	0.694	7.24	[166]
PbS/PbSe	ZnONP	1,2-ethanedithiols	Au	0.60	24.8	0.63	9.4	[169]
PbS/CdS	TiO_2NP	Polysulfide	Cu_2S	0.595	18.81	0.642	7.19	[175]
ZnSe/CdSe	ZnONW	Polysulfide	Cu_2S	0.836	11.96	0.45	4.54	[142]
ZnTe/CdSe	TiO_2NP	Polysulfide	Cu_2S	0.642	19.65	0.57	7.17	[176]
CdS/CdSe/ZnS	TiO_2NP	Polysulfide	Pt	0.55	15.71	0.57	4.91	[177]
CdS/CdSe/ZnSe	TiO_2NP	Polysulfide	Pt	0.58	20.11	0.55	6.39	[177]

(continued)

Table 1.3 (continued)

QDs	ETMs	Electrolyte/HTM	CE	V_{oc} (V)	J_{sc} (mA/cm^2)	FF	PCE (%)	References
CdTe/CdS/CdS	TiO_2NP	Polysulfide	Cu_2S	0.61	20.32	0.51	6.41	[178]
ZnSe/CdSe/ZnSe	ZnOTP	Polysulfide	Cu_2S-GO	0.761	17.3	0.471	6.2	[99]
CdS/CdSe/CdS/ZnS	TiO_2NT	Polysulfide	Cu_2S	0.555	17.7	0.557	5.47	[179]
$CuInS_2$	TiO_2NP	Polysulfide	Cu_2S	0.601	22.82	0.62	8.54	[180]
$CuInSe_2$	TiO_2NP	Polysulfide	Carbon	0.704	21.17	0.621	9.46	[181]
$CdSe_xS_{1-x}$	TiO_2NP	Polysulfide	Cu_2S/GO	0.557	11.2	0.51	3.20	[182]
$CdSe_xTe_{1-x}$	TiO_2NP	Polysulfide	Carbon	0.807	20.67	0.689	11.51	[104]
$Zn_xCd_{1-x}Se$	ZnONC	Polysulfide	Cu_2S	0.65	18.05	0.4	4.74	[97]
Cu–In–Ga–Se	TiO_2NP	Polysulfide	Carbon	0.74	25.01	0.621	11.49	[181]
Cu_2ZnSnS_4	TiO_2NP	Polysulfide	Cu_2S	0.47	17.48	0.40	3.29	[183]
Cu_2ZnSnS_4/CdSe	TiO_2NP	Polysulfide	Cu_2S	0.51	18.68	0.493	4.70	[184]
ZnCuInS	TiO_2NP	Polysulfide	Cu_2S	0.612	22.70	0.615	8.55	[185]
ZnCuInSe	TiO_2NP	Polysulfide	Carbon	0.752	25.97	0.644	12.57	[186]
ZnCuInSe/CdSe	TiO_2NP	Polysulfide	Carbon	0.752	27.39	0.619	12.75	[89]
$CsPbI_3$	TiO_2NP	Spiro-OMeTAD/MoO_x	Au	1.2	14.37	0.78	13.43	[187]
$CH_3NH_3PbI_3$	TiO_2NP	Polyiodide	Pt	0.64	19.29	0.552	6.54	[188]
$CH(NH_2)_2PbI_3$	SnO_2NP	Spiro-OMeTAD	Au	1.1	11.83	0.6442	8.38	[189]

binary QDs in QDSCs. Hu et al. fabricated QDSCs with PCE as high as 8.8% and 7.8% using pure PbS and PbSe, respectively, as the QDs [169]. Nevertheless, their low conduction band edge and electron injection efficiency limited advance in cell's performance. Due to the above-mentioned limitations of the binary QDs, the development of QD sensitisers has revolved to explore multi-compositional QDs to balance the light-harvesting efficiency and electron injection efficiency.

1.4.7 Ternary and Quaternary QD's Light Absorbing Material System

Ternary and quaternary QDs also known as alloyed QDs are another alternative to binary QD due to the feasibility to engineer the physical characteristics, such as band gaps, energy band positions, and photoelectronic properties of QDs without changing the particle size through controlling the compositional ratio between each chalcogenide and metal element [205]. Moreover, alloyed QDs also demonstrated higher chemical stability than their constituents due to their hardened lattice structure and decreased inter-diffusion [206].

Most of the ternary and quaternary QDs used in QDSC have been focused on metal chalcogenide-based materials such as $CdSe_xS_{1-x}$, $CdSe_xTe_{1-x}$, $CuInS_2$ (CIS), $CuInSe_2$ (CIS), Cu–In–Ga–Se (CIGS), Cu_2ZnSnS_4 (CZTS), Zn–Cu–In–S (ZCIS), and Zn–Cu–In–Se (ZCISe). Among them, CdSeS QDs were the first ternary QDs applied as sensitisers in QDSC, delivering a PCE of 0.6% [207]. Nevertheless, the narrow light absorption range of CdSeS QDs has resultant in a fairly poor photovoltaic performance, whereby QDSCs based on multiple absorber layers with various band gaps within the TiO_2 film could only deliver a maximum PCE of 3.2% [182]. This leads to the research of exploring II–VI group QD, the CdSeTe QD sensitiser which displayed broader light absorption range, enhanced chemically stable, and high conduction band edge characteristics, leading to a remarkable certified PCE of 11.16% [104].

Another group of ternary metal chalcogenide QDs which were heavily investigated is I–III–VI group semiconductors, including CIS and CISe owing to their high absorption coefficient and suitable band gap to generate excellent photovoltaic performance [208]. Moreover, they are considered as "green" light absorption materials due to the absence of heavy metal elements, Cd or Pb. Wang et al. demonstrated that the performance of CIS-based QDSCs could be boosted through careful tuning of Cu/In ratios in CIS QDs [180]. As a result, PCE up to 8.54% was achieved when the Cu/In precursor molar ratio is 1/4, which is attributed to the reduced surface defects in the In-rich CIS QDs which effectively inhibited charge recombination at the TiO_2/CIS/electrolyte interfaces. Through the introduction of Zn into CIS QDs, the PCE of ZCIS-based QDSCs was slightly improved to 8.55% which is mainly ascribed to the superior photogenerated electron injection rate from QD into TiO_2 substrate and restraint of charge recombination at the interface [185]. On the other hand, CISe QDs which possess a narrower band gap in comparison to CIS QDs have the capability to increase the light absorption range to near-infrared region (NIR), thus benefiting higher J_{sc} value and the photovoltaic performance. Hyeon et al. showed that a high J_{sc} value of 26.93 mA/cm^2 could be obtained in CISe-based QDSCs by optimising ZnS overlayer thickness, delivering a PCE of 8.1% [209]. Similarly, a QDSC PCE of 9.46% was obtained by Peng et al. when CISe QDs were used as sensitisers and the PCE was further improved to 11.49% when Ga element was alloyed into CISe QDs [181]. The alloyed CIGS QDs were discovered to have optimum electronic structure, which matched well with TiO_2 substrate and favouring the electron extraction process while at the same time, suppressing the intrinsic recombination in CIGS-based QDSCs. Another successful alloying strategy of CISe QDs was reported where Zn element was employed to synthesise ZCISe QDs, yielding a record ZCISe QD-based QDSC PCE of 12.57% [186]. The defect state-related donor–acceptor pair in ZCISe QDs owing to Cu deficiency was believed to be responsible for the enhanced PCE.

1.4.8 Core/Shell QD-Based Light Absorber

Core/shell structure falls under one of the classifications of alloyed semiconductors. QDs with this unique structure have attracted much attention due to the complementary effect of extending the wavelength range of light harvesting, facilitating the charge separation, and limiting the recombination processes in QDSCs. Besides, the shell can act as a protection layer, preventing the core QDs from corrosive organic capping ligands, and thus enhance the chemical, thermal, and photostability of the QDs [85, 210]. Depending on the band gaps and the relative position of electronic energy levels of the core/shell QDs, core/shell QDs can be classified into three categories: type-I, reverse type-I, and type-II band alignment [85]. For type-I, the band gap of the shell material is larger than that of the core material, whereas reverse type-1 is in contrary to type-1. As for type-II, both the valence band (VB) edge and the conduction band (CB) edge of the shell material are located higher or lower than the band gap of core material.

Reverse type-1-structured QDs based on CdS/CdSe QDs were the first core/shell QDs introduced in QDSCs, yielding a PCE of 4.22% [211]. The core/shell structure of CdS/CdSe was found to form a stepwise structure of band edge levels that benefits the electron injection and hole recovery of the QDs through re-organisation of energy levels between the core and shell QDs. This simple yet successful strategy has also led to an extensive usage of direct deposition methods, such SILAR and CBD to make core/shell-structured QDs in QDSCs, which delivered PCE as high as 7.24% [172]. Nevertheless, SILAR and CBD methods which are extensively used in CdS/CdSe-based QDSCs are not feasible to deposit other types of core/shell-structured QDs, such as Te-based core/shell QDs due to limitation in precursor preparation. Moreover, the nucleation and growth of the core/shell QDs in the mesoporous ETM are incontrollable by direct growth methods, leading to the development of pre-synthesised core/shell QD sensitisers.

Most of the type-I- and type-II-structured QDs are entirely or partially pre-synthesised by using capping agents to adjust the nanocrystal size, shape, and light absorption properties. In type-1-structured QDs, the shell is usually used to passivate the surface of the core to improve the optical properties and chemical stability of the core [85]. Jiao et al. demonstrated that type-1-structured PbS/CdS QDs could be obtained through hot injection approach to obtain the core, followed by ion exchange to cover the core with CdS [175]. The authors discovered that the density of trapping state defects on the surface of the PbS QDs has been reduced through the introduction of CdS shell. As a consequence, a significant improvement of cell performance was achieved with PCE of 7.19% in comparison with the pristine PbS-based QDSC (PCE of 1.71%). Similarly, the shell, CdS in the CdSeTe/CdS-based type-1-structured QDs, was revealed to be able to decrease the surface defect density and increase the stability of the core QDs, thus suppressing the charge recombination process in the QDs and QD/TiO_2/electrolyte interfaces [212]. As a result, CdSeTe/CdS-based QDSCs achieved a PCE of 9.48%.

The application of type-II-structured core/shell QDs in QDSCs has attracted a lot attention recently due to their attractive properties and features. The position of CB and VB of the shell could result in significant red shift of the absorption edge, owing to the effect of the exciplex state and spatial separation of electron and hole, facilitating fast electron transfer from the QD sensitiser to the metal-oxide medium [210]. Wang et al. constructed type-II CdTe/CdSe QDSCs which exhibit PCE of 6.76% due to the extended light absorption range to infrared region as well as the fast electron injection and restriction of charge recombination process (Fig. 1.14) [174]. Similar discovery was reported by Yang et al. when employing the CdTe/CdS and CdTe/CdSeS type-II core/shell QDs as a sensitiser in QDSCs. The PCE of the wider light-harvesting range and photoelectronic response improved CdTe/CdSeS-based QDSCs was boosted to 7.24% which is 35% higher than the pure CdTe QDSCs [166].

1.4.9 Co-sensitised QDs

Co-sensitisation strategy was used to harvest broader range of incident light by depositing different light absorber materials with complimentary light absorption range and extinction coefficient in the photoanode [213]. In addition, co-sensitisation could improve the coverage of QDs on TiO_2 surface, protecting the ETM from being directly exposed to the electrolyte, and thus reduce the aggravation of charge recombination and enhance the light-harvesting efficiency [214].

Hu et al. show that the PbS QD-based device performance could be improved through co-sensitising a small amount of PbSe QDs which delivered a PCE of 9.4% as compared to 8.8% for pure PbS-based QDSCs and 7.8% for pure PbSe-based QDSCs

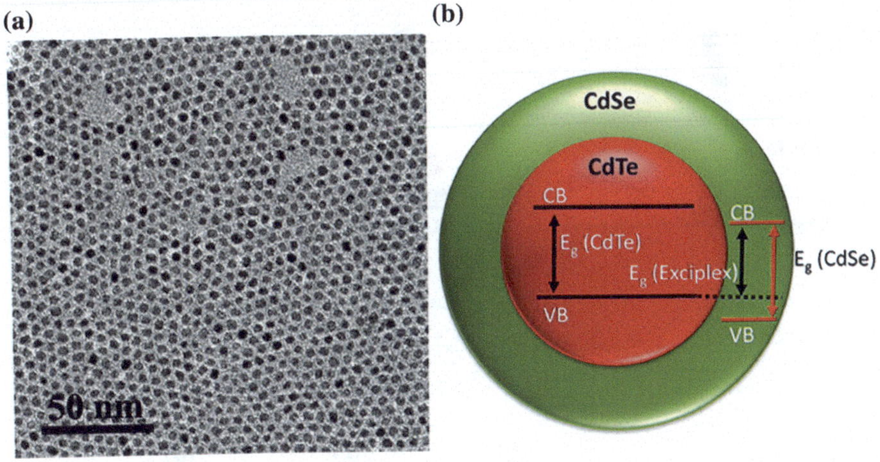

Fig. 1.14 **a** TEM image of 4.9 nm CdTe/CdSe core/shell QDs and **b** cartoon of corresponding core/shell QD indicating the relative bands position (the conduction band and valence band are depicted with a solid black or red line for CdTe and CdSe, respectively. The size of the band gap, Eg, for CdTe, CdSe, and the exciplex state is indicated with arrows) [174]

[169]. The enhancement of device performance was attributed to the improved charge mobility of co-sensitised QDs due to the presence of PbSe, leading to the formation of a larger electronic coupling between neighbouring nanocrystals and thus facilitating faster charge transport. The co-sensitised QDSC performance was further improved to a record high PCE of 12.75% for metal chalcogenide-based QDSCs by Wang et al., through the employment of ZCISe and CdSe QDs as sensitisers (Fig. 1.15) [89]. The synergistic effect of the ZCISe QDs with a wide light absorption range and a high extinction coefficient CdSe QDs was revealed to be responsible for the improved light-harvesting efficiency as well as inhibition of charge recombination. Apart from different materials co-sensitisation, Chen et al. reported that co-sensitising QDSCs with different-sized CdSe (2.5 and 3.5 nm) could also exhibit longer electron lifetime owing to the reduction of charge recombination at the interface as compared to single-sized CdSe-based devices [215].

1.4.10 Organic–Inorganic Hybrid QDs Used in Solar Cells

Organic–inorganic hybrid nanomaterials, especially metal halide perovskite, have been the latest trend in PV with the realisation of both high efficiency and low cost due to their excellent optoelectronic properties (strong absorption and high carrier diffusion length), versatility of structure and materials that can be used, and the low-cost solution processing method for material synthesis [216–218]. The ballistic growth of perovskite solar cells (PSCs) has led to a certified PCE of 25.1%, surpassing

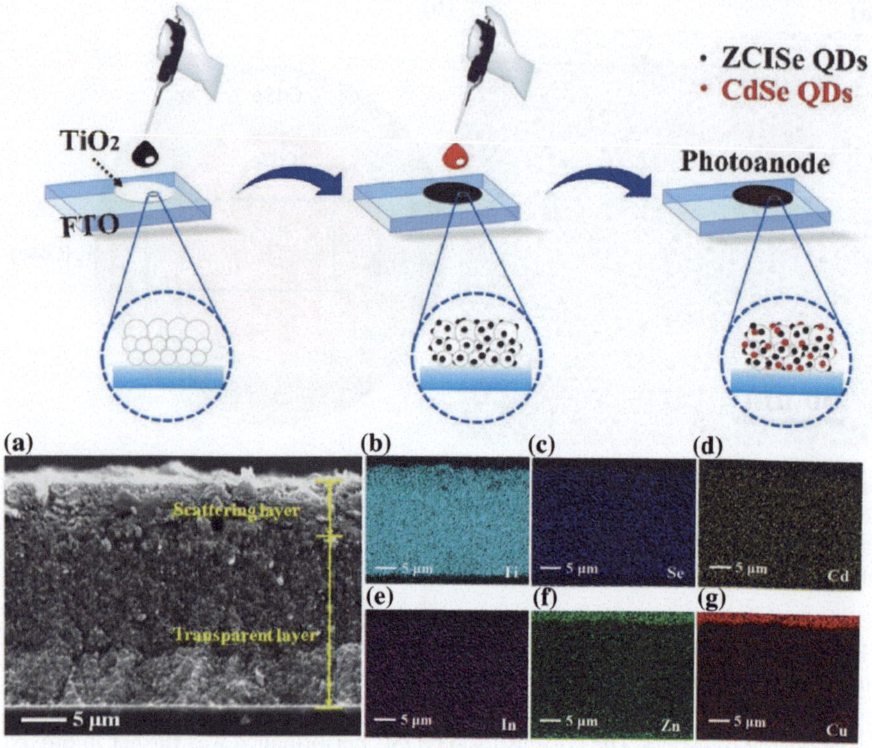

Fig. 1.15 (Top) Schematic illustration for the preparation of co-sensitised photoanodes; (bottom), **a** cross-sectional SEM image and **b–g** elemental mapping images of ZCISe/CdSe QD co-sensitised TiO_2 electrode film [89]

all other emerging PV technologies and is even comparable to the first-generation crystalline silicon solar cells [86].

The first report of perovskite QDSC was revealed by Kojima et al. in 2009 where methylammonium lead triiodide, $CH_3NH_3PbI_3$ ($MAPbI_3$), and methylammonium lead tribromide, $CH_3NH_3PbBr_3$ ($MAPbBr_3$) were applied as the sensitisers in a TiO_2-based device with structure resembling of DSSC, delivering a PCE of 3.8% and 3.13%, respectively [216]. The PCE of the $MAPbI_3$-based QDSC was later increased to 6.54% by Im et al. in 2011 [188]. Nevertheless, the $MAPbI_3$ QD-based QDSCs have poor stability in the presence of the electrolyte due to the gradual dissolution of QDs into the redox electrolyte. Therefore, Xue et al. employed a post-synthetic process for effective control of surface ligand density on formamidinium lead triiodide, $CH(NH_2)_2PbI_3$ ($FAPbI_3$) QDs which improved the ambient and operational stability of the QD films and devices over their bulk form [189]. As a result, the $FAPbI_3$ QD solar cells exhibited a PCE of 8.38%.

The stability of perovskite QDs was further improved by Swarnkar et al. through using $CsPbI_3$, a pure inorganic perovskite material as the light absorber [219]. The

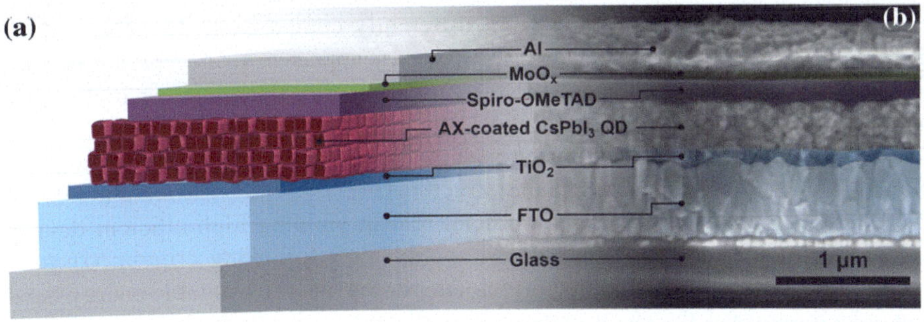

Fig. 1.16 **a** Schematic and **b** SEM cross-sectional images of a perovskite-based QDSC [187]

resultant films exhibited long-range electronic transport and the $CsPbI_3$ QD-based cells delivered a PCE of 10.77%. The same group later introduced an AX post-treatment (where A = FA^+, MA^+, or caesium (Cs^+) and X = I^- or Br^-) to coat the $CsPbI_3$ QD with the AX salt species (Fig. 1.16) [187]. The post-treatment was discovered to greatly improve the electronic coupling between the QDs which boosted the charge carrier mobility of the QDs while preserving the quantum confinement and nanocrystalline characteristic of the QDs. This allowed fabrication of a solar cell with a certified record PCE of 13.43%—the highest efficiency among all QDSCs ever produced at the time of writing.

1.5 Perovskite Solar Cells

In the early stage of PSCs, nano-sized perovskite materials were applied as sensitisers in solar cells with device structure resembling of DSSCs which is gradually developed into p-i-n or n-i-p structured solar cells (Fig. 1.17). Although PSCs evolved from

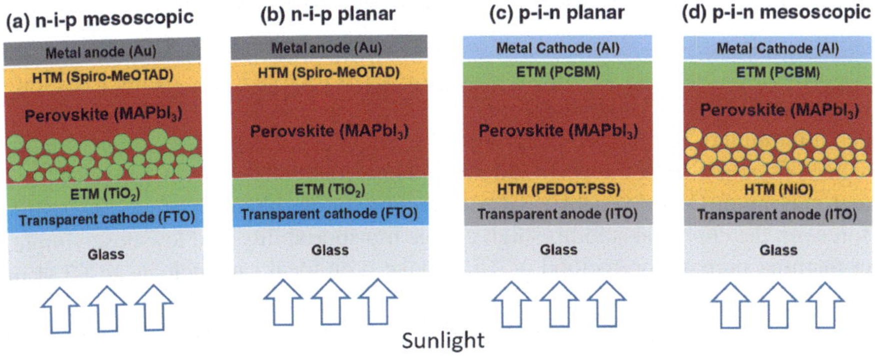

Fig. 1.17 Schematic diagrams of perovskite solar cells in the **a** n-i-p mescoscopic, **b** n-i-p planar, **c** p-i-n (inverted) planar, and **d** p-i-n (inverted) mesoscopic structures [220]

DSSCs, studies suggest that they work differently compared to DSSCs, making them a new type of solar cells. Nevertheless, they appear to overcome the hurdles of DSSCs, in particular the device performance, and offer the greatest paradigm for potential substitutes of Si-based and inorganic thin-film modules.

In PSCs, three typical structures are usually used: (i) mesoporous structure; (ii) planar structure, and (iii) inverted planar structure (Fig. 1.17) [220]. Similar to DSSCs and QDSCs, ETM in PSCs plays a crucial role in yielding high-efficient device. To achieve high performance, the ETM should meet following criteria: (i) good optical transmittance, which reduces incident photon loss; (ii) suitable energy band gap, which improves the electron extraction efficiency; (iii) high electron mobility, which allow fast electron extraction and transport; (iv) ease of production, chemically stable, low toxicity, and low cost, which allow large-scale, mass production. n-type materials, including TiO_2, ZnO, SnO_2, insulative material such as Al_2O_3 and ZrO_2, as well as p-type WO_x, NiO, and Cu_xO, are often used as ETM, scaffold layer or HTM, respectively, in PSCs.

1.5.1 n-Type Photonic-Responsive Materials

PSCs with the conventional n-i-p structure are more extensively investigated as compared to the p-i-n structure, which is partially due to the evolution of PSCs from DSSCs. In a n-i-p structure, the incident light travels through the ETM of PSCs before being absorbed by the light-harvesting material, perovskite. Therefore, the physical and optoelectrical properties of the ETM are important as it influences the light-harvesting efficiency of the devices. Several inorganic n-type semiconducting nanomaterials have been applied as ETMs in the conventional structure PSCs (Table 1.4).

1.5.2 TiO_2-Based Electron Transporting Materials

TiO_2 is the first and thus far, the most popular ETM used in PSCs, owing to its long history in DSSCs. The proper band gap and energy band position where its conduction band minimum (CBM) located slightly lower than that of $MAPbI_3$ facilitate smooth electron injection from the perovskite absorber to the ETL while its lower valence band maximum (VBM) enables TiO_2 exceptional hole-blocking properties [236]. Moreover, the chemical stability, high charge transportability, and low-cost, simple, yet matured preparation method of TiO_2 make it an ideal candidate as an ETM in PSCs.

As aforementioned, ETMs can be used in the form of a compact layer and mesoporous layers. The compact layer is generally less than 100 nm, and the mesoporous layer has thickness in the ranges of hundreds of nm. The electron diffusion length of the perovskite film was reported to be over 1 μm, suggesting that the TiO_2 thickness

Table 1.4 Photovoltaic parameters of PSCs based on different ETMs in n-i-p structure

ETM	Cell structure	V_{oc} (V)	J_{sc} (mA/cm^2)	FF	PCE (%)	References
c-TiO_2	FTO/c-TiO_2/PSK/Spiro-OMeTAD/Au	1.189	22.3	0.806	21.4	[221]
m-TiO_2	FTO/c-TiO_2/m-TiO_2/PSK/PEAI/Spiro-OMeTAD/Au	1.130	25.92	0.8202	24.02	[222]
TiO_2NR	FTO/c-TiO_2/TiO_2NR/PSK/Spiro-OMeTAD/Au	0.945	19.78	0.72	13.45	[223]
TiO_2NW	FTO/TiO_2NW/PSK/Spiro-OMeTAD/Au	0.95	14.8	0.64	9.0	[224]
TiO_2NT	FTO/c-TiO_2/TiO_2NT/PSK/Spiro-OMeTAD/Au	0.886	25.5	0.679	15.34	[225]
TiO_2ND	FTO/TiO_2ND/PSK/Spiro-OMeTAD/Ag	1.08	22.8	0.73	18	[226]
c-ZnO	FTO/c-ZnO/PSK/Spiro-OMeTAD/Ag	1.20	23.51	0.785	21.1	[227]
m-ZnO	FTO/c-ZnO/m-ZnO/PEI/PSK/Spiro-OMeTAD/Ag	1.00	22.75	0.75	20.74	[228]
ZnONR	FTO/ZnONR/PSK/Spiro-OMeTAD/Ag	0.995	21.95	0.74	16.16	[229]
c-SnO_2	ITO/c-SnO_2/PSK/PEAI/Spiro-OMeTAD/Au	1.18	25.2	0.784	23.32	[230]
m-SnO_2	FTO/c-TiO_2/m-SnO_2/PSK/PTAA/Au	1.13	22.58	0.78	19.8	[231]
SnO_2NR	ITO/SnO_2NR/PSK/PTAA/Au	1.03	22	0.753	17.2	[232]
SnO_2NW	FTO/SnO_2NW/PSK/Spiro-OMeTAD/Ag	1.02	21.2	0.655	14.2	[233]
SnO_2NT	FTO/SnO_2NT/PSK/P3HT/Au	0.94	18.38	0.707	12.26	[234]
SnO_2NS	FTO/SnO_2NS/PSK/Spiro-OMeTAD/Au	1.03	21.66	0.689	15.29	[235]

may not be the key factor affecting the electron transport efficiency [237]. In fact, the defects and crystal boundaries of the c-TiO_2 which act as electron traps are responsible for the charge recombination, reducing V_{oc} and device instability. Additionally, since planar PSCs only consist of single compact layer as the ETL and blocking layer, a uniform pinhole-free ETL is crucial for efficient hole-blocking, avoiding intimate contact between the TCO substrate and perovskite. Tan et al. demonstrated hysteresis-free PSCs with PCE up to 21.4% can be achieved, through employing a chlorine-capped TiO_2NP film with a very low trap-state density of ~3×10^{15} cm^{-3} to mitigate interfacial recombination and to improve interface binding [221]. In the meantime, bilayer-structured compact ETL composed of SnO_2 and TiO_2 was also proven to have provided a large free energy difference value between ETL and perovskite as well as defect-free physical contact, resulting in a PCE more than 21% [238, 239].

The ETL of mesoporous PSCs normally contains a mesoporous layer on top of the compact underlayer. The mesoporous layer acts as a scaffold to assist the crystallisation of perovskite with enhanced nucleation at the nanoparticles' surface and expand the contact surface area between perovskite and ETL for efficient photogenerated electron collection [240]. Owing to these reasons, the size of the mesoporous-based TiO_2NPs is important for device performance. If the size is too small, the infiltration of perovskite will be prevented by the close, dense pore whereas if the size is too large, the morphology of perovskite will be affected, therefore increasing the risk of direct connection between the TiO_2NPs and the hole transporting layer (HTL). Apart from the above-mentioned effect, Han et al. revealed that different sizes of TiO_2NPs also affect the charge transfer kinetics at the perovskite/TiO_2 interface [241]. M-TiO_2-based PSCs with a record PCE over 24% is achieved by Kim et al., through careful tuning and selection of appropriate size of mesoporous TiO_2NPs [222].

1D TiO_2 nanostructures including nanorods [223, 243, 244], nanowires [224, 245], nanotubes [225, 246, 247], and nanofibers (Fig. 1.18a, b) [242] possess the merits of higher electron mobility which facilitate fast electron transport over TiO_2NPs. Kim et al. studied the influence of TiO_2NRs length on device performance and discovered that the performance of PSCs degraded with the increase of TiO_2NRs length, due to the disorder arrangement of the nanorods, thus reducing the pore filling [243]. Interestingly, only the charge generation efficiency was found to be affected by different lengths of nanorods while the charge recombination rate remains the same. Similar investigation was carried out by Jiang et al. on TiO_2NWs [245]. The authors suggested that 900-nm-long TiO_2NWs yielded the best PCS performance due to the effective charge transport and additional loading of perovskite materials, allowing higher light absorption and thus the current density. Further modification of TiO_2NWs to 3D branched TiO_2NWs sees the enhanced device performance through efficient electron transport and light scattering (Fig. 1.18c–f) [224]. On the other hand, Gao et al. constructed a TiO_2NT network ETL as light scatterers which is capable of enhancing light absorption, and charge extraction and collection efficiency and thus device performance. Yang et al. further improved the light-harvesting capability through assembling ETL composed of TiO_2NP/TiO_2NT, which showed a substantial increment in photocurrent density compared to pure TiO_2NP-based PCSs [225]. The

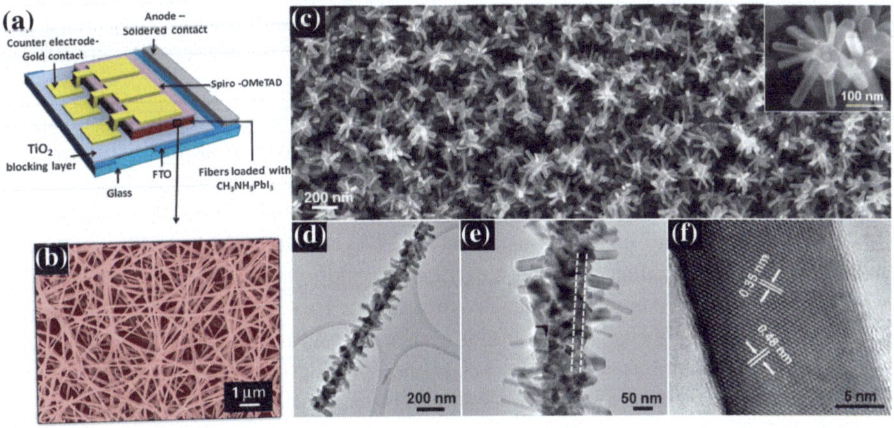

Fig. 1.18 **a** Schematic illustration of the various components of TiO_2NF-based PSCs and **b** SEM of the TiO_2NF film [242]; **c** top view of a 3D TiO_2 NW architecture. Inset is an enlarged SEM image showing the tree-like-branched NR structure; **d** TEM image of a branched TiO_2 NW showing dense and uniform coverage of TiO_2NRs along the entire stem. **e** Higher magnification TEM (HRTEM) image showing the single-crystalline NR branches and the polycrystalline NT trunk. **f** HRTEM image of a TiO_2 NR branch showing the perfect crystal lattice [224]

enhanced charge conduction and light-harvesting efficiency found in the PSCs were attributed to the Mie scattering effect. Similar light-scattering effect was reported by Islam et al. when TiO_2NRs were employed as scaffolding layer and TiO_2NPs as blocking layer [248]. As a result, the PCE of the device was significantly enhanced, due to the improved J_{sc} and reduced charge recombination rate. PSCs based on 3D-nanostructured ETLs, namely nanocones [249], nanohelices [250], and nanodendrite [226], were also fabricated due to their outstanding electron transporting properties as well as light-scattering efficiency, which is beneficial for boosting the device performance. Nevertheless, only the PSCs constructed based on 3D hierarchical structure TiO_2 nanodendrite (TiO_2ND) array were capable of achieving promising device performance with PCE of 18%, owning to the effective electron transporting and hole-blocking properties of the unique nanostructure [226].

1.5.3 ZnO-Based Electron Transporting Materials for PSCs

Even though TiO_2-based PSCs demonstrated promising cell efficient, the preparation of TiO_2 films requires high annealing temperature, up to 500 °C which would limit its prospect in the development of flexible plastic substrate-based PSCs. Additionally, TiO_2–perovskite interfaces are notably unstable under UV light, which is detriment to long-term stability [251]. Therefore, low-temperature processable material such as ZnO is considered as a feasible alternative to TiO_2, owing to its superior optoelectronic properties that offer greater transmittance in the visible to IR regions of the spectrum and lower resistivity, orders of magnitude higher mobility, and a shallower conduction band edge that align more closely with metal halide perovskite [227]. Moreover, ZnO with different morphologies can be easily synthesised through solution method at low temperature due to its anisotropic growth, making it suitable for application in flexible polymer substrates.

ZnO has been successfully demonstrated its function as both blocking layer and ETM in both planar and mesoporous structured PSCs. Particularly in planar PSCs, ZnONPs can be easily deposited on electrically conductive substrate such as FTO and ITO without the mesoporous scaffolding layer. This simple, low-temperature, solution-processable fabrication method is compatible with roll-to-roll processing on flexible substrates which is suitable for large-scale production [252]. Song et al. studied the influence of different ZnO layer thicknesses in affecting PSC performance. It is revealed that ZnO layer with a thickness of 20 nm is the optimum thickness to deliver the best cell performance [253]. Cao et al. and Schutt et al. achieved record high efficiency of 21.1% with ZnO-based planar PSCs, demonstrating the potential of this material [227, 254]. On the contrary, 2D and double-layered ZnO nanostructures are famous for their capability in extending the electron diffusion length and facilitate fast electron transport mesoporous PSCs. Above all, ZnONR is the most investigated nanostructure in ZnO-based mesoporous PSCs due to its boundary-less structure with desirable internal electrical field for charge transport [255]. Son et al. found that both the short-circuit current density and charge collection in ZnONR-based PSCs were better than TiO_2NR-based PSCs, leading to an enhanced PCE of 11% [256]. Mahmood et al. further improved the electron mobility of ZnONRs through nitrogen dopant, leading to an enhancement of PCE of 16.1% (Fig. 1.19) [229]. Meanwhile, Mahmud et al. reported that double-layered ZnO nanostructure with thickness of 45 nm contained the lowest trap states, producing ZnONP-based PSCs with superior charge transport properties [257]. The efficiency of ZnO mesoporous (m-ZnO) devices was further boosted to over 20% through doping ZnO with polymer and employing nanocomposite of m-TiO_2 with ZnO-ZnS [228].

Although ZnO possesses various unique features and properties, the issues of ZnO-perovskite interfacial and thermal instability have led to ZnO as less attractive ETM material compared to TiO_2 and SnO_2. Several reports suggest that the hydroxyl groups present on ZnONPs are the main culprit for the degradation of perovskite [258–260]. Schutt et al. further reveal that the root cause of ZnO-perovskite instability to be the deprotonation of the methylammonium (MA) cation, which lead to the formation of zinc hydroxide [227]. Despite the instability of ZnO-based PSCs, a number of research groups have successfully addressed these stability concerns by "passivating" ZnO surface with magnesium oxide, ethanolamine, poly(ethylenimine), and ZnS which led to increment of ZnO-based cells' PCE over 20% [228, 254, 261, 262]. Nevertheless, the long-term stability of ZnO-based PSCs still remains a concern if MA-based perovskite is employed. Therefore, MA contented perovskite should be

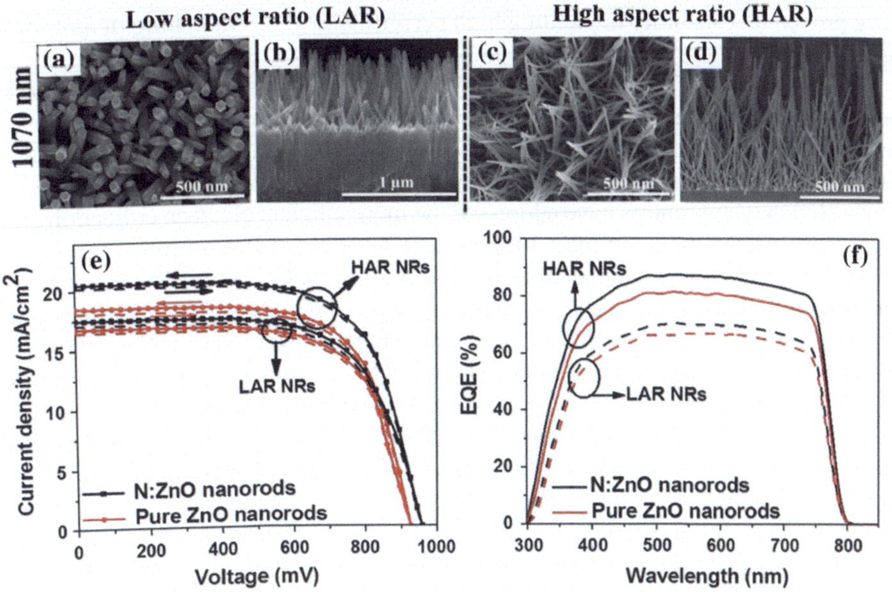

Fig. 1.19 Surface and cross-sectional SEM images of **a**, **b** conventional nitrogen-doped ZnO (N:ZnO) nanorods (rods diameter of 85 nm) with low aspect ratio (LAR) and **c**, **d** modified N:ZnO NRs using PEI as a capping agent during growth (rod diameter of ≈35 nm) resulting in high aspect ratio (HAR) NRs. **e** *J-V* curves both in the forward and reverse scanning directions for the devices based only on optimal 1070-nm-long HAR and LAR for pure ZnO and N:ZnO NRs and **f** EQE spectra of the corresponding devices, respectively [229]

completely avoided and replaced with formamidinium mixed with caesium (FACs)-based perovskite, which has been proven to be able to deliver a record ZnO-based PSC PCE of 21.1% with enhanced long-term stability [227].

1.5.4 SnO_2-Based Electron Transporting Materials in PSCs

Stannic oxide (SnO_2) is another extensively explored metal-oxide materials to be used as compact and mesoporous layers in PSCs. Compared to TiO_2, SnO_2 appears to be a better ETM for PSCs in many aspects including: (1) a deeper conduction band and valence band position, which is beneficial for efficient electron extraction and hole blocking; (2) high electron mobility (240 $cm^2\ V^{-1}\ s^{-1}$), which can enhance the electron transport efficiency and reduce the recombination losses; (3) excellent chemical stability, UV resistivity, good antireflection property, and less photocatalytic activity, which is favourable for long-term stability of the device; (4) Lower crystallisation temperature than TiO_2, which makes it an ideal candidate in the application of flexible solar cells and large-scale fabrication at a lower cost [263]. Owning to these attractive properties, extensive studies on SnO_2 based PSCs, which have proven its capability in yielding high performance PSCs matching that of TiO_2 based PSCs.

The application of SnO_2 as ETMs was first introduced in DSSCs. Although SnO_2 possesses above-mentioned attractive features, it received little attention due to its relatively poor PCE compared with TiO_2- and ZnO-based DSSCs. Nevertheless, SnO_2 has proven its success in delivering high-efficient PSCs with PCE over 21% [264–267]. Several reports have addressed the more efficient electron collection efficiency of compact SnO_2 (c-SnO_2) layers than TiO_2-based PSCs [268–271]. As an example, Dong et al. systematically investigated the SnO_2- and TiO_2-based PSCs and disclosed that the V_{oc} and J_{sc} of the SnO_2-based PSCs are significantly higher as opposed to that of TiO_2-based PSCs, suggesting higher electron collection efficiency of SnO_2 [268]. In addition, the thickness of SnO_2 as compact layer is vital to efficiently avoid direct contact of the perovskite layer with transparent conductive oxide (TCO) layer while, at the same time, allow effective electron extraction. Studies have shown that the optimum c-SnO_2 layer thickness is determined by the preparation conditions [266, 269, 272]. Nevertheless, all reports suggested that SnO_2 nanoparticles (SnO_2NPs) are the most suitable morphology in the compact layer since merely 40-nm-thick c-SnO_2 layer is required to produce a record high-efficient SnO_2-based planar PSCs with PCE of 23.32% [230]. These results have demonstrated the essential in balancing hole-blocking ability and electron transport properties of c-SnO_2 layer through effectively control the layer thickness.

Even though SnO_2-based planar PSCs thus far achieved better PCE than SnO_2-based mesoporous PSCs, the rapid and poorly controlled perovskite crystallisation makes it challenging to commercialise planar PSCs. An extra mesoporous layer of ETM could overcome this problem by altering the surface wettability to improve the uniformity of perovskite over large area [273]. Furthermore, the mesoporous layer could potentially improve the hole-blocking ability of compact layer through filling the pinhole of the compact layer while simultaneously enhance the electron extraction from perovskite layer. These benefits lead to suppression or even complete elimination of the notorious I-V hysteresis. Apparently, the stability of SnO_2-based mesoporous PSCs has proven to be far superior to its planar-based PSCs, counterpart as well as TiO_2-based mesoporous PSCs [59]. Recently, Xiong et al. fabricated PSCs through the incorporation of a thin SnO_2 mesoporous (m-SnO_2) scaffold layer (100 nm) on top of the SnO_2 blocking layer and achieved a hysteresis free and stable PSCs with a high PCE of 19.12% [274]. The authors revealed that the scaffold layer plays a connecting role between the blocking layer and perovskite and forms a cascade energy-level sequence, thus facilitating charge transfer and enhancing charge collection efficiency. Moreover, the PCE of PSCs can be further boosted to 19.80% through the integration of m-SnO_2 scaffold layer on top of c-TiO_2 underlayer, owing to the enhanced charge collection efficiency [231]. Apart from the commonly used SnO_2NPs, some multi-dimensional nanostructured SnO_2 including nanosheets (Fig. 1.20c, d) [235], nanotubes (Fig. 1.20a, b) [234], nanowires [233], and nanorods [232] were also utilised as ETM or mesoporous layer. Nevertheless, the performance of multi-dimensional nanostructured SnO_2 (PCE ~ 16%) is slightly lower than that of SnO_2NPs-based PSCs, suggesting further works needed for improvement.

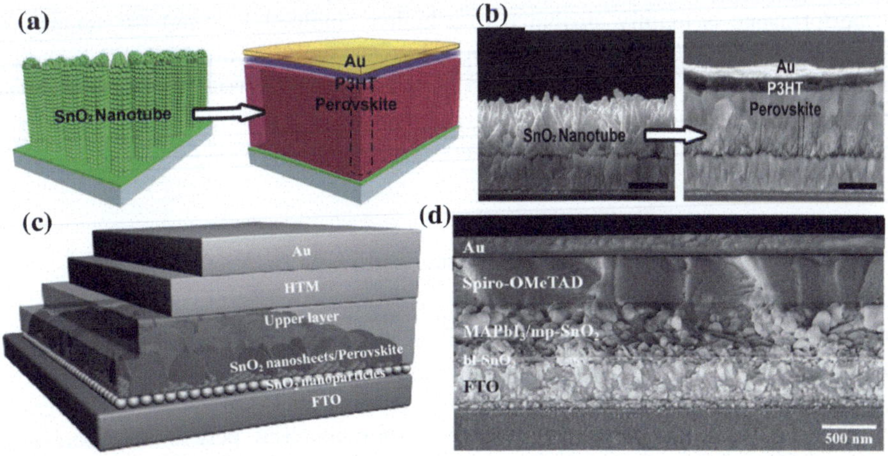

Fig. 1.20 **a** Schematic illustration of PSC fabrication process and **b** the corresponding cross-sectional SEM images of SnO_2NTs and its device [234]; **c** schematic view and **d** cross-sectional SEM image of PSCs with SnO_2 hierarchical ETL [235]

1.5.5 Other Metal-Oxide Scaffold Materials Used in PSCs

Besides the usually used TiO_2, ZnO, and SnO_2 ETMs, a variety of binary metal-oxide materials, for instance ZrO_2 [275–277], Al_2O_3 [278, 279], and SiO_2 [280–282] which have large band gap and are insulate materials in nature, have been used as scaffolding layer in PSCs. PCSs with a PCE of 11.33% were successfully fabricated by Li et al. through integration of 167-nm-thick ZrO_2 layer on top of m-TiO_2 layer, creating a double-layered mesoporous TiO_2/ZrO_2 as scaffold layer [276]. The authors suggested that the additional ZrO_2 layer which acts as spacer layer prevented the transportation of electrons in TiO_2 back to counter-electrode as well as improved the perovskite film quality. Interestingly, Mejía Escobar et al. displayed that PCE as high as 17.9% could be yielded through the implementation of bilayered compact ETM composed of a thin layer of ZrO_2 sandwiched between the TCO and c-TiO_2 [275]. On the other hand, Numata et al. studied the influence of heterogeneous TiO_2/Al_2O_3 composition as a mesoporous scaffold layer on electron transport path and proposed that a 7:3 ratio of TiO_2/Al_2O_3 composite yielded the best PCS performance due to the improved electron injection from perovskite to electron-collecting electrode [279]. Similar phenomenon of enhancement in carrier transfer was revealed when mesoporous SiO_2 layer was integrated into TiO_2 scaffold layer, producing PCSs with PCE of around 12% [280]. Meanwhile, ternary metal oxides including $SrTiO_3$ [283, 284], $BaSnO_3$ [285–287], and Zn_2SnO_4 [288–292] have also demonstrated their capability in delivering high-efficient PSCs. Among them, PSCs fabricated with ~120-nm-thick lanthanum (La)-doped $BaSnO_3$ (LBSO) as the ETL shown the best performance, with an astonishing high PCE of 21.2%, which is more superior to its m-TiO_2-based PSCs counterpart (PCE of 19.7%) [287]. This remarkable achievement in performance is mainly attributed to a higher obtained V_{oc} and FF, which are associated with LBSO extraordinarily high electrical mobility, and higher conduction band minimum and electron density as well as reduced carrier recombination as compared to m-TiO_2-based PSCs. Additionally, LBSO-based devices also showed a greater resistance against photodegradation than the TiO_2-based devices.

1.5.6 p-Type Semiconductor Nanomaterials in PSCs

The extensive research in perovskite field has also led to the rapid development of p-i-n structure PSCs, which is also known as inverted structure PSCs. As opposed to n-i-p structure PSCs, the incident light travels through the HTM in inverted structure PSCs before being harvested by the light absorbing material, perovskite. Therefore, the physical and optoelectronics properties of HTM in this case directly affect the light-harvesting efficiency of the devices. Several inorganic p-type semiconducting nanomaterials have been applied as HTMs in the inverted structure PSCs as listed in Table 1.5. Among them, nickel oxide and copper-based nanomaterials are most widely investigated.

1.5.7 Nickel Oxide

Nickel oxide (NiO_x) is a cubical structured p-type semiconductor with wide band gap, which offers sufficient optical transparency in the visible spectral region that allows solar radiation access to the active layer through it [302]. Additionally, the favourable deep-lying valence band of NiO_x is an ideal electron blocking material that suppresses charge recombination and enhances the hole extraction in PSCs [303]. Furthermore, the chemically stable NiO_x makes it perfect HTM to be applied on TCO substrates. Liu et al. demonstrated that PSCs with PCE of 20.2% could be achieved through a simple, low-temperature annealing process of the undoped NiO_x which enhanced charge extraction capability, better energy alignment, and improved photoluminescence quenching efficiency as compared to sol–gel processed NiO_x [304]. Nevertheless, the low conductivity of NiO_x which results in increased recombination and poor hole extraction has limited the progress of PSC performance [305]. Therefore, the conductivity of the NiO_x film needs to be enhanced either through composition optimisation and doping. Doping NiO_x with various dopants such as Cu [306], Cs [305], Mg [307], Mg:Li [308] has proven to be effective where high-efficient PSCs with PCE over 20% have been achieved. Meanwhile, Abzieher et al. studied the influence of NiO_x layer thickness on the performance of PSCs and reported that 30-nm-thick NiO_x layer yields the best device performance with a PCE of 20.7% due to the low absorbance and high transmittance of incident light [309].

Wang et al. showed that a higher PCE of 9.51% could be obtained through mesoporous NiO_x (m-NiO_x)-based PSCs over planar NiO_x-based PSCs (PCE of 7.9%),

Table 1.5 Photovoltaic parameters of PSCs based on different HTMs in p-i-n structure

ETM	Cell structure	V_{oc} (V)	J_{sc} (mA/cm^2)	FF	PCE (%)	References
c-NiO_x	ITO/c-NiO_2/PSK/PCBM/Ag	1.12	23.18	0.803	20.86	[293]
m-NiO_x	FTO/m-NiO_x/PSK/PCBM/LiF/Al	1.06	20.2	0.813	17.3	[294]
CuSCN	ITO/CuSCN/PSK/PC_{61}BM/PEI/Ag	1.10	20.76	0.73	16.66	[295]
CuI	ITO/PSK-Cu(Tu)I/C_{60}/BCP/Ag	1.13	22.3	0.789	19.9	[296]
CuO	ITO/CuO/PSK/C_{60}/BCP/Ag	0.99	23.2	0.744	17.1	[297]
Cu_2O	ITO/CuO/PSK/PC_{61}BM/Ca/Al	1.07	16.52	0.7551	13.35	[298]
V_2O_5	ITO/V_2O_5/P3CT-K/PSK/PC_{61}BM/ZnO/Ag	1.09	23.24	0.7787	19.7	[299]
MO_x	ITO/PEDOT:PSS-MoO_x/PSK/PC_{61}BM/Bphen/Ag	0.97	21.59	0.7542	15.79	[300]
WO_{3-x}	ITO/WO_{3-x}/PSK/PCBM/LiF/Al	0.87	20.7	0.74	13.3	[301]

due to the efficient light scattering through the extension of light path length [310]. However, the mesoporous NiO_x-based PSCs still inferior to the planar NiO_x-based PSCs, due to the significant light absorption of NiO_x at wavelength below 500 nm as demonstrated in their external quantum efficiency (EQE). Park et al. improved the m-NiO_x-based PSCs to a record high PCE of 17.3% through constructing the (111)-oriented NiO_x nanostructured film which play a role in effective holes extraction and electron leakage prevention [294].

1.5.8 Copper-Based Inorganic Hole Transport Nanomaterials in PSCs

Inorganic semiconducting copper-based HTMs, including copper thiocyanate (CuSCN), copper iodide (CuI), copper oxide (CuO), and copper dioxide (Cu_2O), have been widely used in DSSCs and QDSCs, allowing easy technology transfer to PSCs. The p-type inorganic Cu-based HTMs, are famous for their economical and chemically stable properties while exhibit suitable energy band gap, high hole mobility, high conductivity, and excellent transmittance [311].

CuSCN for instant possesses good transparency in the visible and infrared regions, high mobility of ~0.01–0.1 $cm^2\ V^{-1}\ s^{-1}$ [312], and excellent chemical stability. The employment of CuSCN in inverted planar PSCs demonstrated a high PCE of 16.6% through controlling the surface roughness and contact resistance between the perovskite layer and CuSCN [313]. The efficiency of the CuSCN-based planar PSCs was further improved to 16.66% by Wang et al. and PCE as high as 18.76% was achieved through utilising a composite CuI/CuSCN as HTL [295]. Meanwhile, Chen et al. replaced conventional PEDOT:PSS with a highly conductive, wide band gap semiconductor, CuI in an inverted planar-structured PSCs [314]. The authors observed slight higher performance and enhanced stability of the CuI-based PSCs over PEDOT:PSS-based PSCs due to the hydrophobic nature and ambient stability of CuI. Similar air-stable device phenomenon was reported by Wang et al. with an improved PCE of 14.7% when CuI was employed as HTL [315]. Amazingly, Ye et al. showed that a hysteresis-free device with PCE of 18.1% can be achieved when CuSCN was directly incorporated in the perovskite layer forming perovskite/CuSCN bulk-heterojunction [316]. The integration of CuSCN in perovskite layer was found to have effectively facilitated hole transfer from the perovskite layer to the ITO electrode, passivating hole trap states in perovskite layer, maintain the balance of electron and hole transfer, and eliminate charge accumulation. The same group further boosted the device PCE to a record high 19.9% through modification of CuI to Cu(thiourea)I (Cu(Tu)I) [296]. The modified Cu(Tu)I was found to act as an efficient trap-state passivator, which has identical VBM with perovskite, leading to lower trap-state energy level at a greater extent and eliminate the potential wells for holes at the p-i heterojunctions (Fig. 1.21).

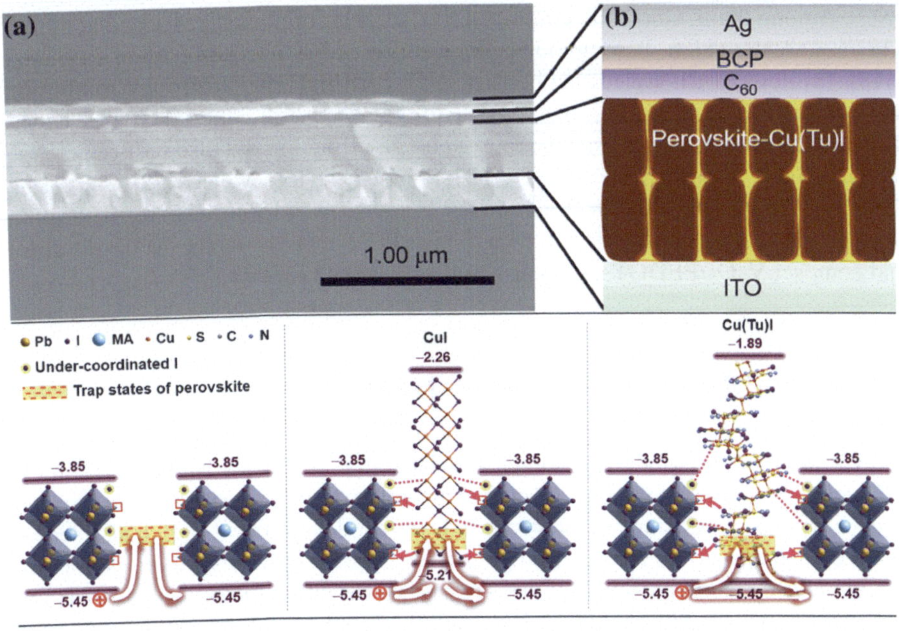

Fig. 1.21 (Top) **a** Cross-sectional SEM image and **b** schematic illustration of the device with Cu(Tu)I as the trap-state passivator; (bottom) schematic illustration of possible mechanism for the trap-state passivation. The size of graphic atoms is adopted randomly according to aesthetics [296]

In contrast to CuSCN and CuI, Cu_2O and CO possess relatively narrower band gap with work function of −5.4 eV [317]. Therefore, it can be used as HTM as well as light absorber in solar cells. Nevertheless, Zuo et al. compared inverted planar PSCs fabricated by using different HTMs, namely PEDOT:PSS, Cu_2O, and CuO, and reported a high PCE of 13.35% and 12.16% obtained from Cu_2O and CuO-based PSCs, respectively, over conventional PEDOT:PSS-based PSCs (11.04%) [298]. The higher device stability and PL quenching results displayed by Cu_2O and CuO layers make them a promising alternative ETMs to replace the unstable PEDOT:PSS. The PCE of CuO_x-based PSCs was further enhanced to 17.1% owning to the fast hole carrier transfer from the perovskite to CuO_x layer and lower contact resistance of the CuO_x layer [297].

1.5.9 Vanadium Oxide

Vanadium oxide (VO_x and V_2O_5) is another HTM that has been used in PSCs, which has a narrow band gap of −2.0 eV and deep work function around −5.2 eV [317]. This means that it can form an efficient hole injection/collection junction with perovskite active layer. To date, only few have adopted vanadium oxide as HTM in

PSCs. Among them, Guo et al. employed VO_x as HTL in planar PSCs and achieved a PCE of 14.5% which shows good ageing stability of 500 h in ambient environment [318]. The device PCE was further boosted to a record high 19.7% through combining V_2O_5 with P3CT-K to form a bilayer-structured HTL [299]. The bilayer HTL was discovered to not only increased the rate of charge extraction but also reduced charge recombination as compared to pure P3CT-K based device. Meanwhile, Guo et al. fabricated layered V_2O_5/PEDOT nanowires, followed by exfoliation to form ultrathin V_2O_5/PEDOT nanobelts and yielded a device PCE of 8.4% using 40 nm uniform and compact V_2O_5/PEDOT nanobelt-based HTL [319].

1.5.10 Molybdenum Oxide-Based HTM

Molybdenum oxide (MoO_3) is another wide band gap, p-type semiconductor. MoO_3 HTL has the capability to reduce the charge recombination by suppressing the exciton quenching and the resistance at photoactive layer/anode interface [302]. It also serves as an optical spacer for improving light absorption. Nevertheless, there is not any report on PSCs based on sole MoO_3 HTL. MoO_3 was either mixed with other HTM or deposited as an additional HTL. For example, MoO_x powders were incorporated into a PEFOT:PSS solution before being deposited as HTL for PSCs [300]. As a result, the perovskite film coverage was improved and a device PCE of over 15% was achieved, due to the presence of MoO_x which provide growing sites for the crystal nucleus of perovskite during annealing process. On the other hand, Hou et al. prepared a MoO_3/PEDOT:PSS bilayer-structured HTL and discovered that the MoO_3 layer could restrain ITO from PEDOT:PSS erosion. Moreover, the hole extraction efficiency of MoO_3/PEDOT:PSS PSCs was found to have increased as compared to pure PEDOT:PSS-based PSCs, thus yielding a device PSC of 14.87% [320].

1.5.11 Tungsten Oxide-Based HTM

Tungsten oxide (WO_x) is a low-cost, nontoxic, and volatile HTM with high work function, which is favourable for effective hole transport and collection [321]. Wan et al. prepared amorphous WO_x ETL at low temperature and revealed that WO_x exhibited comparable light transmittance but higher electrical conductivity in comparison with the conventional TiO_2 ETL [322]. Moreover, WO_x has positive effect on the oriented crystallisation of perovskite, which is beneficial to the photoabsorption, resultant in higher J_{sc} and thus achieved slightly higher average PCE of 8.99% over TiO_2-based PSCs (PCE of 8.79). The performance of WO_x-based planar PSC was further promoted to 13.3% by Masi et al. through fabricating PSCs based on 1D carved rod-shaped nanocrystals of reduced WO_{3-x} [301].

1.6 Summary and Outlook

During the past decade, we have witnessed the swift advancement of material and technical in the quest to obtain record PCE for third-generation solar cells, especially PSCs. Although the PCE of these devices is comparable to other solar cells, the cells' efficiency and long-term stability need to be further improved to be competitive with other existing solar technologies, such as silicon solar cells. Furthermore, "green" and earth-abundant materials should be taken into account in the research of all solar cell components in realising commercialisation of low-cost, environment-friendly products. The shortcomings in these devices will motivate researchers to further develop these kinds of solar technology aiming at commercial application as well as long-term stability. This chapter has introduced the development of photonic-responsive nanomaterials with unique optoelectronic properties and light absorption characteristics, which are well-defined by the particle size and morphology as well as material composition in the field of DSSCs, QDSCs, and PSCs.

References

1. A.B. Michael Schmela, N. Chevillard, M.G. Paredes, M. Heisz, R. Rossi, M. Schmela, Solar-Power Europe, China Photovoltaic Industry Association (CPIA), Dan Whitten & Justin Baca, US Solar Industries Association (SEIA), Japan Photovoltaic Energy Association (JPEA), Faruk Telemcioglu, Günder Turkish Solar Energy Society (GÜNDER), Steve Blume, Smart Energy Council; Rodrigo Lopes Sauaia & Stephanie betz, Brazilian Photovoltaic Solar Energy Association (AbSOLAR), Global Market Outlook 2018–2022, p. 81 (2018)
2. M.A. Green, E.D. Dunlop, D.H. Levi, J. Hohl-Ebinger, M. Yoshita, A.W.Y. Ho-Baillie, Solar cell efficiency tables (version 54). Prog. Photovolt. **27**, 565–575 (2019)
3. A.J. Nozik, M.C. Beard, J.M. Luther, M. Law, R.J. Ellingson, J.C. Johnson, Semiconductor quantum dots and quantum dot arrays and applications of multiple exciton generation to third-generation photovoltaic solar cells. Chem. Rev. **110**, 6873–6890 (2010)
4. H. Gerischer, M.E. Michel-Beyerle, F. Rebentrost, H. Tributsch, Sensitization of charge injection into semiconductors with large band gap. Electrochim. Acta **13**, 1509–1515 (1968)
5. B. O'Regan, M. Grätzel, A low-cost, high-efficiency solar cell based on dye-sensitized colloidal TiO_2 films. Nature **353**, 737–740 (1991)
6. K. Kakiage, Y. Aoyama, T. Yano, K. Oya, J.-I. Fujisawa, M. Hanaya, Highly-efficient dye-sensitized solar cells with collaborative sensitization by silyl-anchor and carboxy-anchor dyes. Chem. Commun. **51**, 15894–15897 (2015)
7. J. Wu, Z. Lan, J. Lin, M. Huang, Y. Huang, L. Fan, G. Luo, Y. Lin, Y. Xie, Y. Wei, Counter electrodes in dye-sensitized solar cells. Chem. Soc. Rev. **46**, 5975–6023 (2017)
8. A. Hagfeldt, M. Graetzel, Light-induced redox reactions in nanocrystalline systems. Chem. Rev. **95**, 49–68 (1995)
9. M.-E. Ragoussi, T. Torres, New generation solar cells: concepts, trends and perspectives. Chem. Commun. **51**, 3957–3972 (2015)
10. A. Hagfeldt, G. Boschloo, L. Sun, L. Kloo, H. Pettersson, Dye-sensitized solar cells. Chem. Rev. **110**, 6595–6663 (2010)
11. M.-E. Yeoh, K.-Y. Chan, Recent advances in photo-anode for dye-sensitized solar cells: a review. Int. J. Energy Res. **41**, 2446–2467 (2017)

12. D. Sengupta, P. Das, B. Mondal, K. Mukherjee, Effects of doping, morphology and film-thickness of photo-anode materials for dye sensitized solar cell application—a review. Renew. Sustain. Energy Rev. **60**, 356–376 (2016)
13. C.-J. Lin, W.-Y. Yu, S.-H. Chien, Transparent electrodes of ordered opened-end TiO_2-nanotube arrays for highly efficient dye-sensitized solar cells. J. Mater. Chem. **20**, 1073–1077 (2010)
14. M. Lv, D. Zheng, M. Ye, J. Xiao, W. Guo, Y. Lai, L. Sun, C. Lin, J. Zuo, Optimized porous rutile TiO_2 nanorod arrays for enhancing the efficiency of dye-sensitized solar cells. Energy Environ. Sci. **6**, 1615–1622 (2013)
15. W.-Q. Wu, Y.-F. Xu, C.-Y. Su, D.-B. Kuang, Ultra-long anatase TiO_2 nanowire arrays with multi-layered configuration on FTO glass for high-efficiency dye-sensitized solar cells. Energy Environ. Sci. **7**, 644–649 (2014)
16. D. Hwang, H. Lee, S.-Y. Jang, S.M. Jo, D. Kim, Y. Seo, D.Y. Kim, Electrospray preparation of hierarchically-structured mesoporous TiO_2 spheres for use in highly efficient dye-sensitized solar cells. ACS Appl. Mater. Interfaces **3**, 2719–2725 (2011)
17. F. Sauvage, D. Chen, P. Comte, F. Huang, L.-P. Heiniger, Y.-B. Cheng, R.A. Caruso, M. Graetzel, Dye-sensitized solar cells employing a single film of mesoporous TiO_2 beads achieve power conversion efficiencies over 10%. ACS Nano **4**, 4420–4425 (2010)
18. S.H. Hwang, J. Yun, J. Jang, Multi-shell porous TiO_2 hollow nanoparticles for enhanced light harvesting in dye-sensitized solar cells. Adv. Func. Mater. **24**, 7619–7626 (2014)
19. M. Saito, S. Fujihara, Large photocurrent generation in dye-sensitized ZnO solar cells. Energy Environ. Sci. **1**, 280–283 (2008)
20. N. Memarian, I. Concina, A. Braga, S.M. Rozati, A. Vomiero, G. Sberveglieri, Hierarchically assembled ZnO nanocrystallites for high-efficiency dye-sensitized solar cells. Angew. Chem. Int. Ed. **123**, 12529–12533 (2011)
21. S. Ameen, M.S. Akhtar, Y.S. Kim, O.B. Yang, H.-S. Shin, Influence of seed layer treatment on low temperature grown ZnO nanotubes: performances in dye sensitized solar cells. Electrochim. Acta **56**, 1111–1116 (2011)
22. A. Rang Arao, V. Dutta, Achievement of 4.7% conversion efficiency in ZnO dye-sensitized solar cells fabricated by spray deposition using hydrothermally synthesized nanoparticles. Nanotechnology **19**, 445712 (2008)
23. C. Xu, J. Wu, U.V. Desai, D. Gao, Multilayer assembly of nanowire arrays for dye-sensitized solar cells. J. Am. Chem. Soc. **133**, 8122–8125 (2011)
24. Z. Dong, X. Lai, J.E. Halpert, N. Yang, L. Yi, J. Zhai, D. Wang, Z. Tang, L. Jiang, Accurate control of multishelled ZnO hollow microspheres for dye-sensitized solar cells with high efficiency. Adv. Mater. **24**, 1046–1049 (2012)
25. C.-Y. Lin, Y.-H. Lai, H.-W. Chen, J.-G. Chen, C.-W. Kung, R. Vittal, K.-C. Ho, Highly efficient dye-sensitized solar cell with a ZnO nanosheet-based photoanode. Energy Environ. Sci. **4**, 3448–3455 (2011)
26. X. Liu, J. Fang, Y. Liu, T.J.F.O.M.S. Lin, Progress in nanostructured photoanodes for dye-sensitized solar cells. Front. Mater. Sci. **10**, 225–237 (2016)
27. J. Zhang, P. Zhou, J. Liu, J. Yu, New understanding of the difference of photocatalytic activity among anatase, rutile and brookite TiO_2. Phys. Chem. Chem. Phys. **16**, 20382–20386 (2014)
28. N.G. Park, J. van de Lagemaat, A.J. Frank, Comparison of dye-sensitized rutile- and anatase-based TiO_2 solar cells. J. Phys. Chem. B **104**, 8989–8994 (2000)
29. D.V. Bavykin, J.M. Friedrich, F.C. Walsh, Protonated titanates and TiO_2 nanostructured materials: synthesis, properties, and applications. Adv. Mater. **18**, 2807–2824 (2006)
30. K. Park, Q. Zhang, D. Myers, G. Cao, Charge transport properties in TiO_2 network with different particle sizes for dye sensitized solar cells. ACS Appl. Mater. Interfaces **5**, 1044–1052 (2013)
31. P. Tahay, M. Babapour Gol Afshani, A. Alavi, Z. Parsa, N. Safari, Interrelationship between TiO_2 nanoparticle size and kind/size of dyes in the mechanism and conversion efficiency of dye sensitized solar cells. Phys. Chem. Chem. Phys. **19**, 11187–11196 (2017)
32. W. Yang, J. Li, Y. Wang, F. Zhu, W. Shi, F. Wan, D. Xu, A facile synthesis of anatase TiO_2 nanosheets-based hierarchical spheres with over 90% 001 facets for dye-sensitized solar cells. Chem. Commun. **47**, 1809–1811 (2011)

33. X. Wu, Z. Chen, G.Q. Lu, L. Wang, Nanosized anatase TiO_2 single crystals with tunable exposed (001) facets for enhanced energy conversion efficiency of dye-sensitized solar cells. Adv. Func. Mater. **21**, 4167–4172 (2011)
34. D.K. Roh, W.S. Chi, H. Jeon, S.J. Kim, J.H. Kim, High efficiency solid-state dye-sensitized solar cells assembled with hierarchical anatase pine tree-like TiO_2 nanotubes. Adv. Func. Mater. **24**, 379–386 (2014)
35. L.-L. Li, C.-Y. Tsai, H.-P. Wu, C.-C. Chen, E.W.-G. Diau, Fabrication of long TiO_2 nanotube arrays in a short time using a hybrid anodic method for highly efficient dye-sensitized solar cells. J. Mater. Chem. **20**, 2753–2758 (2010)
36. S. Ito, N.-L.C. Ha, G. Rothenberger, P. Liska, P. Comte, S.M. Zakeeruddin, P. Péchy, M.K. Nazeeruddin, M. Grätzel, High-efficiency (7.2%) flexible dye-sensitized solar cells with Ti-metal substrate for nanocrystalline-TiO_2 photoanode. Chem. Commun. **38**, 4004–4006 (2006)
37. J. Zhang, Q. Li, S. Li, Y. Wang, C. Ye, P. Ruterana, H. Wang, An efficient photoanode consisting of TiO_2 nanoparticle-filled TiO_2 nanotube arrays for dye sensitized solar cells. J. Power Sources **268**, 941–949 (2014)
38. A.I. Hochbaum, P. Yang, Semiconductor nanowires for energy conversion. Chem. Rev. **110**, 527–546 (2010)
39. H.-Y. Chen, T.-L. Zhang, J. Fan, D.-B. Kuang, C.-Y. Su, Electrospun hierarchical TiO_2 nanorods with high porosity for efficient dye-sensitized solar cells. ACS Appl. Mater. Interfaces **5**, 9205–9211 (2013)
40. K. Fan, J. Yu, W. Ho, Improving photoanodes to obtain highly efficient dye-sensitized solar cells: a brief review. Mater. Horiz. **4**, 319–344 (2017)
41. Y.-C. Park, Y.-J. Chang, B.-G. Kum, E.-H. Kong, J.Y. Son, Y.S. Kwon, T. Park, H.M. Jang, Size-tunable mesoporous spherical TiO_2 as a scattering overlayer in high-performance dye-sensitized solar cells. J. Mater. Chem. **21**, 9582–9586 (2011)
42. S. Mathew, A. Yella, P. Gao, R. Humphry-Baker, B.F.E. Curchod, N. Ashari-Astani, I. Tavernelli, U. Rothlisberger, M.K. Nazeeruddin, M. Grätzel, Dye-sensitized solar cells with 13% efficiency achieved through the molecular engineering of porphyrin sensitizers. Nat. Chem. **6**, 242 (2014)
43. J. Yu, Q. Li, Z. Shu, Dye-sensitized solar cells based on double-layered TiO_2 composite films and enhanced photovoltaic performance. Electrochim. Acta **56**, 6293–6298 (2011)
44. K. Nakayama, T. Kubo, Y. Nishikitani, Electrophoretically deposited TiO_2 nanotube light-scattering layers of dye-sensitized solar cells. Jpn. J. Appl. Phys. **47**, 6610–6614 (2008)
45. G. Wang, W. Xiao, J. Yu, High-efficiency dye-sensitized solar cells based on electrospun TiO_2 multi-layered composite film photoanodes. Energy **86**, 196–203 (2015)
46. L. Yang, W.W.-F. Leung, Application of a bilayer TiO_2 nanofiber photoanode for optimization of dye-sensitized solar cells. Adv. Mater. **23**, 4559–4562 (2011)
47. A.M. Bakhshayesh, M.R. Mohammadi, D.J. Fray, Controlling electron transport rate and recombination process of TiO_2 dye-sensitized solar cells by design of double-layer films with different arrangement modes. Electrochim. Acta **78**, 384–391 (2012)
48. W.-Q. Wu, Y.-F. Xu, H.-S. Rao, C.-Y. Su, D.-B. Kuang, Multistack Integration of three-dimensional hyperbranched anatase titania architectures for high-efficiency dye-sensitized solar cells. J. Am. Chem. Soc. **136**, 6437–6445 (2014)
49. W. Song, Y. Gong, J. Tian, G. Cao, H. Zhao, C. Sun, Novel photoanode for dye-sensitized solar cells with enhanced light-harvesting and electron-collection efficiency. ACS Appl. Mater. Interfaces **8**, 13418–13425 (2016)
50. Z. Dong, H. Ren, C.M. Hessel, J. Wang, R. Yu, Q. Jin, M. Yang, Z. Hu, Y. Chen, Z. Tang, H. Zhao, D. Wang, Quintuple-shelled SnO_2 hollow microspheres with superior light scattering for high-performance dye-sensitized solar cells. Adv. Mater. **26**, 905–909 (2014)
51. J.M. Miranda-Muñoz, S. Carretero-Palacios, A. Jiménez-Solano, Y. Li, G. Lozano, H. Míguez, Efficient bifacial dye-sensitized solar cells through disorder by design. J. Mater. Chem. A **4**, 1953–1961 (2016)
52. B. Tan, Y. Wu, Dye-sensitized solar cells based on anatase TiO_2 nanoparticle/nanowire composites. J. Phys. Chem. B **110**, 15932–15938 (2006)

53. W. Wang, H. Zhang, R. Wang, M. Feng, Y. Chen, Design of a TiO_2 nanosheet/nanoparticle gradient film photoanode and its improved performance for dye-sensitized solar cells. Nanoscale **6**, 2390–2396 (2014)
54. Q. Zhang, D. Myers, J. Lan, S.A. Jenekhe, G. Cao, Applications of light scattering in dye-sensitized solar cells. Phys. Chem. Chem. Phys. **14**, 14982–14998 (2012)
55. Q. Zhang, C.S. Dandeneau, X. Zhou, G. Cao, ZnO nanostructures for dye-sensitized solar cells. Adv. Mater. **21**, 4087–4108 (2009)
56. C.C. Raj, R. Prasanth, A critical review of recent developments in nanomaterials for photoelectrodes in dye sensitized solar cells. J. Power Sources **317**, 120–132 (2016)
57. N. Sakai, R. Usui, T.N. Murakami, Optimum particle size of ZnO for dye-sensitized solar cells. Chem. Lett. **42**, 810–812 (2013)
58. J. Chang, R. Ahmed, H. Wang, H. Liu, R. Li, P. Wang, E.R. Waclawik, ZnO nanocones with high-index $\{10\bar{1}1\}$ facets for enhanced energy conversion efficiency of dye-sensitized solar cells. J. Phys. Chem. C **117**, 13836–13844 (2013)
59. M.S. Akhtar, M.A. Khan, M.S. Jeon, O.B. Yang, Controlled synthesis of various ZnO nanostructured materials by capping agents-assisted hydrothermal method for dye-sensitized solar cells. Electrochim. Acta **53**, 7869–7874 (2008)
60. M. McCune, W. Zhang, Y. Deng, High efficiency dye-sensitized solar cells based on three-dimensional multilayered ZnO nanowire arrays with "caterpillar-like" structure. Nano Lett. **12**, 3656–3662 (2012)
61. J. Han, F. Fan, C. Xu, S. Lin, M. Wei, X. Duan, Z.L. Wang, ZnO nanotube-based dye-sensitized solar cell and its application in self-powered devices. Nanotechnology **21**, 405203 (2010)
62. D. Chu, Y. Masuda, T. Ohji, K. Kato, Formation and photocatalytic application of ZnO nanotubes using aqueous solution. Langmuir **26**, 2811–2815 (2010)
63. Q. Huang, L. Fang, X. Chen, M. Saleem, Effect of polyethyleneimine on the growth of ZnO nanorod arrays and their application in dye-sensitized solar cells. J. Alloy. Compd. **509**, 9456–9459 (2011)
64. S.H. Ko, D. Lee, H.W. Kang, K.H. Nam, J.Y. Yeo, S.J. Hong, C.P. Grigoropoulos, H.J. Sung, Nanoforest of hydrothermally grown hierarchical ZnO nanowires for a high efficiency dye-sensitized solar cell. Nano Lett. **11**, 666–671 (2011)
65. J. Fan, Y. Hao, C. Munuera, M. García-Hernández, F. Güell, E.M.J. Johansson, G. Boschloo, A. Hagfeldt, A. Cabot, Influence of the annealing atmosphere on the performance of ZnO nanowire dye-sensitized solar cells. J. Phys. Chem. C **117**, 16349–16356 (2013)
66. D. Barpuzary, A.S. Patra, J.V. Vaghasiya, B.G. Solanki, S.S. Soni, M. Qureshi, Highly efficient one-dimensional ZnO nanowire-based dye-sensitized solar cell using a metal-free, $D-\pi-A$-type, carbazole derivative with more than 5% power conversion. ACS Appl. Mater. Interfaces **6**, 12629–12639 (2014)
67. M. Navaneethan, J. Archana, M. Arivanandhan, Y. Hayakawa, Functional properties of amine-passivated ZnO nanostructures and dye-sensitized solar cell characteristics. Chem. Eng. J. **213**, 70–77 (2012)
68. C.Y. Jiang, X.W. Sun, G.Q. Lo, D.L. Kwong, J.X. Wang, Improved dye-sensitized solar cells with a ZnO-nanoflower photoanode. Appl. Phys. Lett. **90**, 263501 (2007)
69. W.-C. Chang, L.-Y. Lin, W.-C. Yu, Bifunctional zinc oxide nanoburger aggregates as the dye-adsorption and light-scattering layer for dye-sensitized solar cells. Electrochim. Acta **169**, 456–461 (2015)
70. Y. Shi, C. Zhu, L. Wang, W. Li, K.K. Fung, N. Wang, Asymmetric ZnO panel-like hierarchical architectures with highly interconnected pathways for free-electron transport and photovoltaic improvements. Chem. Eur. J. **19**, 282–287 (2013)
71. Y.-Z. Zheng, H. Ding, Y. Liu, X. Tao, G. Cao, J.-F. Chen, In situ hydrothermal growth of hierarchical ZnO nanourchin for high-efficiency dye-sensitized solar cells. J. Power Sources **254**, 153–160 (2014)
72. A. Sacco, A. Lamberti, R. Gazia, S. Bianco, D. Manfredi, N. Shahzad, F. Cappelluti, S. Ma, E. Tresso, High efficiency dye-sensitized solar cells exploiting sponge-like ZnO nanostructures. Phys. Chem. Chem. Phys. **14**, 16203–16208 (2012)

73. Q. Zhang, T.P. Chou, B. Russo, S.A. Jenekhe, G. Cao, Polydisperse aggregates of ZnO nanocrystallites: a method for energy-conversion-efficiency enhancement in dye-sensitized solar cells. Adv. Func. Mater. **18**, 1654–1660 (2008)
74. D. Wu, Z. Gao, F. Xu, J. Chang, W. Tao, J. He, S. Gao, K. Jiang, Hierarchical ZnO aggregates assembled by orderly aligned nanorods for dye-sensitized solar cells. CrystEngComm **15**, 1210–1217 (2013)
75. C.-W. Kung, H.-W. Chen, C.-Y. Lin, Y.-H. Lai, R. Vittal, K.-C. Ho, Electrochemical synthesis of a double-layer film of ZnO nanosheets/nanoparticles and its application for dye-sensitized solar cells. Prog. Photovolt. **22**, 440–451 (2014)
76. J. Xu, K. Fan, W. Shi, K. Li, T. Peng, Application of ZnO micro-flowers as scattering layer for ZnO-based dye-sensitized solar cells with enhanced conversion efficiency. Sol. Energy **101**, 150–159 (2014)
77. H. Wang, B. Li, J. Gao, M. Tang, H. Feng, J. Li, L. Guo, SnO_2 hollow nanospheres enclosed by single crystalline nanoparticles for highly efficient dye-sensitized solar cells. CrystEngComm **14**, 5177–5181 (2012)
78. Asdim, K. Manseki, T. Sugiura, T. Yoshida, Microwave synthesis of size-controllable SnO_2 nanocrystals for dye-sensitized solar cells. New J. Chem. **38**, 598–603 (2014)
79. A. Le Viet, R. Jose, M.V. Reddy, B.V.R. Chowdari, S. Ramakrishna, Nb_2O_5 photoelectrodes for dye-sensitized solar cells: choice of the polymorph. J. Phys. Chem. C **114**, 21795–21800 (2010)
80. X. Jin, C. Liu, J. Xu, Q. Wang, D. Chen, Size-controlled synthesis of mesoporous Nb_2O_5 microspheres for dye sensitized solar cells. RSC Adv. **4**, 35546–35553 (2014)
81. H. Niu, S. Zhang, Q. Ma, S. Qin, L. Wan, J. Xu, S. Miao, Dye-sensitized solar cells based on flower-shaped α-Fe_2O_3 as a photoanode and reduced graphene oxide–polyaniline composite as a counter electrode. RSC Adv. **3**, 17228–17235 (2013)
82. I. Hod, M. Shalom, Z. Tachan, S. Rühle, A. Zaban, $SrTiO_3$ recombination-inhibiting barrier layer for type II dye-sensitized solar cells. J. Phys. Chem. C **114**, 10015–10018 (2010)
83. B. Tan, E. Toman, Y. Li, Y. Wu, Zinc stannate (Zn_2SnO_4) dye-sensitized solar cells. J. Am. Chem. Soc. **129**, 4162–4163 (2007)
84. D. Hwang, J.-S. Jin, H. Lee, H.-J. Kim, H. Chung, D.Y. Kim, S.-Y. Jang, D. Kim, Hierarchically structured Zn_2SnO_4 nanobeads for high-efficiency dye-sensitized solar cells. Sci. Rep. **4**, 7353 (2014)
85. P. Reiss, M. Protière, L. Li, Core/shell semiconductor nanocrystals. Small **5**, 154–168 (2009)
86. N.R.E. Laboratory, Best Research-Cell Efficiency Chart (2019)
87. M.R. Kim, D. Ma, Quantum-dot-based solar cells: recent advances, strategies, and challenges. J. Phys. Chem. Lett. **6**, 85–99 (2015)
88. Y. Bai, I. Mora-Seró, F. De Angelis, J. Bisquert, P. Wang, Titanium dioxide nanomaterials for photovoltaic applications. Chem. Rev. **114**, 10095–10130 (2014)
89. W. Wang, W. Feng, J. Du, W. Xue, L. Zhang, L. Zhao, Y. Li, X. Zhong, Cosensitized quantum dot solar cells with conversion efficiency over 12%. Adv. Mater. **30**, 1705746 (2018)
90. Q. Zhang, G. Chen, Y. Yang, X. Shen, Y. Zhang, C. Li, R. Yu, Y. Luo, D. Li, Q. Meng, Toward highly efficient CdS/CdSe quantum dots-sensitized solar cells incorporating ordered photoanodes on transparent conductive substrates. Phys. Chem. Chem. Phys. **14**, 6479–6486 (2012)
91. W. Zhang, X. Zeng, H. Wang, R. Fang, Y. Xu, Y. Zhang, W. Chen, High-yield synthesis of "oriented attachment" TiO_2 nanorods as superior building blocks of photoanodes in quantum dot sensitized solar cells. RSC Adv. **6**, 33713–33722 (2016)
92. H.-S. Rao, W.-Q. Wu, Y. Liu, Y.-F. Xu, B.-X. Chen, H.-Y. Chen, D.-B. Kuang, C.-Y. Su, CdS/CdSe co-sensitized vertically aligned anatase TiO_2 nanowire arrays for efficient solar cells. Nano Energy **8**, 1–8 (2014)
93. H. Zhou, L. Li, D. Jiang, Y. Lu, K. Pan, Anatase TiO_2 nanosheets with exposed highly reactive (001) facets as an efficient photoanode for quantum dot-sensitized solar cells. RSC Adv. **6**, 67968–67975 (2016)

94. C. Li, L. Yang, J. Xiao, Y.-C. Wu, M. Søndergaard, Y. Luo, D. Li, Q. Meng, B.B. Iversen, ZnO nanoparticle based highly efficient CdS/CdSe quantum dot-sensitized solar cells. Phys. Chem. Chem. Phys. **15**, 8710–8715 (2013)
95. D. Wu, X. Wang, K. Cao, Y. An, X. Song, N. Liu, F. Xu, Z. Gao, K. Jiang, ZnO nanorods with tunable aspect ratios deriving from oriented-attachment for enhanced performance in quantum-dot sensitized solar cells. Electrochim. Acta **231**, 1–12 (2017)
96. M. Seol, H. Kim, Y. Tak, K. Yong, Novel nanowire array based highly efficient quantum dot sensitized solar cell. Chem. Commun. **46**, 5521–5523 (2010)
97. J. Xu, X. Yang, H. Wang, X. Chen, C. Luan, Z. Xu, Z. Lu, V.A.L. Roy, W. Zhang, C.-S. Lee, Arrays of ZnO/$Zn_xCd_{1-x}Se$ nanocables: band gap engineering and photovoltaic applications. Nano Lett. **11**, 4138–4143 (2011)
98. T.R. Chetia, M.S. Ansari, M. Qureshi, Ethyl cellulose and cetrimonium bromide assisted synthesis of mesoporous, hexagon shaped ZnO nanodisks with exposed $\pm\{0001\}$ polar facets for enhanced photovoltaic performance in quantum dot sensitized solar cells. ACS Appl. Mater. Interfaces **7**, 13266–13279 (2015)
99. K. Yan, L. Zhang, J. Qiu, Y. Qiu, Z. Zhu, J. Wang, S. Yang, A quasi-quantum well sensitized solar cell with accelerated charge separation and collection. J. Am. Chem. Soc. **135**, 9531–9539 (2013)
100. J. Tian, G. Cao, Control of nanostructures and interfaces of metal oxide semiconductors for quantum-dots-sensitized solar cells. J. Phys. Chem. Lett. **6**, 1859–1869 (2015)
101. M. Grätzel, Photoelectrochemical cells. Nature **414**, 338–344 (2001)
102. S. Jiao, J. Du, Z. Du, D. Long, W. Jiang, Z. Pan, Y. Li, X. Zhong, Nitrogen-doped mesoporous carbons as counter electrodes in quantum dot sensitized solar cells with a conversion efficiency exceeding 12%. J. Phys. Chem. Lett. **8**, 559–564 (2017)
103. J. Du, Z. Du, J.-S. Hu, Z. Pan, Q. Shen, J. Sun, D. Long, H. Dong, L. Sun, X. Zhong, L.-J. Wan, Zn–Cu–In–Se quantum dot solar cells with a certified power conversion efficiency of 11.6%. J. Am. Chem. Soc. **138**, 4201–4209 (2016)
104. Z. Du, Z. Pan, F. Fabregat-Santiago, K. Zhao, D. Long, H. Zhang, Y. Zhao, X. Zhong, J.-S. Yu, J. Bisquert, Carbon counter-electrode-based quantum-dot-sensitized solar cells with certified efficiency exceeding 11%. J. Phys. Chem. Lett. **7**, 3103–3111 (2016)
105. L. Etgar, W. Zhang, S. Gabriel, S.G. Hickey, M.K. Nazeeruddin, A. Eychmüller, B. Liu, M. Grätzel, High efficiency quantum dot heterojunction solar cell using anatase (001) TiO_2 nanosheets. Adv. Mater. **24**, 2202–2206 (2012)
106. J. Jia, L. Mu, Y. Lin, X. Zhou, Rutile versus anatase for quantum dot sensitized solar cell. Electrochim. Acta **266**, 103–109 (2018)
107. A. Kongkanand, K. Tvrdy, K. Takechi, M. Kuno, P.V. Kamat, Quantum dot solar cells. Tuning photoresponse through size and shape control of CdSe–TiO_2 architecture. J. Am. Chem. Soc. **130**, 4007–4015 (2008)
108. J. Zhang, J.H. Bang, C. Tang, P.V. Kamat, Tailored TiO_2–$SrTiO_3$ heterostructure nanotube arrays for improved photoelectrochemical performance. ACS Nano **4**, 387–395 (2010)
109. H. Han, P. Sudhagar, T. Song, Y. Jeon, I. Mora-Seró, F. Fabregat-Santiago, J. Bisquert, Y.S. Kang, U. Paik, Three dimensional-TiO_2 nanotube array photoanode architectures assembled on a thin hollow nanofibrous backbone and their performance in quantum dot-sensitized solar cells. Chem. Commun. **49**, 2810–2812 (2013)
110. B. Liu, Y. Sun, X. Wang, L. Zhang, D. Wang, Z. Fu, Y. Lin, T. Xie, Branched hierarchical photoanode of anatase TiO_2 nanotubes on rutile TiO_2 nanorod arrays for efficient quantum dot-sensitized solar cells. J. Mater. Chem. A **3**, 4445–4452 (2015)
111. L. Tao, Y. Xiong, H. Liu, W. Shen, High performance PbS quantum dot sensitized solar cells via electric field assisted in situ chemical deposition on modulated TiO_2 nanotube arrays. Nanoscale **6**, 931–938 (2014)
112. J. Zhang, C. Tang, J.H. Bang, CdS/TiO_2–$SrTiO_3$ heterostructure nanotube arrays for improved solar energy conversion efficiency. Electrochem. Commun. **12**, 1124–1128 (2010)
113. C. Dong, X. Li, J. Qi, First-principles investigation on electronic properties of quantum dot-sensitized solar cells based on anatase TiO_2 nanotubes. J. Phys. Chem. C **115**, 20307–20315 (2011)

114. W. Lee, S.H. Kang, J.-Y. Kim, G.B. Kolekar, Y.-E. Sung, S.-H. Han, TiO_2 nanotubes with a ZnO thin energy barrier for improved current efficiency of CdSe quantum-dot-sensitized solar cells. Nanotechnology **20**, 335706 (2009)
115. H. Huang, L. Pan, C.K. Lim, H. Gong, J. Guo, M.S. Tse, O.K. Tan, Hydrothermal growth of TiO_2 nanorod arrays and in situ conversion to nanotube arrays for highly efficient quantum dot-sensitized solar cells. Small **9**, 3153–3160 (2013)
116. L. Yu, Z. Li, Y. Liu, F. Cheng, S. Sun, Enhanced photoelectrochemical performance of CdSe/Mn-CdS/TiO_2 nanorod arrays solar cell. Appl. Surf. Sci. **309**, 255–262 (2014)
117. Z. Zhang, C. Shi, J. Chen, G. Xiao, L. Li, Combination of short-length TiO_2 nanorod arrays and compact PbS quantum-dot thin films for efficient solid-state quantum-dot-sensitized solar cells. Appl. Surf. Sci. **410**, 8–13 (2017)
118. Y. Chen, Q. Tao, W. Fu, H. Yang, X. Zhou, S. Su, D. Ding, Y. Mu, X. Li, M. Li, Enhanced photoelectric performance of PbS/CdS quantum dot co-sensitized solar cells via hydrogenated TiO_2 nanorod arrays. Chem. Commun. **50**, 9509–9512 (2014)
119. C. Wang, Z. Jiang, L. Wei, Y. Chen, J. Jiao, M. Eastman, H. Liu, Photosensitization of TiO_2 nanorods with CdS quantum dots for photovoltaic applications: a wet-chemical approach. Nano Energy **1**, 440–447 (2012)
120. J. Wan, R. Liu, Y. Tong, S. Chen, Y. Hu, B. Wang, Y. Xu, H. Wang, Hydrothermal etching treatment to rutile TiO_2 nanorod arrays for improving the efficiency of CdS-sensitized TiO_2 solar cells. Nanoscale Res. Lett. **11**, 12 (2016)
121. L. Yu, X. Ren, Z. Yang, Y. Han, Z. Li, The preparation and assembly of CdS_xSe_{1-x} alloyed quantum dots on TiO_2 nanowire arrays for quantum dot-sensitized solar cells. J. Mater. Sci.: Mater. Eelctron. **27**, 7150–7160 (2016)
122. Y.-F. Xu, W.-Q. Wu, H.-S. Rao, H.-Y. Chen, D.-B. Kuang, C.-Y. Su, CdS/CdSe co-sensitized TiO_2 nanowire-coated hollow Spheres exceeding 6% photovoltaic performance. Nano Energy **11**, 621–630 (2015)
123. Z. Peng, Y. Liu, Y. Zhao, K. Chen, Y. Cheng, W. Chen, Incorporation of the TiO_2 nanowire arrays photoanode and Cu2S nanorod arrays counter electrode on the photovoltaic performance of quantum dot sensitized solar cells. Electrochim. Acta **135**, 276–283 (2014)
124. D.R. Baker, P.V. Kamat, Photosensitization of TiO_2 nanostructures with CdS quantum dots: particulate versus tubular support architectures. Adv. Func. Mater. **19**, 805–811 (2009)
125. H.-L. Feng, W.-Q. Wu, H.-S. Rao, L.-B. Li, D.-B. Kuang, C.-Y. Su, Three-dimensional hyperbranched TiO_2/ZnO heterostructured arrays for efficient quantum dot-sensitized solar cells. J. Mater. Chem. A **3**, 14826–14832 (2015)
126. H. Wang, M. Miyauchi, Y. Ishikawa, A. Pyatenko, N. Koshizaki, Y. Li, L. Li, X. Li, Y. Bando, D. Golberg, Single-crystalline rutile TiO_2 hollow spheres: room-temperature synthesis, tailored visible-light-extinction, and effective scattering layer for quantum dot-sensitized solar cells. J. Am. Chem. Soc. **133**, 19102–19109 (2011)
127. R. Zhou, Q. Zhang, E. Uchaker, L. Yang, N. Yin, Y. Chen, M. Yin, G. Cao, Photoanodes with mesoporous TiO_2 beads and nanoparticles for enhanced performance of CdS/CdSe quantum dot co-sensitized solar cells. Electrochim. Acta **135**, 284–292 (2014)
128. Q. Zhang, X. Guo, X. Huang, S. Huang, D. Li, Y. Luo, Q. Shen, T. Toyoda, Q. Meng, Highly efficient CdS/CdSe-sensitized solar cells controlled by the structural properties of compact porous TiO_2 photoelectrodes. Phys. Chem. Chem. Phys. **13**, 4659–4667 (2011)
129. X. Xu, G. Jiang, Q. Wan, J. Shi, G. Xu, L. Miao, Mesoporous titania hollow spheres applied as scattering layers in quantum dots sensitized solar cells. Mater. Chem. Phys. **136**, 1060–1066 (2012)
130. H. Hu, H. Shen, C. Cui, D. Liang, P. Li, S. Xu, W. Tang, Preparation and photoelectrochemical properties of TiO_2 hollow spheres embedded TiO_2/CdS photoanodes for quantum-dot-sensitized solar cells. J. Alloy. Compd. **560**, 1–5 (2013)
131. M. Marandi, E. Rahmani, F. Ahangarani Farahani, Optimization of the photoanode of CdS quantum dot-sensitized solar cells using light-scattering TiO_2 hollow spheres. J. Electron. Mater. **46**, 6769–6783 (2017)

132. J. Xu, Z. Chen, J.A. Zapien, C.-S. Lee, W. Zhang, Surface engineering of ZnO nanostructures for semiconductor-sensitized solar cells. Adv. Mater. **26**, 5337–5367 (2014)
133. J. Tian, Q. Zhang, E. Uchaker, R. Gao, X. Qu, S. Zhang, G. Cao, Architectured ZnO photoelectrode for high efficiency quantum dot sensitized solar cells. Energy Environ. Sci. **6**, 3542–3547 (2013)
134. F.S. Ghoreishi, V. Ahmadi, M. Samadpour, Improved performance of CdS/CdSe quantum dots sensitized solar cell by incorporation of ZnO nanoparticles/reduced graphene oxide nanocomposite as photoelectrode. J. Power Sources **271**, 195–202 (2014)
135. D. Karageorgopoulos, E. Stathatos, E. Vitoratos, Thin ZnO nanocrystalline films for efficient quasi-solid state electrolyte quantum dot sensitized solar cells. J. Power Sources **219**, 9–15 (2012)
136. J. Tian, Q. Zhang, E. Uchaker, Z. Liang, R. Gao, X. Qu, S. Zhang, G. Cao, Constructing ZnO nanorod array photoelectrodes for highly efficient quantum dot sensitized solar cells. J. Mater. Chem. A **1**, 6770–6775 (2013)
137. K.S. Leschkies, R. Divakar, J. Basu, E. Enache-Pommer, J.E. Boercker, C.B. Carter, U.R. Kortshagen, D.J. Norris, E.S. Aydil, Photosensitization of ZnO nanowires with CdSe quantum dots for photovoltaic devices. Nano Lett. **7**, 1793–1798 (2007)
138. M. Seol, E. Ramasamy, J. Lee, K. Yong, Highly efficient and durable quantum dot sensitized ZnO nanowire solar cell using noble-metal-free counter electrode. J. Phys. Chem. C **115**, 22018–22024 (2011)
139. R. Zhang, Q.-P. Luo, H.-Y. Chen, X.-Y. Yu, D.-B. Kuang, C.-Y. Su, CdS/CdSe quantum dot shell decorated vertical ZnO nanowire arrays by spin-coating-based SILAR for photoelectrochemical cells and quantum-dot-sensitized solar cells. ChemPhysChem **13**, 1435–1439 (2012)
140. H. Kim, H. Jeong, T.K. An, C.E. Park, K. Yong, Hybrid-type quantum-dot cosensitized zno nanowire solar cell with enhanced visible-light harvesting. ACS Appl. Mater. Interfaces **5**, 268–275 (2013)
141. W. Lee, S. Kang, T. Hwang, K. Kim, H. Woo, B. Lee, J. Kim, J. Kim, B. Park, Facile conversion synthesis of densely-formed branched ZnO-nanowire arrays for quantum-dot-sensitized solar cells. Electrochim. Acta **167**, 194–200 (2015)
142. J. Xu, X. Yang, Q.-D. Yang, T.-L. Wong, S.-T. Lee, W.-J. Zhang, C.-S. Lee, Arrays of CdSe sensitized ZnO/ZnSe nanocables for efficient solar cells with high open-circuit voltage. J. Mater. Chem. **22**, 13374–13379 (2012)
143. T.R. Chetia, D. Barpuzary, M. Qureshi, Enhanced photovoltaic performance utilizing effective charge transfers and light scattering effects by the combination of mesoporous, hollow 3D-ZnO along with 1D-ZnO in CdS quantum dot sensitized solar cells. Phys. Chem. Chem. Phys. **16**, 9625–9633 (2014)
144. J. Tian, L. Lv, X. Wang, C. Fei, X. Liu, Z. Zhao, Y. Wang, G. Cao, Microsphere light-scattering layer assembled by ZnO nanosheets for the construction of high efficiency (>5%) quantum dots sensitized solar cells. J. Phys. Chem. C **118**, 16611–16617 (2014)
145. Z. Zhu, J. Qiu, K. Yan, S. Yang, Building high-efficiency CdS/CdSe-sensitized solar cells with a hierarchically branched double-layer architecture. ACS Appl. Mater. Interfaces **5**, 4000–4005 (2013)
146. L. Yu, Z. Li, Synthesis of $Zn_xCd_{1-x}Se$@ZnO hollow spheres in different sizes for quantum dots sensitized solar cells application. Nanomaterials (Basel) **9**, 132 (2019)
147. J. Xiao, Q. Huang, J. Xu, C. Li, G. Chen, Y. Luo, D. Li, Q. Meng, CdS/CdSe co-sensitized solar cells based on a new SnO_2 photoanode with a three-dimensionally interconnected ordered porous structure. J. Phys. Chem. C **118**, 4007–4015 (2014)
148. S. Greenwald, S. Rühle, M. Shalom, S. Yahav, A. Zaban, Unpredicted electron injection in CdS/CdSe quantum dot sensitized ZrO_2 solar cells. Phys. Chem. Chem. Phys. **13**, 19302–19306 (2011)
149. K. Meng, P.K. Surolia, K.R. Thampi, BaTiO3 photoelectrodes for CdS quantum dot sensitized solar cells. J. Mater. Chem. A **2**, 10231–10238 (2014)

150. A. Pimachev, U. Poudyal, V. Proshchenko, W. Wang, Y. Dahnovsky, Large enhancement in photocurrent by Mn doping in CdSe/ZTO quantum dot sensitized solar cells. Phys. Chem. Chem. Phys. **18**, 26771–26776 (2016)
151. R. Mastria, A. Rizzo, Mastering heterostructured colloidal nanocrystal properties for light-emitting diodes and solar cells. J. Mater. Chem. C **4**, 6430–6446 (2016)
152. H.J. Yun, T. Paik, B. Diroll, M.E. Edley, J.B. Baxter, C.B. Murray, Nanocrystal size-dependent efficiency of quantum dot sensitized solar cells in the strongly coupled CdSe nanocrystals/TiO_2 system. ACS Appl. Mater. Interfaces **8**, 14692–14700 (2016)
153. C. Xia, W. Wu, T. Yu, X. Xie, C. van Oversteeg, H.C. Gerritsen, C. de Mello Donega, Size-dependent band-gap and molar absorption coefficients of colloidal $CuInS_2$ quantum dots. ACS Nano **12**, 8350–8361 (2018)
154. A.M. Smith, S. Nie, Semiconductor nanocrystals: structure, properties, and band gap engineering. Acc. Chem. Res. **43**, 190–200 (2010)
155. R.T. Ross, A.J. Nozik, Efficiency of hot-carrier solar energy converters. J. Appl. Phys. **53**, 3813–3818 (1982)
156. R.D. Schaller, V.M. Agranovich, V.I. Klimov, High-efficiency carrier multiplication through direct photogeneration of multi-excitons via virtual single-exciton states. Nat. Phys. **1**, 189–194 (2005)
157. K. Tvrdy, P.A. Frantsuzov, P.V. Kamat, Photoinduced electron transfer from semiconductor quantum dots to metal oxide nanoparticles. Proc. Natl. Acad. Sci. **108**, 29 (2011)
158. Z. Peng, Y. Liu, W. Shu, K. Chen, W. Chen, Synthesis of various sized $CuInS_2$ quantum dots and their photovoltaic properties as sensitizers for TiO_2 photoanodes. Eur. J. Inorg. Chem. **2012**, 5239–5244 (2012)
159. J. Yang, J.-Y. Kim, J.H. Yu, T.-Y. Ahn, H. Lee, T.-S. Choi, Y.-W. Kim, J. Joo, M.J. Ko, T. Hyeon, Copper–indium–selenide quantum dot-sensitized solar cells. Phys. Chem. Chem. Phys. **15**, 20517–20525 (2013)
160. D.H. Jara, S.J. Yoon, K.G. Stamplecoskie, P.V. Kamat, Size-dependent photovoltaic performance of $CuInS_2$ quantum dot-sensitized solar cells. Chem. Mater. **26**, 7221–7228 (2014)
161. W.E. Buhro, V.L. Colvin, Shape matters. Nat. Mater. **2**, 138–139 (2003)
162. L.-S. Li, J. Hu, W. Yang, A.P. Alivisatos, Band gap variation of size- and shape-controlled colloidal CdSe quantum rods. Nano Lett. **1**, 349–351 (2001)
163. R.K. Chava, M. Kang, Ag2S quantum dot sensitized zinc oxide photoanodes for environment friendly photovoltaic devices. Mater. Lett. **199**, 188–191 (2017)
164. P.N. Kumar, A. Kolay, M. Deepa, S.M. Shivaprasad, A.K. Srivastava, Stability, scale-up, and performance of quantum dot solar cells with carbonate-treated titanium oxide films. ACS Appl. Mater. Interfaces **9**, 25278–25290 (2017)
165. G. Jiang, Z. Pan, Z. Ren, J. Du, C. Yang, W. Wang, X. Zhong, Poly(vinyl pyrrolidone): a superior and general additive in polysulfide electrolytes for high efficiency quantum dot sensitized solar cells. J. Mater. Chem. A **4**, 11416–11421 (2016)
166. J. Yang, X. Zhong, CdTe based quantum dot sensitized solar cells with efficiency exceeding 7% fabricated from quantum dots prepared in aqueous media. J. Mater. Chem. A **4**, 16553–16561 (2016)
167. J. Duan, Q. Tang, B. He, L. Yu, Efficient In2S3 quantum dot–sensitized solar cells: a promising power conversion efficiency of 13.0%. Electrochim. Acta **139**, 381–385 (2014)
168. S. Yang, P. Zhao, X. Zhao, L. Qu, X. Lai, InP and Sn:InP based quantum dot sensitized solar cells. J. Mater. Chem. A **3**, 21922–21929 (2015)
169. L. Hu, S. Huang, R. Patterson, J.E. Halpert, Enhanced mobility in PbS quantum dot films via PbSe quantum dot mixing for optoelectronic applications. J. Mater. Chem. A **7**, 4497–4502 (2019)
170. Y.C. Choi, D.U. Lee, J.H. Noh, E.K. Kim, S.I. Seok, Highly improved Sb2S3 sensitized-inorganic–organic heterojunction solar cells and quantification of traps by deep-level transient spectroscopy. Adv. Mater. **24**, 3587–3592 (2014)

171. A. Tubtimtae, M.-W. Lee, G.-J. Wang, Ag2Se quantum-dot sensitized solar cells for full solar spectrum light harvesting. J. Power Sources **196**, 6603–6608 (2011)
172. F. Huang, L. Zhang, Q. Zhang, J. Hou, H. Wang, H. Wang, S. Peng, J. Liu, G. Cao, High efficiency CdS/CdSe quantum dot sensitized solar cells with two ZnSe layers. ACS Appl. Mater. Interfaces **8**, 34482–34489 (2016)
173. H. Zhang, K. Cheng, Y.M. Hou, Z. Fang, Z.X. Pan, W.J. Wu, J.L. Hua, X.H. Zhong, Efficient CdSe quantum dot-sensitized solar cells prepared by a postsynthesis assembly approach. Chem. Commun. **48**, 11235–11237 (2012)
174. J. Wang, I. Mora-Seró, Z. Pan, K. Zhao, H. Zhang, Y. Feng, G. Yang, X. Zhong, J. Bisquert, Core/shell colloidal quantum dot exciplex states for the development of highly efficient quantum-dot-sensitized solar cells. J. Am. Chem. Soc. **135**, 15913–15922 (2013)
175. S. Jiao, J. Wang, Q. Shen, Y. Li, X. Zhong, Surface engineering of PbS quantum dot sensitized solar cells with a conversion efficiency exceeding 7%. J. Mater. Chem. A **4**, 7214–7221 (2016)
176. S. Jiao, Q. Shen, I. Mora-Seró, J. Wang, Z. Pan, K. Zhao, Y. Kuga, X. Zhong, J. Bisquert, Band engineering in core/shell ZnTe/CdSe for photovoltage and efficiency enhancement in exciplex quantum dot sensitized solar cells. ACS Nano **9**, 908–915 (2015)
177. F. Huang, Q. Zhang, B. Xu, J. Hou, Y. Wang, R.C. Massé, S. Peng, J. Liu, G. Cao, A comparison of ZnS and ZnSe passivation layers on CdS/CdSe co-sensitized quantum dot solar cells. J. Mater. Chem. A **4**, 14773–14780 (2016)
178. A. Sahasrabudhe, S. Bhattacharyya, Dual sensitization strategy for high-performance core/shell/quasi-shell quantum dot solar cells. Chem. Mater. **27**, 4848–4859 (2015)
179. L. Mu, C. Liu, J. Jia, X. Zhou, Y. Lin, Dual post-treatment: a strategy towards high efficiency quantum dot sensitized solar cells. J. Mater. Chem. A **1**, 8353–8357 (2013)
180. G. Wang, H. Wei, J. Shi, Y. Xu, H. Wu, Y. Luo, D. Li, Q. Meng, Significantly enhanced energy conversion efficiency of $CuInS_2$ quantum dot sensitized solar cells by controlling surface defects. Nano Energy **35**, 17–25 (2017)
181. W. Peng, J. Du, Z. Pan, N. Nakazawa, J. Sun, Z. Du, G. Shen, J. Yu, J.-S. Hu, Q. Shen, X. Zhong, Alloying strategy in Cu–In–Ga–Se quantum dots for high efficiency quantum dot sensitized solar cells. ACS Appl. Mater. Interfaces **9**, 5328–5336 (2017)
182. P.K. Santra, P.V. Kamat, Tandem-layered quantum dot solar cells: tuning the photovoltaic response with luminescent ternary cadmium chalcogenides. J. Am. Chem. Soc. **135**, 877–885 (2013)
183. B. Bai, D. Kou, W. Zhou, Z. Zhou, S. Wu, Application of quaternary Cu_2ZnSnS_4 quantum dot-sensitized solar cells based on the hydrolysis approach. Green Chem. **17**, 4377–4382 (2015)
184. B. Bai, D. Kou, W. Zhou, Z. Zhou, Q. Tian, Y. Meng, S. Wu, Quaternary Cu_2ZnSnS_4 quantum dot-sensitized solar cells: synthesis, passivation and ligand exchange. J. Power Sources **318**, 35–40 (2016)
185. L. Yue, H. Rao, J. Du, Z. Pan, J. Yu, X. Zhong, Comparative advantages of Zn–Cu–In–S alloy QDs in the construction of quantum dot-sensitized solar cells. RSC Adv. **8**, 3637–3645 (2018)
186. L. Zhang, Z. Pan, W. Wang, J. Du, Z. Ren, Q. Shen, X. Zhong, Copper deficient Zn–Cu–In–Se quantum dot sensitized solar cells for high efficiency. J. Mater. Chem. A **5**, 21442–21451 (2017)
187. E.M. Sanehira, A.R. Marshall, J.A. Christians, S.P. Harvey, P.N. Ciesielski, L.M. Wheeler, P. Schulz, L.Y. Lin, M.C. Beard, J.M. Luther, Enhanced mobility $CsPbI_3$ quantum dot arrays for record-efficiency, high-voltage photovoltaic cells. Sci. Adv. **3**, eaao4204 (2017)
188. J.-H. Im, C.-R. Lee, J.-W. Lee, S.-W. Park, N.-G. Park, 6.5% efficient perovskite quantum-dot-sensitized solar cell. Nanoscale **3**, 4088–4093 (2011)
189. J. Xue, J.-W. Lee, Z. Dai, R. Wang, S. Nuryyeva, M.E. Liao, S.-Y. Chang, L. Meng, D. Meng, P. Sun, O. Lin, M.S. Goorsky, Y. Yang, Surface ligand management for stable $FAPbI_3$ perovskite quantum dot solar cells. Joule **2**, 1866–1878 (2018)
190. A. Tubtimtae, K.-L. Wu, H.-Y. Tung, M.-W. Lee, G.J. Wang, Ag_2S quantum dot-sensitized solar cells. Electrochem. Commun. **12**, 1158–1160 (2010)

191. W.-T. Sun, Y. Yu, H.-Y. Pan, X.-F. Gao, Q. Chen, L.-M. Peng, CdS quantum dots sensitized TiO_2 nanotube-array photoelectrodes. J. Am. Chem. Soc. **130**, 1124–1125 (2008)
192. Y.-L. Lee, B.-M. Huang, H.-T. Chien, Highly efficient CdSe-sensitized TiO_2 photoelectrode for quantum-dot-sensitized solar cell applications. Chem. Mater. **20**, 6903–6905 (2008)
193. M.A. Becker, J.G. Radich, B.A. Bunker, P.V. Kamat, How does a SILAR CdSe film grow? Tuning the deposition steps to suppress interfacial charge recombination in solar cells. J. Phys. Chem. Lett. **5**, 1575–1582 (2014)
194. Z. Yang, H.-T. Chang, CdHgTe and CdTe quantum dot solar cells displaying an energy conversion efficiency exceeding 2%. Sol. Energy Mater. Sol. Cells **94**, 2046–2051 (2010)
195. B.S. Kwak, Y. Im, M. Kang, Design of a free-ruthenium In2S3 crystalline photosensitized solar cell. Int. J. Photoenergy **2014**, 8 (2014)
196. J.-W. Lee, D.-Y. Son, T.K. Ahn, H.-W. Shin, I.Y. Kim, S.-J. Hwang, M.J. Ko, S. Sul, H. Han, N.-G. Park, Quantum-dot-sensitized solar cell with unprecedentedly high photocurrent. Sci. Rep. **3**, 1050 (2013)
197. L.-Y. Chang, R.R. Lunt, P.R. Brown, V. Bulović, M.G. Bawendi, Low-temperature solution-processed solar cells based on PbS colloidal quantum dot/CdS heterojunctions. Nano Lett. **13**, 994–999 (2013)
198. V. González-Pedro, C. Sima, G. Marzari, P.P. Boix, S. Giménez, Q. Shen, T. Dittrich, I. Mora-Seró, High performance PbS quantum dot sensitized solar cells exceeding 4% efficiency: the role of metal precursors in the electron injection and charge separation. Phys. Chem. Chem. Phys. **15**, 13835–13843 (2013)
199. X. Zhang, Y. Zhang, H. Wu, L. Yan, Z. Wang, J. Zhao, W.W. Yu, A.L. Rogach, PbSe quantum dot films with enhanced electron mobility employed in hybrid polymer/nanocrystal solar cells. RSC Adv. **6**, 17029–17035 (2016)
200. J. Zhang, J. Gao, C.P. Church, E.M. Miller, J.M. Luther, V.I. Klimov, M.C. Beard, PbSe quantum dot solar cells with more than 6% efficiency fabricated in ambient atmosphere. Nano Lett. **14**, 6010–6015 (2014)
201. S.H. Im, C.-S. Lim, J.A. Chang, Y.H. Lee, N. Maiti, H.-J. Kim, M.K. Nazeeruddin, M. Grätzel, S.I. Seok, Toward interaction of sensitizer and functional moieties in hole-transporting materials for efficient semiconductor-sensitized solar cells. Nano Lett. **11**, 4789–4793 (2011)
202. J.A. Chang, S.H. Im, Y.H. Lee, H.-J. Kim, C.-S. Lim, J.H. Heo, S.I. Seok, Panchromatic photon-harvesting by hole-conducting materials in inorganic-organic heterojunction sensitized-solar cell through the formation of nanostructured electron channels. Nano Lett. **12**, 1863–1867 (2012)
203. O.S. Hutter, L.J. Phillips, K. Durose, J.D. Major, 6.6% efficient antimony selenide solar cells using grain structure control and an organic contact layer. Sol. Energy Mater. Sol. Cells **188**, 177–181 (2018)
204. C. Chen, L. Wang, L. Gao, D. Nam, D. Li, K. Li, Y. Zhao, C. Ge, H. Cheong, H. Liu, H. Song, J. Tang, 6.5% certified efficiency Sb_2Se_3 solar cells using PbS colloidal quantum dot film as hole-transporting layer. ACS Energy Lett. **2**, 2125–2132 (2017)
205. R.E. Bailey, S. Nie, Alloyed semiconductor quantum dots: tuning the optical properties without changing the particle size. J. Am. Chem. Soc. **125**, 7100–7106 (2003)
206. M.D. Regulacio, M.-Y. Han, Composition-tunable alloyed semiconductor nanocrystals. Acc. Chem. Res. **43**, 621–630 (2010)
207. A. Kumar, K.-T. Li, A.R. Madaria, C.J.N.R. Zhou, Sensitization of hydrothermally grown single crystalline TiO_2 nanowire array with CdSeS nanocrystals for photovoltaic applications. Nano Res. **4**, 1181–1190 (2011)
208. M.G. Panthani, V. Akhavan, B. Goodfellow, J.P. Schmidtke, L. Dunn, A. Dodabalapur, P.F. Barbara, B.A. Korgel, Synthesis of $CuInS_2$, $CuInSe_2$, and $Cu(In_xGa_{1-x})Se_2$ (CIGS) nanocrystal "inks" for printable photovoltaics. J. Am. Chem. Soc. **130**, 16770–16777 (2008)
209. J.-Y. Kim, J. Yang, J.H. Yu, W. Baek, C.-H. Lee, H.J. Son, T. Hyeon, M.J. Ko, Highly efficient copper–indium–selenide quantum dot solar cells: suppression of carrier recombination by controlled ZnS overlayers. ACS Nano **9**, 11286–11295 (2015)

210. R. Ghosh Chaudhuri, S. Paria, Core/shell nanoparticles: classes, properties, synthesis mechanisms, characterization, and applications. Chem. Rev. **112**, 2373–2433 (2012)
211. Y.-L. Lee, Y.-S. Lo, Highly efficient quantum-dot-sensitized solar cell based on co-sensitization of CdS/CdSe. Adv. Func. Mater. **19**, 604–609 (2009)
212. J. Yang, J. Wang, K. Zhao, T. Izuishi, Y. Li, Q. Shen, X. Zhong, CdSeTe/CdS type-I core/shell quantum dot sensitized solar cells with efficiency over 9%. J. Phys. Chem. C **119**, 28800–28808 (2015)
213. P.V. Kamat, Boosting the efficiency of quantum dot sensitized solar cells through modulation of interfacial charge transfer. Acc. Chem. Res. **45**, 1906–1915 (2012)
214. Z. Pan, H. Rao, I. Mora-Seró, J. Bisquert, X. Zhong, Quantum dot-sensitized solar cells. Chem. Soc. Rev. **47**, 7659–7702 (2018)
215. J. Chen, W. Lei, W.Q. Deng, Reduced charge recombination in a co-sensitized quantum dot solar cell with two different sizes of CdSe quantum dot. Nanoscale **3**, 674–677 (2011)
216. A. Kojima, K. Teshima, Y. Shirai, T. Miyasaka, Organometal halide perovskites as visible-light sensitizers for photovoltaic cells. J. Am. Chem. Soc. **131**, 6050–6051 (2009)
217. M. Grätzel, The light and shade of perovskite solar cells. Nat. Mater. **13**, 838 (2014)
218. N.-G. Park, Perovskite solar cells: an emerging photovoltaic technology. Mater. Today **18**, 65–72 (2015)
219. A. Swarnkar, A.R. Marshall, E.M. Sanehira, B.D. Chernomordik, D.T. Moore, J.A. Christians, T. Chakrabarti, J.M. Luther, Quantum dot–induced phase stabilization of α-$CsPbI_3$ perovskite for high-efficiency photovoltaics. Science **354**, 92–95 (2016)
220. Z. Song, S.C. Watthage, A.B. Phillips, M.J. Heben, Pathways toward high-performance perovskite solar cells: review of recent advances in organo-metal halide perovskites for photovoltaic applications. J. Photonics Energy **6**, 1–23, 23 (2016)
221. H. Tan, A. Jain, O. Voznyy, X. Lan, F.P.G. de Arquer, J.Z. Fan, R. Quintero-Bermudez, M. Yuan, B. Zhang, Y. Zhao, F. Fan, P. Li, L.N. Quan, Y. Zhao, Z.-H. Lu, Z. Yang, S. Hoogland, E.H. Sargent, Efficient and stable solution-processed planar perovskite solar cells via contact passivation. Science **355**, 722 (2017)
222. M. Kim, G.-H. Kim, T.K. Lee, I.W. Choi, H.W. Choi, Y. Jo, Y.J. Yoon, J.W. Kim, J. Lee, D. Huh, H. Lee, S.K. Kwak, J.Y. Kim, D.S. Kim, Methylammonium chloride induces intermediate phase stabilization for efficient perovskite solar cells. Joule (2019)
223. S.S. Mali, C.S. Shim, H.K. Park, J. Heo, P.S. Patil, C.K. Hong, Ultrathin atomic layer deposited TiO_2 for surface passivation of hydrothermally grown 1D TiO_2 nanorod arrays for efficient solid-state perovskite solar cells. Chem. Mater. **27**, 1541–1551 (2015)
224. Y. Yu, J. Li, D. Geng, J. Wang, L. Zhang, T.L. Andrew, M.S. Arnold, X. Wang, Development of lead iodide perovskite solar cells using three-dimensional titanium dioxide nanowire architectures. ACS Nano **9**, 564–572 (2015)
225. H.-Y. Yang, W.-Y. Rho, S.K. Lee, S.H. Kim, Y.-B. Hahn, TiO_2 nanoparticles/nanotubes for efficient light harvesting in perovskite solar cells. Nanomaterials (Basel) **9**, 326 (2019)
226. C. Chen, S. Wu, J. Wang, S. Chen, T. Peng, R. Li, Improved photovoltaic performance of perovskite solar cells based on three-dimensional rutile TiO_2 nanodendrite array film. Nanoscale **10**, 20836–20843 (2018)
227. K. Schutt, P.K. Nayak, A.J. Ramadan, B. Wenger, Y.-H. Lin, H.J. Snaith, Overcoming zinc oxide interface instability with a methylammonium-free perovskite for high-performance solar cells. Adv. Funct. Mater 1900466 (2019)
228. K. Mahmood, M.T. Mehran, F. Rehman, M.S. Zafar, S.W. Ahmad, R.-H. Song, Electrosprayed polymer-hybridized multidoped ZnO mesoscopic nanocrystals yield highly efficient and stable perovskite solar cells. ACS Omega **3**, 9648–9657 (2018)
229. K. Mahmood, B.S. Swain, A. Amassian, 16.1% efficient hysteresis-free mesostructured perovskite solar cells based on synergistically improved ZnO nanorod arrays. Adv. Energy Mater. **5**, 1500568 (2015)
230. Q. Jiang, Y. Zhao, X. Zhang, X. Yang, Y. Chen, Z. Chu, Q. Ye, X. Li, Z. Yin, J. You, Surface passivation of perovskite film for efficient solar cells. Nat. Photonics **13**, 460–466 (2019)

231. Y. Lee, S. Paek, K.T. Cho, E. Oveisi, P. Gao, S. Lee, J.-S. Park, Y. Zhang, R. Humphry-Baker, A.M. Asiri, M.K. Nazeeruddin, Enhanced charge collection with passivation of the tin oxide layer in planar perovskite solar cells. J. Mater. Chem. A **5**, 12729–12734 (2017)
232. E.J. Yeom, S.S. Shin, W.S. Yang, S.J. Lee, W. Yin, D. Kim, J.H. Noh, T.K. Ahn, S.I. Seok, Controllable synthesis of single crystalline Sn-based oxides and their application in perovskite solar cells. J. Mater. Chem. A **5**, 79–86 (2017)
233. G.S. Han, H.S. Chung, D.H. Kim, B.J. Kim, J.-W. Lee, N.-G. Park, I.S. Cho, J.-K. Lee, S. Lee, H.S. Jung, Epitaxial 1D electron transport layers for high-performance perovskite solar cells. Nanoscale **7**, 15284–15290 (2015)
234. C. Gao, S. Yuan, B. Cao, J. Yu, SnO_2 nanotube arrays grown via an in situ template-etching strategy for effective and stable perovskite solar cells. Chem. Eng. J. **325**, 378–385 (2017)
235. Q. Liu, M.-C. Qin, W.-J. Ke, X.-L. Zheng, Z. Chen, P.-L. Qin, L.-B. Xiong, H.-W. Lei, J.-W. Wan, J. Wen, G. Yang, J.-J. Ma, Z.-Y. Zhang, G.-J. Fang, Enhanced stability of perovskite solar cells with low-temperature hydrothermally grown SnO_2 electron transport layers. Adv. Func. Mater. **26**, 6069–6075 (2016)
236. J. Lian, B. Lu, F. Niu, P. Zeng, X. Zhan, Electron-transport materials in perovskite solar cells. Small Methods **2**, 1800082 (2018)
237. C.S. Ponseca, T.J. Savenije, M. Abdellah, K. Zheng, A. Yartsev, T. Pascher, T. Harlang, P. Chabera, T. Pullerits, A. Stepanov, J.-P. Wolf, V. Sundström, Organometal halide perovskite solar cell materials rationalized: ultrafast charge generation, high and microsecond-long balanced mobilities, and slow recombination. J. Am. Chem. Soc. **136**, 5189–5192 (2014)
238. S. Song, G. Kang, L. Pyeon, C. Lim, G.-Y. Lee, T. Park, J. Choi, Systematically optimized bilayered electron transport layer for highly efficient planar perovskite solar cells ($\eta = 21.1\%$). ACS Energy Lett. **2**, 2667–2673 (2017)
239. M.M. Tavakoli, P. Yadav, R. Tavakoli, J. Kong, Surface engineering of TiO_2 ETL for highly efficient and hysteresis-less planar perovskite solar cell (21.4%) with enhanced open-circuit voltage and stability. Adv. Energy Mater. **8**, 1800794 (2018)
240. H. Liu, Z. Huang, S. Wei, L. Zheng, L. Xiao, Q. Gong, Nano-structured electron transporting materials for perovskite solar cells. Nanoscale **8**, 6209–6221 (2016)
241. Y. Yang, K. Ri, A. Mei, L. Liu, M. Hu, T. Liu, X. Li, H. Han, The size effect of TiO_2 nanoparticles on a printable mesoscopic perovskite solar cell. J. Mater. Chem. A **3**, 9103–9107 (2015)
242. S. Dharani, H.K. Mulmudi, N. Yantara, P.T. Thu Trang, N.G. Park, M. Graetzel, S. Mhaisalkar, N. Mathews, P.P. Boix, High efficiency electrospun TiO_2 nanofiber based hybrid organic–inorganic perovskite solar cell. Nanoscale **6**, 1675–1679 (2014)
243. H.-S. Kim, J.-W. Lee, N. Yantara, P.P. Boix, S.A. Kulkarni, S. Mhaisalkar, M. Grätzel, N.-G. Park, High efficiency solid-state sensitized solar cell-based on submicrometer rutile TiO_2 nanorod and $CH_3NH_3PbI_3$ perovskite sensitizer. Nano Lett. **13**, 2412–2417 (2013)
244. A. Fakharuddin, F. Di Giacomo, I. Ahmed, Q. Wali, T.M. Brown, R. Jose, Role of morphology and crystallinity of nanorod and planar electron transport layers on the performance and long term durability of perovskite solar cells. J. Power Sources **283**, 61–67 (2015)
245. Q. Jiang, X. Sheng, Y. Li, X. Feng, T. Xu, Rutile TiO_2 nanowire-based perovskite solar cells. Chem. Commun. **50**, 14720–14723 (2014)
246. X. Wang, Z. Li, W. Xu, S.A. Kulkarni, S.K. Batabyal, S. Zhang, A. Cao, L.H. Wong, TiO2 nanotube arrays based flexible perovskite solar cells with transparent carbon nanotube electrode. Nano Energy **11**, 728–735 (2015)
247. X. Liang, Y. Cheng, X. Xu, R. Dong, D. Li, Z. Zhou, R. Wei, G. Dong, S.-W. Tsang, J.C. Ho, Enhanced performance of perovskite solar cells based on vertical TiO_2 nanotube arrays with full filling of $CH_3NH_3PbI_3$. Appl. Surf. Sci. **451**, 250–257 (2018)
248. N. Islam, M. Yang, K. Zhu, Z. Fan, Mesoporous scaffolds based on TiO_2 nanorods and nanoparticles for efficient hybrid perovskite solar cells. J. Mater. Chem. A **3**, 24315–24321 (2015)
249. D. Zhong, B. Cai, X. Wang, Z. Yang, Y. Xing, S. Miao, W.-H. Zhang, C. Li, Synthesis of oriented TiO_2 nanocones with fast charge transfer for perovskite solar cells. Nano Energy **11**, 409–418 (2015)

250. J.-W. Lee, S.H. Lee, H.-S. Ko, J. Kwon, J.H. Park, S.M. Kang, N. Ahn, M. Choi, J.K. Kim, N.-G. Park, Opto-electronic properties of TiO_2 nanohelices with embedded $HC(NH_2)_2PbI_3$ perovskite solar cells. J. Mater. Chem. A **3**, 9179–9186 (2015)
251. T. Leijtens, G.E. Eperon, S. Pathak, A. Abate, M.M. Lee, H.J. Snaith, Overcoming ultra-violet light instability of sensitized TiO_2 with meso-superstructured organometal tri-halide perovskite solar cells. Nat. Commun. **4**, 2885 (2013)
252. M. Liu, M.B. Johnston, H.J. Snaith, Efficient planar heterojunction perovskite solar cells by vapour deposition. Nature **501**, 395 (2013)
253. J. Song, J. Bian, E. Zheng, X.-F. Wang, W. Tian, T. Miyasaka, Efficient and environmentally stable perovskite solar cells based on ZnO electron collection layer. Chem. Lett. **44**, 610–612 (2015)
254. J. Cao, B. Wu, R. Chen, Y. Wu, Y. Hui, B.-W. Mao, N. Zheng, Efficient, hysteresis-free, and stable perovskite solar cells with ZnO as electron-transport layer: effect of surface passivation. Adv. Mater. **30**, 1705596 (2018)
255. K. Mahmood, B.S. Swain, A. Amassian, Double-layered ZnO nanostructures for efficient perovskite solar cells. Nanoscale **6**, 14674–14678 (2014)
256. D.-Y. Son, J.-H. Im, H.-S. Kim, N.-G. Park, 11% efficient perovskite solar cell based on ZnO nanorods: an effective charge collection system. J. Phys. Chem. C **118**, 16567–16573 (2014)
257. M.A. Mahmud, N.K. Elumalai, M.B. Upama, D. Wang, K.H. Chan, M. Wright, C. Xu, F. Haque, A. Uddin, Low temperature processed ZnO thin film as electron transport layer for efficient perovskite solar cells. Sol. Energy Mater. Sol. Cells **159**, 251–264 (2017)
258. Y. Dkhissi, S. Meyer, D. Chen, H.C. Weerasinghe, L. Spiccia, Y.-B. Cheng, R.A. Caruso, Stability comparison of perovskite solar cells based on zinc oxide and titania on polymer substrates. Chemsuschem **9**, 687–695 (2016)
259. Y. Cheng, Q.-D. Yang, J. Xiao, Q. Xue, H.-W. Li, Z. Guan, H.-L. Yip, S.-W. Tsang, Decomposition of organometal halide perovskite films on zinc oxide nanoparticles. ACS Appl. Mater. Interfaces **7**, 19986–19993 (2015)
260. J. Yang, B.D. Siempelkamp, E. Mosconi, F. De Angelis, T.L. Kelly, Origin of the thermal instability in $CH_3NH_3PbI_3$ thin films deposited on ZnO. Chem. Mater. **27**, 4229–4236 (2015)
261. X. Zhao, H. Shen, Y. Zhang, X. Li, X. Zhao, M. Tai, J. Li, J. Li, X. Li, H. Lin, Aluminum-doped zinc oxide as highly stable electron collection layer for perovskite solar cells. ACS Appl. Mater. Interfaces **8**, 7826–7833 (2016)
262. R. Chen, J. Cao, Y. Duan, Y. Hui, T.T. Chuong, D. Ou, F. Han, F. Cheng, X. Huang, B. Wu, N. Zheng, High-efficiency, hysteresis-less, UV-stable perovskite solar cells with cascade ZnO–ZnS electron transport layer. J. Am. Chem. Soc. **141**, 541–547 (2019)
263. L. Xiong, Y. Guo, J. Wen, H. Liu, G. Yang, P. Qin, G. Fang, Review on the application of SnO_2 in perovskite solar cells. Adv. Func. Mater. **28**, 1802757 (2018)
264. Z. Liu, K. Deng, J. Hu, L. Li, Coagulated SnO_2 colloids for high-performance planar perovskite solar cells with negligible hysteresis and improved stability. Angew. Chem. Int. Ed. (2019)
265. Q. Jiang, Z. Chu, P. Wang, X. Yang, H. Liu, Y. Wang, Z. Yin, J. Wu, X. Zhang, J. You, Planar-structure perovskite solar cells with efficiency beyond 21%. Adv. Mater. **29**, 1703852 (2017)
266. Q. Jiang, L. Zhang, H. Wang, X. Yang, J. Meng, H. Liu, Z. Yin, J. Wu, X. Zhang, J. You, Enhanced electron extraction using SnO_2 for high-efficiency planar-structure $HC(NH_2)_2PbI_3$-based perovskite solar cells. Nat. Energy **2**, 16177 (2016)
267. Y. Luan, X. Yi, P. Mao, Y. Wei, J. Zhuang, N. Chen, T. Lin, C. Li, J. Wang, High-performance planar perovskite solar cells with negligible hysteresis using 2,2,2-trifluoroethanol-incorporated SnO_2. iScience **16**, 433–441 (2019)
268. Q. Dong, Y. Shi, C. Zhang, Y. Wu, L. Wang, Energetically favored formation of SnO_2 nanocrystals as electron transfer layer in perovskite solar cells with high efficiency exceeding 19%. Nano Energy **40**, 336–344 (2017)
269. W. Ke, G. Fang, Q. Liu, L. Xiong, P. Qin, H. Tao, J. Wang, H. Lei, B. Li, J. Wan, G. Yang, Y. Yan, Low-temperature solution-processed tin oxide as an alternative electron transporting layer for efficient perovskite solar cells. J. Am. Chem. Soc. **137**, 6730–6733 (2015)

270. J. Barbé, M.L. Tietze, M. Neophytou, B. Murali, E. Alarousu, A.E. Labban, M. Abulikemu, W. Yue, O.F. Mohammed, I. McCulloch, A. Amassian, S. Del Gobbo, Amorphous tin oxide as a low-temperature-processed electron-transport layer for organic and hybrid perovskite solar cells. ACS Appl. Mater. Interfaces **9**, 11828–11836 (2017)
271. P. Pinpithak, H.-W. Chen, A. Kulkarni, Y. Sanehira, M. Ikegami, T. Miyasaka, Low-temperature and ambient air processes of amorphous SnO_x-based mixed halide perovskite planar solar cell. Chem. Lett. **46**, 382–384 (2017)
272. C. Wang, D. Zhao, C.R. Grice, W. Liao, Y. Yu, A. Cimaroli, N. Shrestha, P.J. Roland, J. Chen, Z. Yu, P. Liu, N. Cheng, R.J. Ellingson, X. Zhao, Y. Yan, Low-temperature plasma-enhanced atomic layer deposition of tin oxide electron selective layers for highly efficient planar perovskite solar cells. J. Mater. Chem. A **4**, 12080–12087 (2016)
273. B. Roose, J.-P.C. Baena, K.C. Gödel, M. Graetzel, A. Hagfeldt, U. Steiner, A. Abate, Mesoporous SnO_2 electron selective contact enables UV-stable perovskite solar cells. Nano Energy **30**, 517–522 (2016)
274. L. Xiong, M. Qin, C. Chen, J. Wen, G. Yang, Y. Guo, J. Ma, Q. Zhang, P. Qin, S. Li, G. Fang, Fully high-temperature-processed SnO_2 as blocking layer and scaffold for efficient, stable, and hysteresis-free mesoporous perovskite solar cells. Adv. Func. Mater. **28**, 1706276 (2018)
275. M.A. Mejía Escobar, S. Pathak, J. Liu, H.J. Snaith, F. Jaramillo, ZrO_2/TiO_2 electron collection layer for efficient meso-superstructured hybrid perovskite solar cells. ACS Appl. Mater. Interfaces **9**, 2342–2349 (2017)
276. Y. Li, L. Zhao, S. Wei, M. Xiao, B. Dong, L. Wan, S. Wang, Effect of ZrO_2 film thickness on the photoelectric properties of mixed-cation perovskite solar cells. Appl. Surf. Sci. **439**, 506–515 (2018)
277. M. Che, L. Zhu, Y.L. Zhao, D.S. Yao, X.Q. Gu, J. Song, Y.H. Qiang, Enhancing current density of perovskite solar cells using TiO_2-ZrO_2 composite scaffold layer. Mater. Sci. Semicond. Process. **56**, 29–36 (2016)
278. H. Si, Q. Liao, Z. Zhang, Y. Li, X. Yang, G. Zhang, Z. Kang, Y. Zhang, An innovative design of perovskite solar cells with Al_2O_3 inserting at ZnO/perovskite interface for improving the performance and stability. Nano Energy **22**, 223–231 (2016)
279. Y. Numata, Y. Sanehira, T. Miyasaka, Impacts of Heterogeneous TiO_2 and Al_2O_3 composite mesoporous scaffold on formamidinium lead trihalide perovskite solar cells. ACS Appl. Mater. Interfaces **8**, 4608–4615 (2016)
280. N. Cheng, P. Liu, S. Bai, Z. Yu, W. Liu, S.-S. Guo, X.-Z. Zhao, Application of mesoporous SiO_2 layer as an insulating layer in high performance hole transport material free $CH_3NH_3PbI_3$ perovskite solar cells. J. Power Sources **321**, 71–75 (2016)
281. S.H. Hwang, J. Roh, J. Lee, J. Ryu, J. Yun, J. Jang, Size-controlled SiO_2 nanoparticles as scaffold layers in thin-film perovskite solar cells. J. Mater. Chem. A **2**, 16429–16433 (2014)
282. F. Qi, C. Wang, N. Cheng, P. Liu, Y. Xiao, F. Li, X. Sun, W. Liu, S. Guo, X.-Z. Zhao, Improving the performance through SPR effect by employing Au@SiO_2 core-shell nanoparticles incorporated TiO_2 scaffold in efficient hole transport material free perovskite solar cells. Electrochim. Acta **282**, 10–15 (2018)
283. A. Bera, K. Wu, A. Sheikh, E. Alarousu, O.F. Mohammed, T. Wu, Perovskite oxide $SrTiO_3$ as an efficient electron transporter for hybrid perovskite solar cells. J. Phys. Chem. C **118**, 28494–28501 (2014)
284. Y. Okamoto, R. Fukui, M. Fukazawa, Y. Suzuki, $SrTiO_3/TiO_2$ composite electron transport layer for perovskite solar cells. Mater. Lett. **187**, 111–113 (2017)
285. L. Zhu, Z. Shao, J. Ye, X. Zhang, X. Pan, S. Dai, Mesoporous $BaSnO_3$ layer based perovskite solar cells. Chem. Commun. **52**, 970–973 (2016)
286. L. Zhu, J. Ye, X. Zhang, H. Zheng, G. Liu, X. Pan, S. Dai, Performance enhancement of perovskite solar cells using a La-doped $BaSnO_3$ electron transport layer. J. Mater. Chem. A **5**, 3675–3682 (2017)
287. S.S. Shin, E.J. Yeom, W.S. Yang, S. Hur, M.G. Kim, J. Im, J. Seo, J.H. Noh, S.I. Seok, Colloidally prepared La-doped $BaSnO_3$ electrodes for efficient, photostable perovskite solar cells. Science **356**, 167 (2017)

288. L.S. Oh, D.H. Kim, J.A. Lee, S.S. Shin, J.-W. Lee, I.J. Park, M.J. Ko, N.-G. Park, S.G. Pyo, K.S. Hong, J.Y. Kim, Zn_2SnO_4-based photoelectrodes for organolead halide perovskite solar cells. J. Phys. Chem. C **118**, 22991–22994 (2014)
289. A. Bera, A.D. Sheikh, M.A. Haque, R. Bose, E. Alarousu, O.F. Mohammed, T. Wu, Fast crystallization and improved stability of perovskite solar cells with Zn_2SnO_4 electron transporting layer: interface matters. ACS Appl. Mater. Interfaces **7**, 28404–28411 (2015)
290. W.-Q. Wu, D. Chen, F. Li, Y.-B. Cheng, R.A. Caruso, Solution-processed Zn_2SnO_4 electron transporting layer for efficient planar perovskite solar cells. Mater. Today Energy **7**, 260–266 (2018)
291. S.S. Shin, W.S. Yang, J.H. Noh, J.H. Suk, N.J. Jeon, J.H. Park, J.S. Kim, W.M. Seong, S.I. Seok, High-performance flexible perovskite solar cells exploiting Zn_2SnO_4 prepared in solution below 100 °C. Nat. Commun. **6**, 7410 (2015)
292. M. Tai, X. Zhao, H. Shen, Y. Guo, M. Zhang, Y. Zhou, X. Li, Z. Yao, X. Yin, J. Han, X. Li, H. Lin, Ultrathin Zn_2SnO_4 (ZTO) passivated ZnO nanocone arrays for efficient and stable perovskite solar cells. Chem. Eng. J. **361**, 60–66 (2019)
293. W. Chen, Y. Zhou, L. Wang, Y. Wu, B. Tu, B. Yu, F. Liu, H.-W. Tam, G. Wang, A.B. Djurišić, L. Huang, Z. He, Molecule-doped nickel oxide: verified charge transfer and planar inverted mixed cation perovskite solar cell. Adv. Mater. **30**, 1800515 (2018)
294. J.H. Park, J. Seo, S. Park, S.S. Shin, Y.C. Kim, N.J. Jeon, H.-W. Shin, T.K. Ahn, J.H. Noh, S.C. Yoon, C.S. Hwang, S.I. Seok, Efficient $CH_3NH_3PbI_3$ perovskite solar cells employing nanostructured p-type NiO electrode formed by a pulsed laser deposition. Adv. Mater. **27**, 4013–4019 (2015)
295. H. Wang, Z. Yu, J. Lai, X. Song, X. Yang, A. Hagfeldt, L. Sun, One plus one greater than two: high-performance inverted planar perovskite solar cells based on a composite CuI/CuSCN hole-transporting layer. J. Mater. Chem. A **6**, 21435–21444 (2018)
296. S. Ye, H. Rao, Z. Zhao, L. Zhang, H. Bao, W. Sun, Y. Li, F. Gu, J. Wang, Z. Liu, Z. Bian, C. Huang, A breakthrough efficiency of 19.9% obtained in inverted perovskite solar cells by using an efficient trap state passivator Cu(thiourea)I. J. Am. Chem. Soc. **139**, 7504–7512 (2017)
297. W. Sun, Y. Li, S. Ye, H. Rao, W. Yan, H. Peng, Y. Li, Z. Liu, S. Wang, Z. Chen, L. Xiao, Z. Bian, C. Huang, High-performance inverted planar heterojunction perovskite solar cells based on a solution-processed CuO_x hole transport layer. Nanoscale **8**, 10806–10813 (2016)
298. C. Zuo, L. Ding, Solution-processed Cu_2O and CuO as hole transport materials for efficient perovskite solar cells. Small **11**, 5528–5532 (2015)
299. C. Duan, M. Zhao, C. Zhao, Y. Wang, J. Li, W. Han, Q. Hu, L. Yao, H. Jian, F. Lu, T. Jiu, Inverted $CH_3NH_3PbI_3$ perovskite solar cells based on solution-processed V_2O_5 film combined with P3CT salt as hole transport layer. Mater. Today Energy **9**, 487–495 (2018)
300. Y. Jiang, C. Li, H. Liu, R. Qin, H. Ma, Poly(3,4-ethylenedioxythiophene):poly(styrene sulfonate)(PEDOT:PSS)–molybdenum oxide composite films as hole conductors for efficient planar perovskite solar cells. J. Mater. Chem. A **4**, 9958–9966 (2016)
301. S. Masi, R. Mastria, R. Scarfiello, S. Carallo, C. Nobile, S. Gambino, T. Sibillano, C. Giannini, S. Colella, A. Listorti, P.D. Cozzoli, A. Rizzo, Room-temperature processed films of colloidal carved rod-shaped nanocrystals of reduced tungsten oxide as interlayers for perovskite solar cells. Phys. Chem. Chem. Phys. **20**, 11396–11404 (2018)
302. P. Tonui, S.O. Oseni, G. Sharma, Q. Yan, G. Tessema Mola, Perovskites photovoltaic solar cells: an overview of current status. Renew. Sustain. Energy Rev. **91**, 1025–1044 (2018)
303. J. Cui, F. Meng, H. Zhang, K. Cao, H. Yuan, Y. Cheng, F. Huang, M. Wang, $CH_3NH_3PbI_3$-based planar solar cells with magnetron-sputtered nickel oxide. ACS Appl. Mater. Interfaces **6**, 22862–22870 (2014)
304. Z. Liu, J. Chang, Z. Lin, L. Zhou, Z. Yang, D. Chen, C. Zhang, S. Liu, Y. Hao, High-performance planar perovskite solar cells using low temperature, solution–combustion-based nickel oxide hole transporting layer with efficiency exceeding 20%. Adv. Energy Mater. **8**, 1703432 (2018)

305. W. Chen, F.-Z. Liu, X.-Y. Feng, A.B. Djurišić, W.K. Chan, Z.-B. He, Cesium doped NiO_x as an efficient hole extraction layer for inverted planar perovskite solar cells. Adv. Energy Mater. **7**, 1700722 (2017)
306. W. Chen, Y. Wu, J. Fan, A.B. Djurišić, F. Liu, H.W. Tam, A. Ng, C. Surya, W.K. Chan, D. Wang, Z.-B. He, Understanding the doping effect on NiO: toward high-performance inverted perovskite solar cells. Adv. Energy Mater. **8**, 1703519 (2018)
307. G. Li, Y. Jiang, S. Deng, A. Tam, P. Xu, M. Wong, H.-S. Kwok, Overcoming the limitations of sputtered nickel oxide for high-efficiency and large-area perovskite solar cells. Adv. Sci. **4**, 1700463 (2017)
308. Y. Wu, F. Xie, H. Chen, X. Yang, H. Su, M. Cai, Z. Zhou, T. Noda, L. Han, Thermally stable $MAPbI_3$ perovskite solar cells with efficiency of 19.19% and area over 1 cm^2 achieved by additive engineering. Adv. Mater. **29**, 1701073 (2017)
309. T. Abzieher, S. Moghadamzadeh, F. Schackmar, H. Eggers, F. Sutterlüti, A. Farooq, D. Kojda, K. Habicht, R. Schmager, A. Mertens, R. Azmi, L. Klohr, J.A. Schwenzer, M. Hetterich, U. Lemmer, B.S. Richards, M. Powalla, U.W. Paetzold, Electron-beam-evaporated nickel oxide hole transport layers for perovskite-based photovoltaics. Adv. Energy Mater. **9**, 1802995 (2019)
310. K.-C. Wang, J.-Y. Jeng, P.-S. Shen, Y.-C. Chang, E.W.-G. Diau, C.-H. Tsai, T.-Y. Chao, H.-C. Hsu, P.-Y. Lin, P. Chen, T.-F. Guo, T.-C. Wen, p-type mesoscopic nickel oxide/organometallic perovskite heterojunction solar cells. Sci. Rep. **4**, 4756 (2014)
311. R. Singh, P.K. Singh, B. Bhattacharya, H.-W. Rhee, Review of current progress in inorganic hole-transport materials for perovskite solar cells. Appl. Mater. Today **14**, 175–200 (2019)
312. P. Pattanasattayavong, G.O.N. Ndjawa, K. Zhao, K.W. Chou, N. Yaacobi-Gross, B.C. O'Regan, A. Amassian, T.D. Anthopoulos, Electric field-induced hole transport in copper(i) thiocyanate (CuSCN) thin-films processed from solution at room temperature. Chem. Commun. **49**, 4154–4156 (2013)
313. S. Ye, W. Sun, Y. Li, W. Yan, H. Peng, Z. Bian, Z. Liu, C. Huang, CuSCN-based inverted planar perovskite solar cell with an average PCE of 15.6%. Nano Lett. **15**, 3723–3728 (2015)
314. W.-Y. Chen, L.-L. Deng, S.-M. Dai, X. Wang, C.-B. Tian, X.-X. Zhan, S.-Y. Xie, R.-B. Huang, L.-S. Zheng, Low-cost solution-processed copper iodide as an alternative to PEDOT:PSS hole transport layer for efficient and stable inverted planar heterojunction perovskite solar cells. J. Mater. Chem. A **3**, 19353–19359 (2015)
315. H. Wang, Z. Yu, X. Jiang, J. Li, B. Cai, X. Yang, L. Sun, Efficient and stable inverted planar perovskite solar cells employing cui as hole-transporting layer prepared by solid-gas transformation. Energy Technol. **5**, 1836–1843 (2017)
316. S. Ye, H. Rao, W. Yan, Y. Li, W. Sun, H. Peng, Z. Liu, Z. Bian, Y. Li, C. Huang, A strategy to simplify the preparation process of perovskite solar cells by co-deposition of a hole-conductor and a perovskite layer. Adv. Mater. **28**, 9648–9654 (2016)
317. W. Yan, S. Ye, Y. Li, W. Sun, H. Rao, Z. Liu, Z. Bian, C. Huang, Hole-transporting materials in inverted planar perovskite solar cells. Adv. Energy Mater. **6**, 1600474 (2016)
318. Q. Guo, C. Wang, J. Li, Y. Bai, F. Wang, L. Liu, B. Zhang, T. Hayat, A. Alsaedi, Z.A. Tan, Low-temperature solution-processed vanadium oxide as hole transport layer for efficient and stable perovskite solar cells. Phys. Chem. Chem. Phys. **20**, 21746–21754 (2018)
319. C.X. Guo, K. Sun, J. Ouyang, X. Lu, Layered V_2O_5/PEDOT nanowires and ultrathin nanobelts fabricated with a silk reelinglike process. Chem. Mater. **27**, 5813–5819 (2015)
320. F. Hou, Z. Su, F. Jin, X. Yan, L. Wang, H. Zhao, J. Zhu, B. Chu, W. Li, Efficient and stable planar heterojunction perovskite solar cells with an MoO_3/PEDOT:PSS hole transporting layer. Nanoscale **7**, 9427–9432 (2015)
321. C. Tao, S. Ruan, G. Xie, X. Kong, L. Shen, F. Meng, C. Liu, X. Zhang, W. Dong, W. Chen, Role of tungsten oxide in inverted polymer solar cells. Appl. Phys. Lett. **94**, 043311 (2009)
322. K. Wang, Y. Shi, Q. Dong, Y. Li, S. Wang, X. Yu, M. Wu, T. Ma, Low-temperature and solution-processed amorphous WOX as electron-selective layer for perovskite solar cells. J. Phys. Chem. Lett. **6**, 755–759 (2015)

Chapter 2
Microwave-Responsive Nanomaterials for Catalysis

Tuo Ji and Jiahua Zhu

Abstract Microwave heating is a powerful and non-conventional energy source for heterogeneous catalytic reactions, which has attracted considerable attention during the past decades. With the presence of microwave-responsive catalysts, microwave can selectively heat the designed catalyst surface, and expedite the reaction rate at catalyst/solvent interface. This chapter strives to extensively review the recent work on microwave-responsive catalysts and their roles in heterogeneous catalytic reactions. The fundamental mechanism of microwave heating is illustrated to explain its functions in the catalytic reactions. The working principle of microwave-responsive catalyst and related evaluation methods are discussed in this chapter. Additionally, the advantages of microwave-responsive catalysts for specific reactions have been categorized and reviewed. At last, a few different strategies to enhance microwave thermal effects have been summarized. It is concluded that developing microwave-responsive catalysts is a practical method to expedite reaction rate, enhance energy-efficiency, and improve product quality. Therefore, designing microwave-responsive catalysts could be an effective strategy for highly efficient reactions and future industrial-scale applications.

Keywords Microwave · Catalyst · Nanomaterial · Energy efficiency · Heterogeneous reaction

2.1 Introduction

Microwave heating has been considered as a powerful energy source for material synthesis [1], organic synthesis [2], fuel engineering [3], green chemistry [4], and biomass conversion [5]. Since the first publication of microwave heating in chemical reactions, studies in this area grow exponentially and have been extensively reviewed.

T. Ji · J. Zhu (✉)
Department of Chemical and Biomolecular Engineering, The University of Akron, Akron, OH 44325, USA
e-mail: jzhu1@uakron.edu

Z. Sun and T. Liao (eds.), *Responsive Nanomaterials for Sustainable Applications*, Springer Series in Materials Science 297,
https://doi.org/10.1007/978-3-030-39994-8_2

The major advantages of microwave in reactions have been well recognized as rapid reaction rate, cleaner products, and milder reaction conditions [6–8].

Compared to homogeneous reaction, microwave heating could have additional benefits in heterogeneous reaction, such as promoting selective heating, expediting interfacial reaction, and improving mass transfer [9, 10]. These characteristics would create a unique reaction environment at catalyst/solution interface. Hence, the design of microwave-responsive catalysts has been recognized as an effective approach to achieve the goal in microwave heating heterogeneous reactions. Such catalysts not only provide catalytically active sites, but also responsible for in situ heat generation under microwave.

The focus of this chapter is to present the latest research on microwave-responsive catalysts and their roles in heterogeneous catalytic reactions. The fundamental mechanisms of microwave heating are summarized. The working principle of microwave-responsive catalyst and related evaluation methods are discussed. The discussion is then expanded to different types of reactions including liquid-phase, gas-phase reactions, as well as biomass pyrolysis. At last, different strategies to enhance microwave thermal effect are discussed.

2.2 The Principle of Microwave Heating

2.2.1 *Microwave Heating*

Microwave is one kind of electromagnetic wave, which consists of two perpendicular components—electric and magnetic fields. The wavelength of microwave falls in the range of 1 mm to 1 m, corresponding to a frequency range of 300–0.3 GHz (between the infrared and radio frequencies). In general, radar operates in the frequency range of 1–30 GHz, while telecommunication uses the rest frequency band. To avoid interference with telecommunication devices, household and industrial microwave ovens are operated at either 915 MHz or 2.45 GHz, which are reserved by the Federal Communications Commission (FCC) [11]. Most of the domestic microwave ovens and laboratory-scale microwave reactors use the frequency of 2.45 GHz.

Conventional heating involves the process of conduction and convection, where the heat is usually supplied from an external heating source (such as oil bath, steam, or resistive heater) and then propagates to the targeted area via heat transfer media. The efficiency of such heating process is usually low because the heat transfer depends on many factors including the thermal conductivity of the heat transfer media, convection currents, and heating volume [10, 12]. Different from conventional heating, microwave heating is more likely an energy conversion process rather than heat transfer, where electromagnetic energy can be converted to thermal energy with appropriate design. The most appealing feature of microwave in chemical reactions is volumetric heating, where uniform temperature field can be created and target temperature can be reached within a short period of time [13].

2.2.2 *The Mechanism of Microwave Heating*

Microwave heating hinges on the ability of the materials to absorb and transform microwave energy into heat. Based on the different interaction modes, materials can be categorized as conductors (e.g., metals, graphite), insulators (e.g., quartz glass, porcelain, ceramics), and dielectrics (e.g., water, carbon, Fe_3O_4).

When microwave irradiates on a conductor, most of them are reflected from its surface. In contrast, microwave can penetrate through the insulator without loss. These two types of materials are not suitable for heat generation. Dielectric materials can absorb microwave and generate heat (called dielectric heating). Dielectrics have two important features: (1) few free charge carriers and (2) exhibit dipole moment or ability to generate dipole moment [14]. Dipole moment is defined as two equal and opposite charges separated by a finite distance, which is widely existed in the dielectrics [15]. With the presence of an external electric field, non-polar molecules can be induced and form a temporary dipole moment. When electric field and dipoles are misaligned, dipole rotation would occur to align themselves to the electric field. Such alignment occurs at a high frequency of 2.45×10^9 times per second at microwave frequency of 2.45 GHz. The high-frequency rotation generates a huge amount of friction heat inside the dielectrics, and this is exactly how microwave can be used to generate heat. It is also worth mentioning that both rotated molecules and oscillatory migration of ions in the system can generate heat under microwave radiation.

The ability of dielectrics to convert microwave energy is related to their dielectric properties, which depends on the complex permittivity as described in (2.1):

$$\varepsilon = \varepsilon' - j \cdot \varepsilon'' \qquad (2.1)$$

where the real part of permittivity (ε') signifies the ability to store electric energy. The imaginary part, termed as dielectric loss (ε''), signifies the ability to convert electric energy. The ability of dielectrics to convert microwave into heat energy can be assessed by $\tan \delta$, (2.2):

$$\tan \delta = \varepsilon'' / \varepsilon' \qquad (2.2)$$

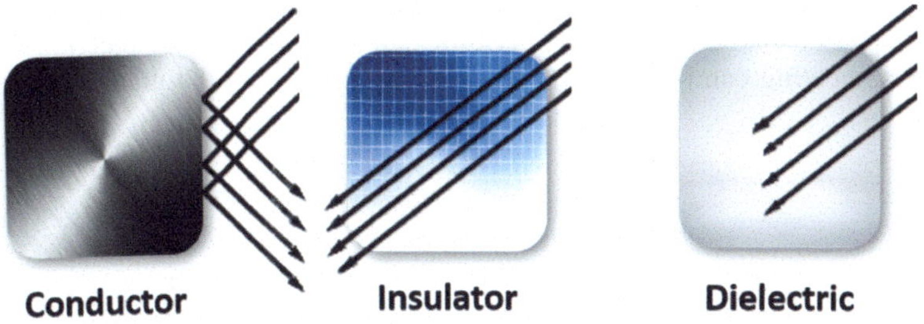

Fig. 2.1 Interaction of microwave with conductor, insulator, and dielectric materials

where tan δ is the dissipation factor of dielectrics. A larger value of tan δ means a higher conversion efficiency from microwave to heat [16]. Besides the intrinsic material properties, other factors also influence the dielectric properties such as system temperature, microwave frequency, and state of materials such as purity, chemical state, and structure [17]. In general, materials with high real permittivity (ε') are often used in dielectric capacitors, while dielectrics with high dissipation factor (tan δ) can be used as microwave absorption materials.

The conversion of microwave energy to heat can be calculated by (2.3, 2.4) [18]:

$$Q_0 = \sigma E^2 \tag{2.3}$$

$$\sigma = 2\pi\varepsilon_0\varepsilon'' f \tag{2.4}$$

where Q_0 is dissipated heat in the electric field (W/m^3), E is electric field intensity, σ is dielectric conductivity, f is the frequency of energy source (Hz), and ε_0 is the permittivity of vacuum or free space (8.85×10^{12} F/m). If the materials are induced by oscillated magnetic fields, the magnetic loss can be described in a similar way, (2.5):

$$\mu = \mu' - j \cdot \mu'' \tag{2.5}$$

where the real part of permeability (μ') signifies the ability to store magnetic energy, and imaginary part (μ'') signifies the ability to convert magnetic energy. Therefore, the total dissipated energy can be calculated by (2.6) [19]:

$$Q = 2\pi\varepsilon_0\varepsilon'' f E^2 + 2\pi\mu_0\mu'' f H^2 \tag{2.6}$$

where H is magnetic field intensity, and μ_0 is the permeability of vacuum or free space ($\mu_0 = 4\pi \times 10^{-7}$ H/m).

Another factor influencing microwave heating is penetration depth D_p, which is given by (2.7) [20]

$$D_p = \frac{1}{2\pi f}\left(\frac{2}{\mu'\mu_0\varepsilon_0\mu'\kappa'}\right)^{1/2}\left[\left(1 + \tan^2\delta\right)^{\frac{1}{2}} - 1\right]^{-1/2} \tag{2.7}$$

At frequency near the microwave regime, D_p is in the order of meters. Uniform volumetric heating can be achieved when D_p is larger than the material or system dimensions.

2.2.3 Microwave Heating in Catalytic Reactions

Microwave heating has been studied in a wide range of reactions, including redox [21], pyrolysis [22], dehydration [23], esterification [24], cycloaddition [25], and decarboxylation [26]. The promising outcomes of microwave heating are higher yield, milder reaction condition, shorter reaction time, and sometimes to trigger reactions that cannot occur in conventional heating mode. All these beneficial features arouse great interests from scientists with an aim to explain the reaction mechanisms under microwave irradiation [27]. According to the literature in the past decades, so-called non-thermal effect (microwave effect) is widely used to explain the microwave-assisted reactions [10]. That is said, when a reaction is under a highly polarized field, microwave could impact the mobility and diffusion of reactants/products and increase the probability of reaction. However, more recent studies have shown solid evidence of thermal effect in microwave reactions. For instance, higher local temperature has been detected in a microwave heated reaction, which is 13–26 °C higher than normal boiling point of solvent [28].

Microwave in heterogeneous catalytic reactions has attracted great attention recently [30, 31]. Though more complicated, the interaction of microwave with liquid, solid, and solid/liquid interface has triggered more efficient mass transfer and faster reaction rate. Figure 2.2 shows the temperature distribution of Magtrieve™ in toluene under microwave irritation. The temperature of Magtrieve™ reached 140 °C within 2 min, which is higher than the bulk toluene phase [29]. In a case study of 1-octanol oxidation reaction, such intensified local heating promoted the reaction yield to 67% within 5 min, which is higher than the yield with conventional heating.

Selective heating of heterogeneous catalysts has been experimentally demonstrated as "hot spots" in a few earlier studies. This is considered solid evidence of thermal effect since the temperature of catalysts is obviously higher than the surrounding medium. Zhang et al. estimated the temperature of hot spots to be 100–200 °C higher than the bulk phase and the size to be about 100 μm in the H_2S

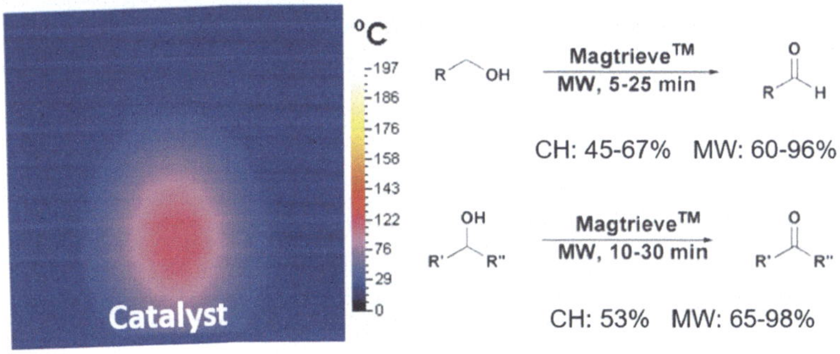

Fig. 2.2 Temperature profile of Magtrieve™ in a toluene solution after 2 min microwave irradiation [29]

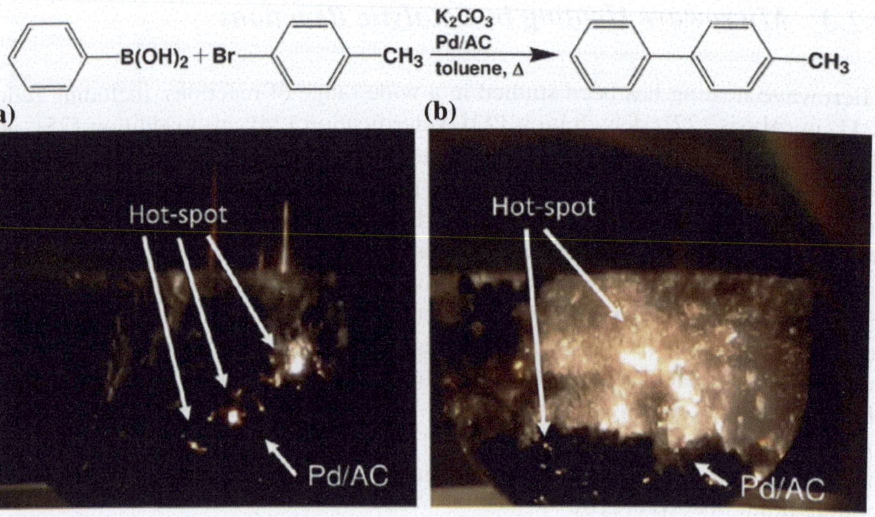

Fig. 2.3 High-speed camera photographs of hot-spot generation on the Pd/AC catalysts surface in the Suzuki–Miyaura coupling reaction: **a** 40 s and **b** 120 s irradiation [32]

decomposition reaction [33]. Satoshi et al. studied Suzuki–Miyaura coupling reaction by using Pd/activated carbon (AC) catalyst and recorded the formation of hot spots on catalysts surface in real time by using a high-speed camera [32]. As shown in Fig. 2.3a, microscopic electric discharges (whitish-orange arcing) are generated on the catalyst surface when microwave irradiates for 40 s. With 120 s irradiation, hot spots are widely populated on the catalyst surface, Fig. 2.3b. Researchers believe that hot spots positively contribute to a fast reaction in microwave [34, 35]. The overheated reactants near hot spots facilitated free radicals formation and triggered faster reactions. In the aqueous-phase oxidation reaction, for instance, hot spots break down H_2O molecules into hydroxyl (·OH) and hydrogen (·H) radicals [36]. To achieve such selective heating and hot-spot generation exactly on heterogeneous catalyst surface, scientists developed microwave-responsive catalysts where the heat generation and reactive sites can be spatially aligned to maximize the energy efficiency of heterogeneous reactions [37].

2.3 Microwave-Responsive Catalysts

2.3.1 The Principle of Microwave-Responsive Catalysts

Microwave-responsive catalysts are defined as a family of catalysts those can effectively absorb microwave and convert microwave to heat. These catalysts should satisfy the basic requirements of conventional catalysts such as rich reactive sites, good thermal and chemical stabilities. Besides, catalysts should be dielectrically active to absorb microwave as well as convert microwave to heat. There are a few different approaches to design and fabricate microwave-responsive catalysts, which are detailed in the following three sections.

(1) *Integrating conventional catalyst with microwave receptor*. A straightforward approach to design microwave-responsive catalysts is constructing a core–shell structure with microwave absorber core and catalyst shell. For example, carbon nanotube (CNT)/TiO_2 core/shell structured catalysts were used in microwave-assisted saccharides (glucose, fructose, etc.) dehydration reaction [37]. High-resolution transmission electron microscopy (HRTEM) image in the inset of Fig. 2.4a confirms the core–shell structure of CNT/TiO_2. Upon microwave irradiation, the internal CNT core would be heated and provided localized thermal energy to accelerate the reaction at the TiO_2 shell. Results showed that TiO_2/CNT catalyst pushes reaction temperature to 162 °C, which is 30% higher than pure TiO_2 system (125 °C) at the same power level of 15 W. As a result, the highest energy efficiency of 4.2 mol KJ^{-1} L^{-1} is obtained by TiO_2/CNT catalyst, which is six times higher than pure TiO_2 and four times higher than CNTs.

(2) *Using dielectrics as catalyst supports*. Dielectric materials like carbon, conductive polymers, and some metal oxides can be used as catalyst supports. For example, carbon-based solid-acid catalyst is one type of microwave-responsive catalysts, which has been used in microwave-assisted acid hydrolysis or pyrolysis reactions [38–40]. In such catalysts, the acid groups on the carbon supports can act as active sites and carbon support itself can be directly heated up by microwave. Moreover, metal particles (Fe, Cu, Pt, and Pd) can be doped on carbon materials and used as catalysts for microwave-assisted redox reactions. For example, Varisli et al. developed Fe/mesoporous carbon catalysts for

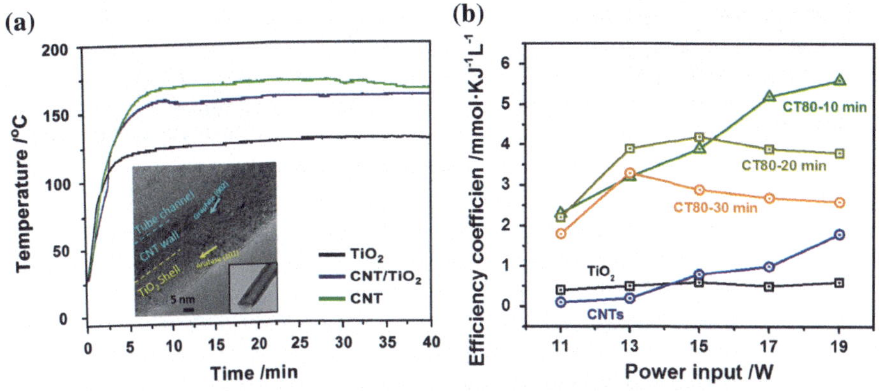

Fig. 2.4 **a** System temperature profiles by using TiO_2, CNTs/TiO_2, and CNTs catalysts. The inset is the HRTEM image of CNTs/TiO_2. Reaction condition: [fructose] = 0.55 mmol, [catalyst] = 50 mg, input power = 15 W. **b** Energy efficiency of HMF production at different power inputs [37]

microwave-assisted ammonia decomposition to produce hydrogen [41]. Mesoporous carbon acts as catalyst supports and microwave receptors. The enhanced thermal effect promoted a complete ammonia conversion reaction at 450 °C, which a much higher reaction temperature of 600 °C is required in conventional heating reactor.

(3) *Creating dipoles at the catalyst surface*. Recently, scientists devoted great efforts to modify catalyst surface or use crystal defect engineering methods to enhance the microwave response of conventional catalysts. Xia et al. found that hydrogenated TiO_2 nanocrystals can respond to electromagnetic wave at 2–18 GHz due to the formation of an amorphous TiO_2 layer on the surface after hydrogenation [43]. Interfacial polarization is explained as the main reason for the induced microwave-responsive behavior, where the crystal defects play an important role at the interface that triggers microwave absorption at the grain boundary. Such phenomenon was also observed in sulfated TiO_2 nanomaterials [42]. When ion-exchanged titanates ($H_2Ti_nO_{2n+1}$) were directly heated at 400 °C, a large amount of Ti-O-SO_4^{2-} acid sites were formed between the crystal layers of titanates. As shown in Figure 2.5, the sulfated TiO_2 formed Bronsted acid sites on O^- and Lewis acid sites on Ti^+, which accelerated charge transfer and form electrical dipole moment on the surface. As a result, such polarized surface becomes very active in the electromagnetic field that helps to generate a significant amount of heat. The results reveal that energy efficiency can be enhanced by nine times by using sulfated TiO_2 catalyst [5.6 mmol $(kJ\ L)^{-1}$] when compared to TiO_2 solid acid [0.6 mmol $(kJ\ L)^{-1}$] in the fructose dehydration reaction.

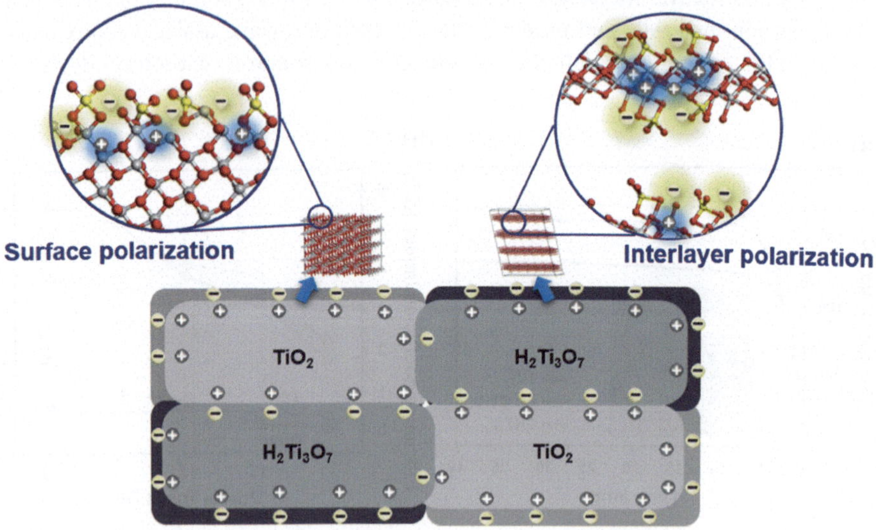

Fig. 2.5 Proposed mechanism of polarization on TiO_2 and $H_2Ti_3O_7$ [42]

2.3.2 Catalytic Performance Evaluation

Under constant power input, the microwave absorption ability of catalysts can be assessed by maximum system temperature and heating rate. To quantify the microwave absorption properties, catalysts can be measured by open-ended coaxial probe method. The catalyst powders are mixed with paraffin wax in an ether solution, then dried by evaporating the solvent. The dried samples are collected and compressed into a toroidal shape. The relative complex permittivity and permeability of the wax-based composites can be measured on a vector network analyzer. Based on these results, microwave absorption properties can be quantified.

Besides the activity, selectivity, and stability, energy efficiency is another factor should be considered in the microwave heating catalytic reactions. Energy efficiency can be calculated by (2.8):

$$\eta = \frac{n_p}{P \cdot t \cdot n_c} \tag{2.8}$$

where η is the energy efficiency factor (J^{-1}), P is the input power of microwave, t is elapsed time, and n_p and n_c are the molar mass of the target product and catalyst, respectively. η directly reflects the energy needed to produce a certain amount of product.

2.3.3 Materials for Microwave-Responsive Catalysts

Carbon materials (activated carbon, CNTs, carbon fiber, mesoporous carbon, graphene, etc.) are the primary selection of most microwave-responsive catalysts, due to their good microwave absorption properties ($0.1 < \tan\delta < 2.95$) and high thermal conductivity (1–3000 W/m K) [44, 45]. Carbon-based microwave-responsive catalysts have been widely used in various reactions such as biomass conversion [46], phenol oxidation [35], NO_x reduction [47], and SO_2 oxidation [48].

Though carbon materials are good candidate in general, they can be dramatically different depending on the specific type of carbon. For example, activated carbon is a type of amorphous carbon with rich porous structure and larger surface area that can be a good catalyst support, while its lower dielectric loss (compared to CNT and graphene-based material [49]) makes it not suitable for microwave receptor. Usually, carbon with high graphitization degree comes with high thermal conductivity and excellent microwave absorption [50]. However, high graphitization degree may sacrifice surface properties for catalyst development. Therefore, the specific demands of carbon materials should be evaluated based on practical application conditions.

Conducting polymers such as polyacetylene (PA), polyaniline (PANI), polypyrrole (PPy), poly(3,4-ethylene dioxythiophene) (PEDOT) are another family of materials suitable for microwave-responsive catalysts [51]. The unique features of

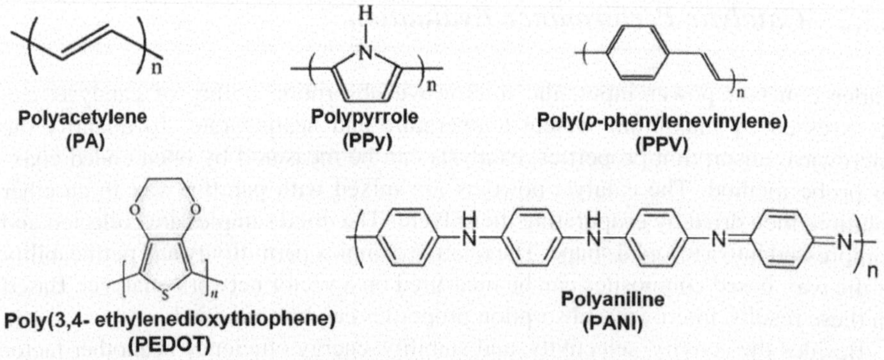

Fig. 2.6 Molecular structure of representative conducting polymers

reversible doping/dedoping, tunable chemical and electrochemical properties, and facile synthesis make them promising materials in catalysis [52, 53]. For example, PANI has been used as catalyst support to load noble metal catalysts [54, 55]. Besides, its abundant acid sites and convenient doping of different acids make it an excellent solid-acid catalyst in hydrolysis reactions [56]. In addition, its high electrical conductivity and dielectric loss make them suitable microwave receptors [57].

2.4 The State of Art of Microwave-Responsive Catalysts in Different Reactions

Microwave-responsive catalysts can be used in liquid-phase, gas-phase, and solid-solid-phase reactions. In this section, the state of art of microwave-responsive catalysts in these reactions is discussed.

2.4.1 Liquid-Phase Organic Synthesis

In liquid-phase microwave-assisted organic synthesis, polar solvents are often used since they are usually good microwave absorbers [59]. Common polar solvents include water, dimethyl sulfoxide (DMSO), ethylene glycol, acetonitrile, N,N-dimethylformamide (DMF), acetic acid, dichloromethane, and acetone.

The temperature window of reaction depends on the maximum temperature of the solvent can reach. The main limitation of using solvent is that the dielectric constant of most solvents decreases with increasing temperature, as shown in Fig. 2.7. For example, the dielectric constant of water is 78 at 25 °C, while it drops down to 20 at 300 °C [60]. Such decrease in dielectric constant means that more energy

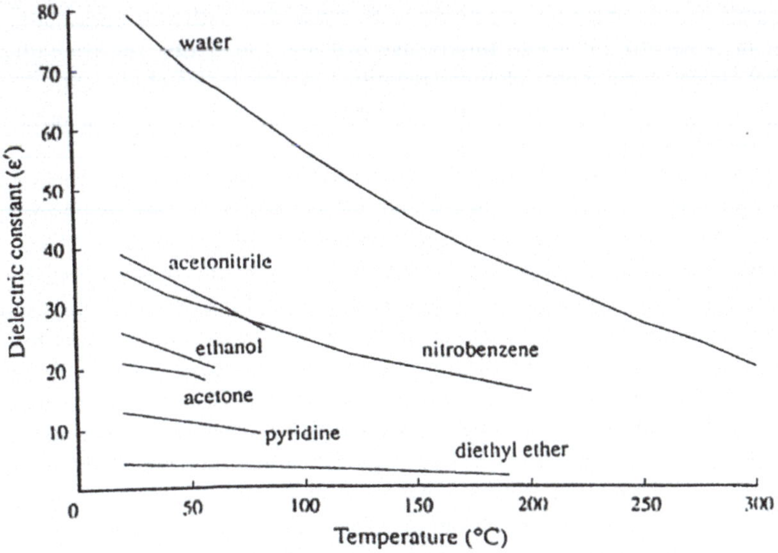

Fig. 2.7 Curves of dielectric constants versus temperature for liquids [58]

input is required to maintain the heating rate or equilibrium temperature as system temperature goes up.

Microwave-responsive catalysts have been demonstrated promising performance in liquid-phase oxidation reactions. For example, Crosswhite et al. prepared a series of nanoscale magnetic spinel oxides with composition of MCr_2O_4 (M = Cu, Co, Fe) and used as catalysts for methanol oxidation [61]. Figure 2.8a compares the heating profiles of pure methanol/water (1:1, v:v) solution and solutions containing spinel catalysts at 5 W. The relatively higher temperature of the heterogeneous system indicates a better microwave response possibly due to the superior microwave absorption

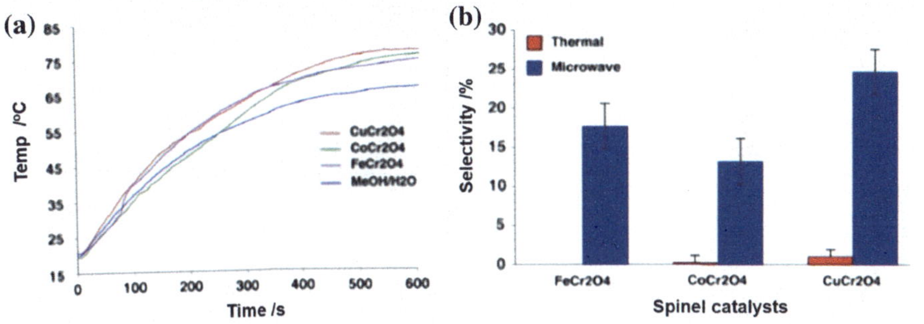

Fig. 2.8 **a** Heating curves and **b** selectivity of formaldehyde for the three spinel catalysts after 80 min of reaction time (6 mL of 1:1 $MeOH:H_2O$ v:v, 166 mg of catalyst) at 60 °C under (blue) microwave and (red) conventional heating [61]

of the catalysts. The reaction selectivity is as low as 2% with conventional heating, while it increases to 25% with microwave heating, Fig. 2.8b. These results indicated that microwave not only changed reaction kinetics, it also changed the reaction pathway that leads to a new reaction equilibrium and maximized the production of targeted products.

Garcia-Costa et al. studied the effect of catalytic support on microwave-assisted phenol oxidation reaction [35]. They found that the rate of phenol oxidation is significantly increased by using Fe/activated carbon (Fe/AC) than Fe/γ-Al_2O_3 under microwave irradiation. Microwave generated hot spots on the surface of AC support, which promoted the formation of $HO_X\cdot$ free radical and thus phenol oxidation. Liu et al. proposed a similar radical mechanism in organic dye oxidation, where hot spots facilitate the pyrolysis of H_2O molecules into hydroxyl radical ($\cdot OH$) and hydrogen radical ($\cdot H$) [36]. O_2 molecules absorbed on catalyst were reduced to superoxide radical ($\cdot O_2^-$). Both $\cdot O_2^-$ and $\cdot OH$ are known to be strong oxidative species for the organic dye oxidation. Different from free radical mechanism, Zhang et al. suggested a hole (h^+) formation process in $NiCo_2O_4$ that is responsible for the oxidation of organics [62]. In this study, photoluminescence spectroscopy was used to monitor the formation of h^+ under microwave irradiation. As illustrated in Fig. 2.9, microwave irradiation created hot spots (temperature > 1473 K) on the $NiCo_2O_4$ surface, which accelerated the movement of electrons (e^-) and induced the separation of e^-/h^+ pairs through thermal excitation.

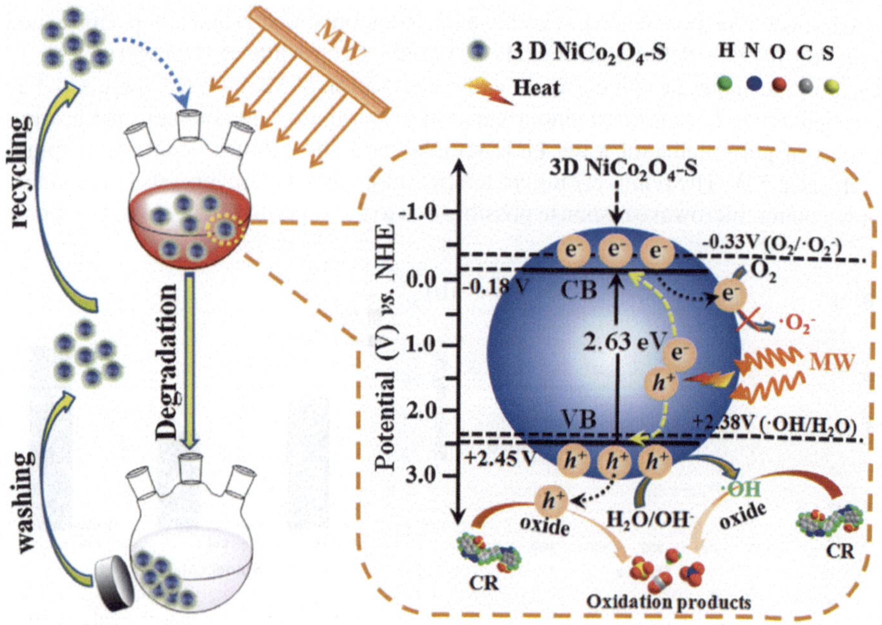

Fig. 2.9 Scheme of h^+ ions formation under microwave irradiation and their effect on degradation of organic dye [62]

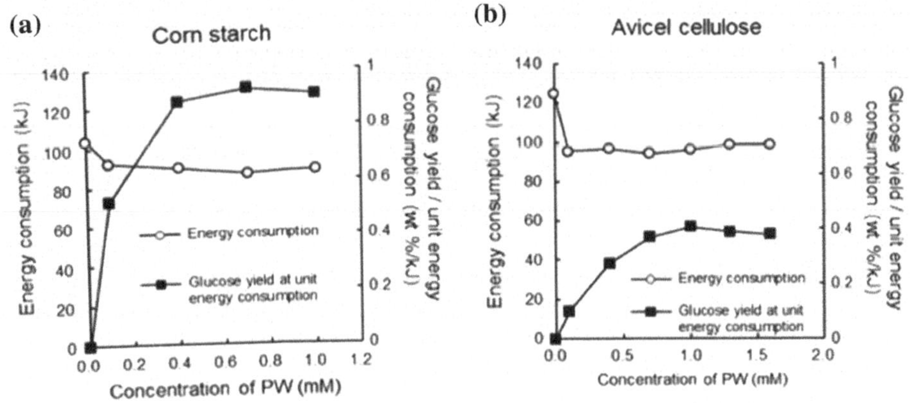

Fig. 2.10 Relations of the POM concentration versus the energy consumption and on efficiency (the glucose yield per unit energy) for **a** corn starch hydrolyzed at 160 °C and **b** Avicel cellulose hydrolyzed at 200 °C

Apart from the supported catalysts, some materials acquire both catalytic activity and microwave absorption property. Polyoxometalate (POM) is one kind of metallic clusters, which has a typical Keggin-type structure ($[XM4O_{40}]^-$) consisting of heteroatoms (X) of P or Si, addenda atoms (M) of Mo or W and oxygen atoms (terminal, edge-bridging, and corner-bridging) [63]. Benefited from the strong Bronsted acidity, oxidative activity, and recyclability, they are often used in heterogeneous catalytic reactions [64, 65]. Tsubaki et al. synthesized a series of POM clusters and used as acid catalysts for the microwave-assisted hydrolysis of corn starch and crystalline cellulose [66]. They found that POMs increased the dielectric loss of the reaction system, leading to a higher energy efficiency of the reactions. As shown in Fig. 2.10, the addition of POM clusters resulted in 17 and 23% reduction of energy consumption for starch and Avicel cellulose hydrolysis, respectively.

PANI can be also used as a microwave-responsive solid-acid catalyst. Not only abundant acid sites and good microwave response, but the acid sites on the backbone can also be easily redoped by a simple acid-washing process. Therefore, the reusability of such catalysts can be greatly improved, which also addressed the water poisoning challenge of solid acids in real practices [67]. Moreover, PANI can be easily polymerized on other supports to facilitate its separation from the reaction mixture [23].

2.4.2 *Gas-Phase Reaction*

Unlike liquid medium, gas molecules can be hardly heated up with microwave irradiation, since the distance between gas molecules is too far to generate effective molecular friction and collision [68]. Hence, catalysts should be responsible for heat generation in gas-phase reactions. Hot-spot formation in gas-phase reaction has been considered the main reason for enhanced conversion of studied reactions including reforming reaction [69, 70], NH_3 decomposition [41], dehydrogenation [71, 72], and hydrodechlorination [73].

Fidalgo et al. studied microwave-assisted dry reforming of CH_4 and CO_2 by using AC catalysts [74]. Results revealed that higher conversion of CO_2 and CH_4 can be achieved in the microwave oven than in the electric furnace, Fig. 2.11. Microwave generated "micro-plasmas" on catalyst surface, resulting in a higher temperature of AC catalyst bed than surrounding atmosphere. In a separate study, it was found that the porosity and surface groups of AC could greatly impact the microwave absorption and activity in CO_2 reforming reaction [75]. However, the existence of oxygen surface groups would obstruct the micro-plasma generation and thus lower the heating capability of AC catalyst.

Tarasov et al. also studied CO_2 reforming under microwave heating using Ni/TiO_2 [76]. Compared to the conventional heating, microwave heating showed obvious advantages in higher propane and CO_2 conversion as well as less coke formation. X-ray absorption spectroscopy (EXAFS) in Fig. 2.12 revealed a partial reduction of Ni^{2+} cation to Ni^0 under microwave irradiation, which could be responsible for the change of catalytic environment and reaction selectivity.

It is worth noticing that metal particles are not only catalytically active for reactions, but also active for graphitized carbon formation during reforming reaction. Odedairo et al. studied the effect of Cr/Fe/Ta doping on the performance of Ni/CeO_2 catalyst in dry reforming of CH_4 [77]. A few layers of graphene were formed on

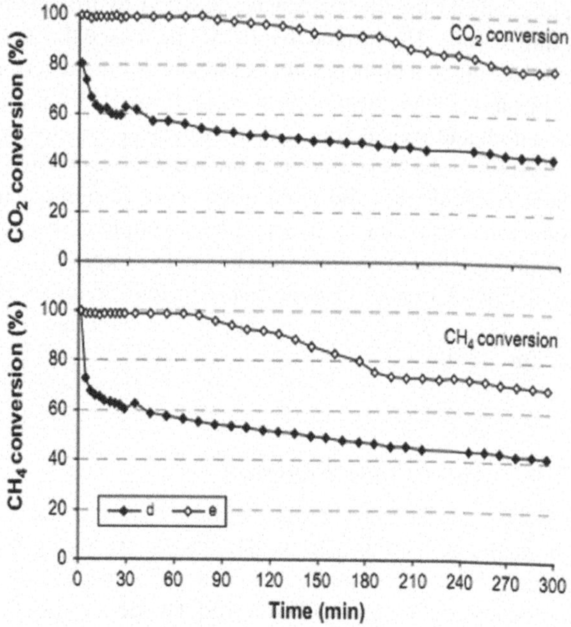

Fig. 2.11 Effect of the heating mode on the dry reforming reaction. Test "d" carried out in the electric furnace and test "e" carried out in the microwave oven

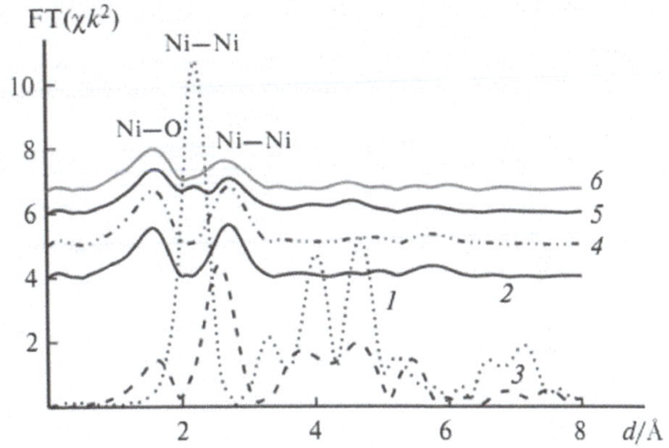

Fig. 2.12 EXAFS spectra of the nickel reference compounds [1, Ni foil, 2, $Ni_2(OH)_2CO_3$, 3, NiO] and Ni catalyst (5%)/TiO_2 (4, initial; 5, after microwaving, 310 °C, H_2; 6, after conventional thermal treatment, 350 °C, H_2) [76]

metal particle surface, Fig. 2.13, which further improves the catalyst performance of the reforming reaction.

Apart from reforming reactions, several studies focused on microwave-assisted decomposition reactions. For example, Deng et al. used pyrolysis residue of sewage sludge (PRSS) as microwave-responsive catalyst in methane decomposition reaction [78]. PRSS has been demonstrated a better microwave absorption than commercial AC, Fig. 2.14, probably due to the inorganic components as revealed by elemental analysis (18 wt% C, 36 wt% O and about 20 wt% of K, Al, Fe, Ca, Mg). It is also found that the gas atmosphere influences the microwave absorption as well. The stable system temperature follows the sequence of $T_{\text{nitrogen}} > T_{\text{hydrogen}} > T_{\text{methane}}$, where nitrogen is considered helpful for the formation of hot spots and higher reaction temperature. Guler et al. developed molybdenum/mesoporous carbon (Mo/C) catalysts

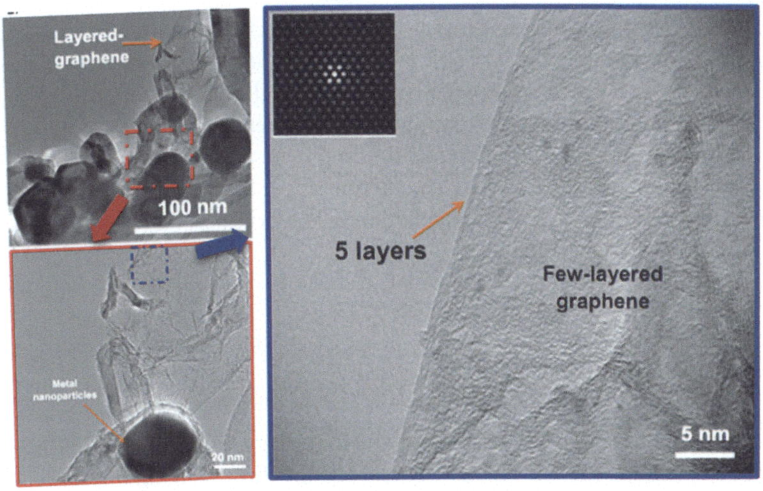

Fig. 2.13 TEM images of layered graphene formed on Ta/Ni/CeO_2 catalyst after 14 h [77]

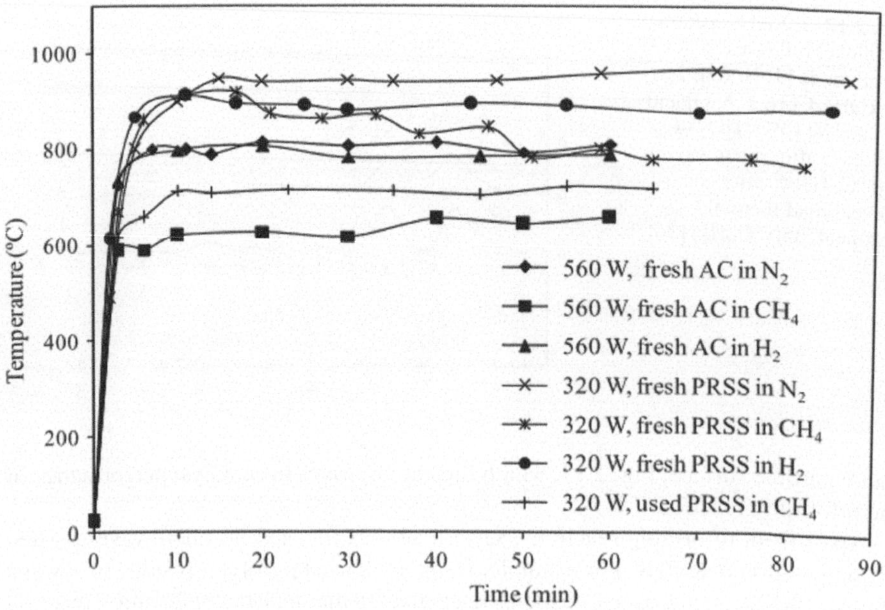

Fig. 2.14 Temperature curves of PRSS and AC under microwave in the different atmosphere (VHSV = $0.3\ \text{L}\ \text{g}^{-1}\ \text{h}^{-1}$)

and used them in microwave-assisted ammonia decomposition reaction [34]. During the reaction, molybdenum carbide was formed, which is highly active for ammonia decomposition. Complete conversion was achieved at 400 °C with microwave heating, while only 49% conversion can be achieved at 600 °C by conventional heating.

In other reactions, Suttisawa et al. demonstrated a tetralin dehydrogenation method by using Pt/AC catalyst [79]. Microwave heating increased dehydrogenation conversion by 19% as compared to conventional heating. Pillai et al. studied microwave-assisted hydrodechlorination of chlorinated benzenes [73]. They found that microwave irradiation is conducive to the fast desorption of the products from catalyst surface and catalyst activity can be remained. In terms of energy efficiency, a significant reduction of power consumption is also observed in the microwave reaction system.

2.4.3 *Solid Biomass Pyrolysis*

Pyrolysis of biomass is a promising technique for energy refinery and waste management [80, 81]. Char/coke, bio-oil, and syngas are the three main products of a pyrolysis reaction. The properties of products largely depend on the biomass properties and pyrolysis reaction conditions. The temperature is the key factor that influences product composition and quality. Like in other reactions, microwave pyrolysis can accelerate the heating rate, reduce the energy consumption, and control the product fraction [82, 83]. However, due to poor microwave response of most biomass species, catalysts are often added to increase the microwave absorption and control the property of pyrolysis products [84]. Omoriyekomwan et al. studied microwave pyrolysis of palm kernel shell by using AC and lignite char as catalysts with an aim to produce phenol-rich bio-oil [46]. The introduction of AC in the reaction is believed to suppress secondary pyrolysis reaction and minimize the production of by-products. Microwave has been demonstrated helpful to increase the phenol production in bio-oil, where the concentration of phenol and total phenolics in bio-oil reached to 64.58 and 71.24%, respectively.

Doping metal particles on AC support have been demonstrated successful to improve microwave heating performance. Zhang et al. developed a series of rice-husk char (RHC)-supported metallic (Ni, Fe and Cu) catalyst for syngas production [85]. Compared to pure RHC, as shown in Fig. 2.15a, the maximum temperature increased by 24.27%, 15.71%, and 9.56% after incorporating Fe, Ni, and Cu components, respectively. Meanwhile, the maximum heating rate increased by 87.21%, 61.76%, and 60.55%, respectively. Metal particles induce corona discharge phenomena during the microwave heating, which can boost hot-spot formation and improve the microwave thermal effects [86]. Meanwhile, the metal particles can accelerate the cracking/reforming reactions and increase the yield of syngas, Fig. 2.15b. Dong et al. studied microwave-assisted pyrolysis of bamboo with AC supported iron (III) catalyst [87]. Compared to pure bamboo, the maximal reaction temperature was increased by 44.5% after incorporating 10 wt% catalyst. Further increase in catalyst loading decreased the maximum temperature, while gas yield was significantly increased from 25.62 to 65.16%. Moreover, the gas quality was also improved with syngas composition increased from 66.24 to 81.14 vol.% together with an increased H_2/CO ratio from 0.78 to 1.04.

Mohamed et al. investigated the pyrolysis of switchgrass by adding different species of K_3PO_4, clinoptilolite, bentonite, SiC, and AC [88]. Switchgrass mixed

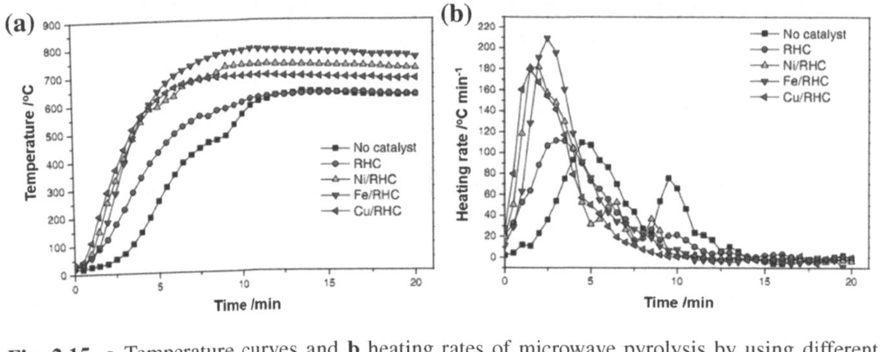

Fig. 2.15 **a** Temperature curves and **b** heating rates of microwave pyrolysis by using different catalysts [85]

with 10 wt% K_3PO_4 + 10 wt% bentonite reached 400 °C in 2.8 min by microwave heating, compared with 28.8 min by conventional heating, shown in Fig. 2.16a. Figure 2.16b shows the pyrolysis product distribution with the addition of SiC, AC and combinations of K_3PO_4, bentonite, or clinoptilolite. These results indicated that SiC and AC favored gas production, while K_3PO_4/bentonite increased bio-oil yield.

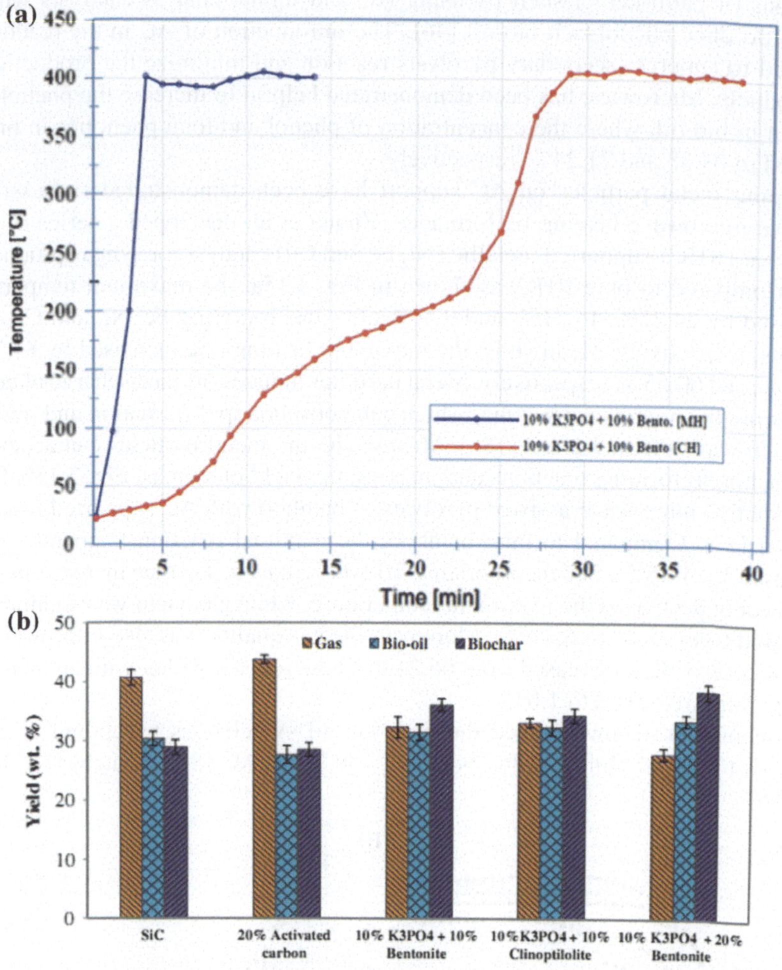

Fig. 2.16 **a** Comparison of temperature profile with microwave heating and conventional heating, **b** pyrolysis products distribution of switchgrass with 20 wt% SiC, 10 wt% K_3PO_4 + 10 wt% bentonite, 10 wt% K_3PO_4 + 20 wt% bentonite, 10 wt% K_3PO_4 + 10 wt% clinoptilolite, and 20 wt% of activated carbon [88]

2.5 Strategies to Enhance the Microwave Thermal Effect

Microwave-responsive catalysts have been demonstrated successful in a wide range of catalytic reactions mainly due to the spatial alignment of reactive sites and hot spots. Strategies discussed in this section attempt to enhance the thermal effect via material design. The outcome of such strategies could further expedite the reaction rate and improve the energy efficiency in a wider spectrum of reactions.

2.5.1 Integrating Magnetic Loss Materials

To further improve the microwave thermal effect, magnetic loss phase can be coupled with dielectric loss phase, where dual absorption peaks can be realized to maximize the generation of hot spots. Common magnetic loss materials are Fe_3O_4, cobalt (Co), γ-Fe_2O_3, Fe, etc. [89–92], which can be introduced in catalyst during synthesis or via a post-doping process [93, 94]. These materials can absorb microwaves from both electric and magnetic components. Besides, the introduction of magnetic loss materials would improve the electron conductivity of catalyst, which could further improve its microwave response [95]. However, most magnetic loss materials cannot exert the best performance at 2.45 GHz, since largest absorption is often observed in the frequency range of 4–14 GHz [96]. However, successful cases were also reported to improve catalyst performance by incorporating magnetic loss materials. Liu et al. synthesized Fe_3O_4/CNT catalyst for catalytic degradation of organics [36]. As shown in Fig. 2.17, the reflection loss reached a minimum of -19.8 dB at about 2.6 GHz for the Fe_3O_4/CNTs, indicating a better microwave absorption at 2.45 GHz than that of the individual components. As such, the hot spot formation can be promoted as well as the oxidation of antibiotics such as chlortetracycline, tetracycline, and oxytetracycline.

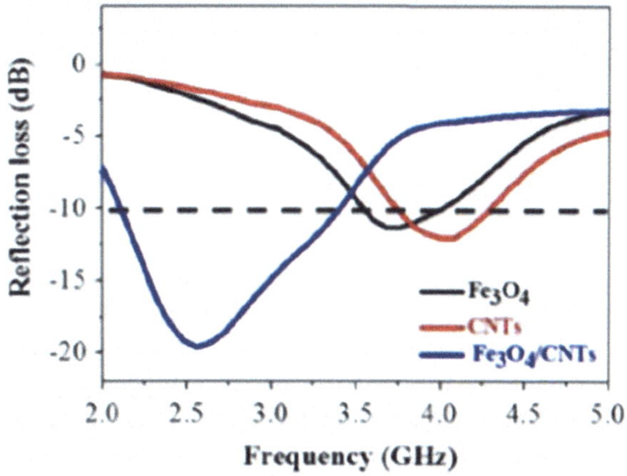

Fig. 2.17 Reflection loss of CNTs, Fe_3O_4, and Fe_3O_4/CNTs

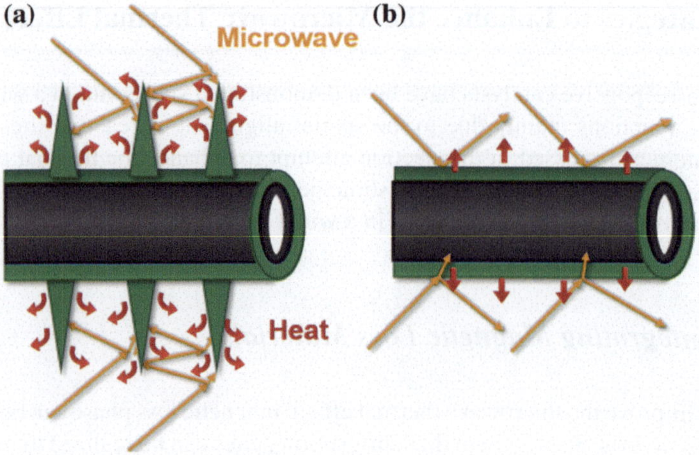

Fig. 2.18 Scheme of microwave transmission and reflection and heat dissipation on **a** rough surface and **b** smooth surface [23]

2.5.2 Morphology Control

Morphology is another factor that affects microwave–material interaction. Considerable research has been done to understand the effect of material morphology on microwave response behavior [94, 97], which is often explained by the change of electromagnetic impedance match [98]. Usually, nanomaterials with complex and rough surface structure are favored for microwave absorption [23]. Since scattering and multiple absorption of the incident microwave could occur on rough surfaces. In a core–shell structured CNT/PANI catalyst with flake-patterned PANI grew radially from the center of shell, microwave absorption can be improved with enhanced reflection. The effect of surface structure on the microwave transmission/reflection is schematically presented in Fig. 2.18, where most of the electromagnetic wave can be translated into the material on a rough surface while reflected on a smooth surface. As a result, higher system temperature can be achieved with rough CNT/PANI (158 °C) as compared to the smooth one (153 °C).

2.5.3 Heteroatoms Doping

Heteroatom-doped materials, especially carbon materials, have drawn prominent interest in catalysts [99], electrochemistry [100], energy storage [101], and sensors [102]. The advantage of heteroatoms doping including: (1) (N, S, B, etc.) dopants with higher electronegativity result in higher positive charge density on adjacent carbon atoms and thus greater polarity; (2) dopants bring more electrons to the delocalized π bonds of carbon and thus larger electrical conductivity; (3) doping process generates more point defects (voids, vacancies, Stone–Wales defects) and line defects (grain boundaries, dislocation) in the carbon skeleton, which breaks the electron–hole symmetry. In general, heteroatom doping is beneficial in terms of surface polarization and hot-spot formation.

As mentioned in Sect. 2.3.1, sulfated TiO_2 can form surface polarization to enhance microwave response. The sulfate groups at catalyst surface not only serve as polarization center to generate heat under the microwave, but also act as catalytic acid sites for saccharides conversion reaction. The surface structure and properties of TiO_2 can be further promoted by incorporating other n-type dopants, such as vanadium- (V), niobium (Nb), and tantalum (Ta) [104]. Nb^{5+} ($r = 0.70$ Å) has a similar ionic radius as Ti^{4+} ($r = 0.68$ Å), and thus, the lattice mismatch between TiO_2 and Nb_2O_5 is negligible. In the literature, manipulation electronic structure of titania by incorporating Nb atoms has been extensively discussed [105]. The multivalent cation doping increases the electrical conductivity of TiO_2. Microwave absorption study showed the equilibrium temperature increased from 146 °C of TiO_2 to 161 °C of Nb-doped TiO_2 at 15 W. Mechanism analysis of microwave absorption found that sulfate acid sites polarized the surrounding H_2O molecules at the interface, which accelerated fructose dehydration to HMF. Moreover, the introduction of Nb-enhanced electron transfer in the matrix and the ratio of Bronsted acid, which further promoted the polarization effect at the interface, Fig. 2.19. As a result, the highest rate constant ($0.58\ min^{-1}$) and HMF yield (40.5 mol%) were achieved with the Nb-doped TiO_2 catalyst.

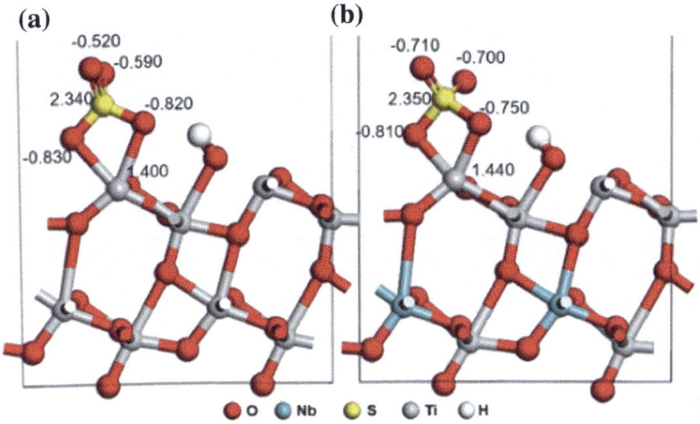

Fig. 2.19 Simulation of sulfonic groups on **a** anatase (101) **b** Nb-doped anatase (101) [103]

2.6 Summary and Future Perspectives

Microwave-responsive catalysts integrate the catalyst function and microwave absorption into one single material, which has been demonstrated effective to expedite reaction rate, enhance energy efficiency, and improve product quality. With the presence of microwave-responsive catalysts, microwaves can selectively heat the catalyst in a heterogeneous reaction and thus expedite mass transfer and reaction at catalyst/solvent interface. Dielectric loss materials such as carbon and conductive polymers have been explored as microwave-responsive catalysts. Taking advantages of thermal effects, these catalysts exhibited promising performance in liquid-phase organic synthesis, gas-phase reaction, and solid biomass pyrolysis. To further improve the performance of microwave-responsive catalysts, several strategies such as integrating magnetic loss components, morphology control, and heteroatoms doping, were explored. Until now, the advantages of microwave-responsive catalysts have been realized, while the potential of such catalysts has not yet been fully explored. The complex mechanism of thermal effects (e.g., hot spots, superheating) is yet to be explored. More experimental techniques, computational simulation tools, and fundamental theories need to be developed.

Considerable attention has been given to microwave-assisted catalytic reactions nowadays. However, most of the existing research focuses on the microwave energy input for chemical reactions, while the development of microwave-responsive catalyst is still at its early stage. According to Google Scholar, only less than 100 studies were reported for the design of microwave-responsive catalysts (up to 2018). In terms of catalyst design, the major limitation is the fixed microwave frequency at 2.45 GHz. In the future, microwave reactors with tunable frequency can be designed and the operating frequency can be optimized for specific reactions. In addition, energy consumption of microwave-assisted reaction is rarely discussed in existing reports. Even though microwave heating could accelerate the reaction rate, the generation of microwave itself also consumes large amount of energy. Therefore, it is suggested that the energy efficiency of reactions should be included in future reports, which is critical to establish an economic factor for industrial-scale applications.

At last, many previous works demonstrated the critical role of hot spots in the catalytic reactions. Hot spots are considered responsible for free radical generation, improved mass transfer, altered reaction pathway, etc. However, it remains a great challenge to fully understand the formation of hot spots and their fundamental role in reactions mainly due to the technical gap between characterization at macroscale and reaction at nanoscale [106]. To address this challenge, characterization techniques should be further developed to capture the experimental phenomenon at nanoscale.

References

1. S. Faraji, F.N. Ani, Microwave-assisted synthesis of metal oxide/hydroxide composite electrodes for high power supercapacitors—a review. J. Power Sources **263**, 338–360 (2014)

2. D. Dallinger, C.O. Kappe, Microwave-assisted synthesis in water as solvent. Chem. Rev. **107**(6), 2563–2591 (2007)
3. C. Si, J. Wu, Y. Wang, Y. Zhang, X. Shang, Drying of low-rank coals: a review of fluidized bed technologies. Drying Technol. **33**(3), 277–287 (2015)
4. F.J. Barba, Z. Zhu, M. Koubaa, A.S. Sant'Ana, V. Orlien, Green alternative methods for the extraction of antioxidant bioactive compounds from winery wastes and by-products: a review. Trends Food Sci. Technol. **49**, 96–109 (2016)
5. F. Mushtaq, R. Mat, F.N. Ani, A review on microwave assisted pyrolysis of coal and biomass for fuel production. Renew. Sustain. Energy Rev. **39**, 555–574 (2014)
6. S. Nomanbhay, M. Ong, A review of microwave-assisted reactions for biodiesel production. Bioengineering **4**(2), 57 (2017)
7. F. Motasemi, M.T. Afzal, A review on the microwave-assisted pyrolysis technique. Renew. Sustain. Energy Rev. **28**, 317–330 (2013)
8. R.N. Baig, R.S. Varma, Alternative energy input: mechanochemical, microwave and ultrasound-assisted organic synthesis. Chem. Soc. Rev. **41**(4), 1559–1584 (2012)
9. N. Remya, J.-G. Lin, Current status of microwave application in wastewater treatment—a review. Chem. Eng. J. **166**(3), 797–813 (2011)
10. A. de la Hoz, A. Diaz-Ortiz, A. Moreno, Microwaves in organic synthesis. Thermal and non-thermal microwave effects. Chem. Soc. Rev. **34**(2), 164–178 (2005)
11. A.K. Datta, P.M. Davidson, Microwave and radio frequency processing. J. Food Sci. **65**, 32–41 (2000)
12. M. Oghbaei, O. Mirzaee, Microwave versus conventional sintering: a review of fundamentals, advantages and applications. J. Alloys Compd. **494**(1–2), 175–189 (2010)
13. E. Thostenson, T.-W. Chou, Microwave processing: fundamentals and applications. Compos. A **30**(9), 1055–1071 (1999)
14. J.M. Osepchuk, A history of microwave heating applications. IEEE Trans. Microw. Theory Tech. **32**(9), 1200–1224 (1984)
15. R. Kelly, N. Rowson, Microwave reduction of oxidised ilmenite concentrates. Miner. Eng. **8**(11), 1427–1438 (1995)
16. H. Kingston, L. Jassie, Microwave energy for acid decomposition at elevated temperatures and pressures using biological and botanical samples. Anal. Chem. **58**(12), 2534–2541 (1986)
17. F. Içier, T. Baysal, Dielectrical properties of food materials-1: factors affecting and industrial uses. Crit. Rev. Food Sci. Nutr. **44**(6), 465–471 (2004)
18. R.V. Decareau, *Microwaves in the Food Processing Industry* (Academic Press, London, 1985)
19. D.D. Dinčov, K.A. Parrott, K.A. Pericleous, Heat and mass transfer in two-phase porous materials under intensive microwave heating. J. Food Eng. **65**(3), 403–412 (2004)
20. A.A. Metaxas, R.J. Meredith, *Industrial Microwave Heating* (IET, 1983)
21. C. González-Arellano, J.M. Campelo, D.J. Macquarrie, J.M. Marinas, A.A. Romero, R. Luque, Efficient microwave oxidation of alcohols using low-loaded supported metallic iron nanoparticles. ChemSusChem Chem. Sustain. Energy Mater. **1**(8–9), 746–750 (2008)
22. S.S. Lam, H.A. Chase, A review on waste to energy processes using microwave pyrolysis. Energies **5**(10), 4209–4232 (2012)
23. T. Ji, R. Tu, L. Mu, X. Lu, J. Zhu, Structurally tuning microwave absorption of core/shell structured CNT/polyaniline catalysts for energy efficient saccharide-HMF conversion. Appl. Catal. B **220**, 581–588 (2018)
24. T. Lieu, S. Yusup, M. Moniruzzaman, Kinetic study on microwave-assisted esterification of free fatty acids derived from Ceiba pentandra seed oil. Bioresour. Technol. **211**, 248–256 (2016)
25. R. Babu, S.-H. Kim, A.C. Kathalikkattil, R.R. Kuruppathparambil, D.W. Kim, S.J. Cho, D.-W. Park, Aqueous microwave-assisted synthesis of non-interpenetrated metal-organic framework for room temperature cycloaddition of CO_2 and epoxides. Appl. Catal. A **544**, 126–136 (2017)
26. R. Bernini, E. Mincione, M. Barontini, G. Provenzano, L. Setti, Obtaining 4-vinylphenols by decarboxylation of natural 4-hydroxycinnamic acids under microwave irradiation. Tetrahedron **63**(39), 9663–9667 (2007)

27. J. Jacob, L. Chia, F. Boey, Thermal and non-thermal interaction of microwave radiation with materials. J. Mater. Sci. **30**(21), 5321–5327 (1995)
28. D.R. Baghurst, D.M.P. Mingos, Superheating effects associated with microwave dielectric heating. J. Chem. Soc. Chem. Commun. (9), 674–677 (1992)
29. D. Bogdal, M. Lukasiewicz, J. Pielichowski, A. Miciak, S. Bednarz, Microwave-assisted oxidation of alcohols using Magtrieve^{TM}. Tetrahedron **59**(5), 649–653 (2003)
30. H. Will, P. Scholz, B. Ondruschka, Microwave-assisted heterogeneous gas-phase catalysis. Chem. Eng. Technol. Ind. Chem. Plant Equipment Process Eng. Biotechnol. **27**(2), 113–122 (2004)
31. N. Sharma, U.K. Sharma, E.V. Van der Eycken, Microwave-assisted organic synthesis: overview of recent applications. Green Techn. Org. Synth. Med. Chem. 441–468 (2018)
32. S. Horikoshi, A. Osawa, M. Abe, N. Serpone, On the generation of hot-spots by microwave electric and magnetic fields and their impact on a microwave-assisted heterogeneous reaction in the presence of metallic Pd nanoparticles on an activated carbon support. J. Phys. Chem. C **115**(46), 23030–23035 (2011)
33. X. Zhang, D.O. Hayward, D.M.P. Mingos, Apparent equilibrium shifts and hot-spot formation for catalytic reactions induced by microwave dielectric heating. Chem. Commun. **11**, 975–976 (1999)
34. M. Guler, T. Dogu, D. Varisli, Hydrogen production over molybdenum loaded mesoporous carbon catalysts in microwave heated reactor system. Appl. Catal. B **219**, 173–182 (2017)
35. A.L. Garcia-Costa, J.A. Zazo, J.J. Rodriguez, J.A. Casas, Microwave-assisted catalytic wet peroxide oxidation. Comparison of Fe catalysts supported on activated carbon and gamma-alumina. Appl. Catal. B **218**, 637–642 (2017)
36. S.Y. Liu, L.F. Mei, X.L. Liang, L.B. Liao, G.C. Lv, S.F. Ma, S.Y. Lu, A. Abdelkader, K. Xi, Anchoring Fe_3O_4 nanoparticles on carbon nanotubes for microwave-induced catalytic degradation of antibiotics. ACS Appl. Mater. Interfaces **10**(35), 29467–29475 (2018)
37. T. Ji, R. Tu, L. Mu, X. Lu, J. Zhu, Enhancing energy efficiency in saccharide–HMF conversion with core/shell structured microwave responsive catalysts. ACS Sustain. Chem. Eng. **5**(5), 4352–4358 (2017)
38. T. Chen, C. Fan, One-pot generation of mesoporous carbon supported nanocrystalline $H_3PW_{12}O_{40}$ heteropoly acid with high performance in microwave esterification of acetic acid and isoamyl alcohol. J. Porous Mater. **20**(5), 1225–1230 (2013)
39. S. Horikoshi, T. Minagawa, S. Tsubaki, A. Onda, N. Serpone, Is selective heating of the sulfonic acid catalyst $ACSO_3H$ by microwave radiation crucial in the acid hydrolysis of cellulose to glucose in aqueous media? Catalysts **7**(8), 231 (2017)
40. Y. Wu, Z. Fu, D. Yin, Q. Xu, F. Liu, C. Lu, L. Mao, Microwave-assisted hydrolysis of crystalline cellulose catalyzed by biomass char sulfonic acids. Green Chem. **12**(4), 696–700 (2010)
41. D. Varisli, C. Korkusuz, T. Dogu, Microwave-assisted ammonia decomposition reaction over iron incorporated mesoporous carbon catalysts. Appl. Catal. B **201**, 370–380 (2017)
42. T. Ji, R. Tu, L. Li, L. Mu, C. Liu, X. Lu, J. Zhu, Localizing microwave heat by surface polarization of titanate nanostructures for enhanced catalytic reaction efficiency. Appl. Catal. B **227**, 266–275 (2018)
43. T. Xia, C. Zhang, N.A. Oyler, X. Chen, Hydrogenated TiO_2 nanocrystals: a novel microwave absorbing material. Adv. Mater. **25**(47), 6905–6910 (2013)
44. J.A. Menendez, A. Arenillas, B. Fidalgo, Y. Fernandez, L. Zubizarreta, E.G. Calvo, J.M. Bermudez, Microwave heating processes involving carbon materials. Fuel Process. Technol. **91**(1), 1–8 (2010)
45. N. Mehra, L. Mu, T. Ji, J. Zhu, Chapter 3—Thermal conduction in polymer composites, in *Polymer-Based Multifunctional Nanocomposites and Their Applications*, ed. by K. Song, C. Liu, J.Z. Guo (Elsevier, Amsterdam, 2019), pp. 77–110
46. J.E. Omoriyekomwan, A. Tahmasebi, J.L. Yu, Production of phenol-rich bio-oil during catalytic fixed-bed and microwave pyrolysis of palm kernel shell. Bioresour. Technol. **207**, 188–196 (2016)

47. X. Zhang, D.O. Hayward, Applications of microwave dielectric heating in environment-related heterogeneous gas-phase catalytic systems. Inorg. Chim. Acta **359**(11), 3421–3433 (2006)
48. C.Y. Cha, D.S. Kim, Microwave induced reactions of sulfur dioxide and nitrogen oxides in char and anthracite bed. Carbon **39**(8), 1159–1166 (2001)
49. J. Lee, J. Kim, T. Hyeon, Recent progress in the synthesis of porous carbon materials. Adv. Mater. **18**(16), 2073–2094 (2006)
50. M. Biercuk, M.C. Llaguno, M. Radosavljevic, J. Hyun, A.T. Johnson, J.E. Fischer, Carbon nanotube composites for thermal management. Appl. Phys. Lett. **80**(15), 2767–2769 (2002)
51. T.K. Das, S. Prusty, Review on conducting polymers and their applications. Polym. Plast. Technol. Eng. **51**(14), 1487–1500 (2012)
52. A.D. Chowdhury, N. Agnihotri, A. De, Hydrolysis of sodium borohydride using Ru–Co-PEDOT nanocomposites as catalyst. Chem. Eng. J. **264**, 531–537 (2015)
53. B. Mu, A. Wang, One-pot fabrication of multifunctional superparamagnetic attapulgite/Fe_3O_4/polyaniline nanocomposites served as an adsorbent and catalyst support. J. Mater. Chem. A **3**(1), 281–289 (2015)
54. R.K. Pandey, V. Lakshminarayanan, Electro-oxidation of formic acid, methanol, and ethanol on electrodeposited Pd-polyaniline nanofiber films in acidic and alkaline medium. J. Phys. Chem. C **113**(52), 21596–21603 (2009)
55. S. Guo, S. Dong, E. Wang, Polyaniline/Pt hybrid nanofibers: high-efficiency nanoelectrocatalysts for electrochemical devices. Small **5**(16), 1869–1876 (2009)
56. R. Yuan, H. Wang, T. Ji, L. Mu, L. Chen, Y. Zhu, J. Zhu, Superhydrophobic polyaniline hollow spheres with mesoporous brain-like convex-fold shell textures. J. Mater. Chem. A **3**(38), 19299–19303 (2015)
57. P. Saini, V. Choudhary, B. Singh, R. Mathur, S. Dhawan, Polyaniline–MWCNT nanocomposites for microwave absorption and EMI shielding. Mater. Chem. Phys. **113**(2), 919–926 (2009)
58. J.A. Dean, *Lange's Handbook of Chemistry* (McGraw-Hill Inc., New York; London, 1999)
59. A.M. Rodríguez, P. Prieto, A. de la Hoz, Á. Díaz-Ortiz, D.R. Martín, J.I. García, Influence of polarity and activation energy in microwave–assisted organic synthesis (MAOS). ChemistryOpen **4**(3), 308–317 (2015)
60. C.R. Strauss, R.W. Trainor, Developments in microwave-assisted organic chemistry. Aust. J. Chem. **48**(10), 1665–1692 (1995)
61. M. Crosswhite, J. Hunt, T. Southworth, K. Serniak, A. Ferrari, A.E. Stiegman, Development of magnetic nanoparticles as microwave-specific catalysts for the rapid, low-temperature synthesis of formalin solutions. ACS Catal. **3**(6), 1318–1323 (2013)
62. X.Y. Liu, D. Xu, D.F. Zhang, G.Z. Zhang, L. Zhang, Superior performance of 3D Co-Ni bimetallic oxides for catalytic degradation of organic dye: Investigation on the effect of catalyst morphology and catalytic mechanism. Appl. Catal. B **186**, 193–203 (2016)
63. I.V. Kozhevnikov, Catalysis by heteropoly acids and multicomponent polyoxometalates in liquid-phase reactions. Chem. Rev. **98**(1), 171–198 (1998)
64. T. Yamase, M.T. Pope, *Polyoxometalate Chemistry for Nano-composite Design* (Springer Science & Business Media, 2006)
65. H. Lv, Y.V. Geletii, C. Zhao, J.W. Vickers, G. Zhu, Z. Luo, J. Song, T. Lian, D.G. Musaev, C.L. Hill, Polyoxometalate water oxidation catalysts and the production of green fuel. Chem. Soc. Rev. **41**(22), 7572–7589 (2012)
66. S. Tsubaki, K. Oono, T. Ueda, A. Onda, K. Yanagisawa, T. Mitani, J. Azuma, Microwave-assisted hydrolysis of polysaccharides over polyoxometalate clusters. Bioresour. Technol. **144**, 67–73 (2013)
67. T. Okuhara, Water-tolerant solid acid catalysts. Chem. Rev. **102**(10), 3641–3666 (2002)
68. P. Lidström, J. Tierney, B. Watheyb, J. Westmana, Microwave assisted organic synthesis: a review. Tetrahedron **57**, 9225–9283 (2001)
69. X. Zhang, C.S.-M. Lee, D.M.P. Mingos, D.O. Hayward, Carbon dioxide reforming of methane with Pt catalysts using microwave dielectric heating. Catal. Lett. **88**(3–4), 129–139 (2003)

70. T. Durka, G.D. Stefanidis, T. Van Gerven, A.I. Stankiewicz, Microwave-activated methanol steam reforming for hydrogen production. Int. J. Hydrogen Energy **36**(20), 12843–12852 (2011)
71. S. Horikoshi, A. Osawa, S. Sakamoto, N. Serpone, Control of microwave-generated hot spots Part IV Control of hot spots on a heterogeneous microwave-absorber catalyst surface by a hybrid internal/external heating method. Chem. Eng. Process. Process Intensif. **69**, 52–56 (2013)
72. B. Nigrovski, U. Zavyalova, P. Scholz, K. Pollok, M. Müller, B. Ondruschka, Microwave-assisted catalytic oxidative dehydrogenation of ethylbenzene on iron oxide loaded carbon nanotubes. Carbon **46**(13), 1678–1686 (2008)
73. U.R. Pillai, E. Sahle-Demessie, R.S. Varma, Hydrodechlorination of chlorinated benzenes in a continuous microwave reactor. Green Chem. **6**(6), 295–298 (2004)
74. B. Fidalgo, A. Dominguez, J.J. Pis, J.A. Menendez, Microwave-assisted dry reforming of methane. Int. J. Hydrogen Energy **33**(16), 4337–4344 (2008)
75. B. Fidalgo, A. Arenillas, J.A. Menendez, Influence of porosity and surface groups on the catalytic activity of carbon materials for the microwave-assisted CO_2 reforming of CH_4. Fuel **89**(12), 4002–4007 (2010)
76. A.L. Tarasov, O.P. Tkachenko, O.A. Kirichenko, L.M. Kustov, Microwave-activated carbon dioxide reforming of propane over Ni/TiO_2 catalysts. Russ. Chem. Bull. **65**(12), 2820–2824 (2016)
77. T. Odedairo, J. Ma, J.L. Chen, S.B. Wang, Z.H. Zhu, Influences of doping Cr/Fe/Ta on the performance of Ni/CeO_2 catalyst under microwave irradiation in dry reforming of CH_4. J. Solid State Chem. **233**, 166–177 (2016)
78. W.Y. Deng, Y.X. Su, S.G. Liu, H.G. Shen, Microwave-assisted methane decomposition over pyrolysis residue of sewage sludge for hydrogen production. Int. J. Hydrogen Energy **39**(17), 9169–9179 (2014)
79. Y. Suttisawat, S. Horikoshi, H. Sakai, M. Abe, Hydrogen production from tetralin over microwave-accelerated Pt-supported activated carbon. Int. J. Hydrogen Energy **35**(12), 6179–6183 (2010)
80. Y. Wan, P. Chen, B. Zhang, C. Yang, Y. Liu, X. Lin, R. Ruan, Microwave-assisted pyrolysis of biomass: catalysts to improve product selectivity. J. Anal. Appl. Pyrolysis **86**(1), 161–167 (2009)
81. A.V. Bridgwater, Review of fast pyrolysis of biomass and product upgrading. Biomass Bioenergy **38**, 68–94 (2012)
82. X. Zhao, M. Wang, H. Liu, L. Li, C. Ma, Z. Song, A microwave reactor for characterization of pyrolyzed biomass. Bioresour. Technol. **104**, 673–678 (2012)
83. C. Yin, Microwave-assisted pyrolysis of biomass for liquid biofuels production. Bioresour. Technol. **120**, 273–284 (2012)
84. Y. Fernández, A. Arenillas, J.Á. Menéndez, Microwave heating applied to pyrolysis. In *Advances in Induction and Microwave Heating of Mineral and Organic Materials* (InTech, 2011)
85. S.P. Zhang, Q. Dong, L. Zhang, Y.Q. Xiong, High quality syngas production from microwave pyrolysis of rice husk with char-supported metallic catalysts. Bioresour. Technol. **191**, 17–23 (2015)
86. P. Lahijani, Z.A. Zainal, A.R. Mohamed, M. Mohammadi, Microwave-enhanced CO_2 gasification of oil palm shell char. Bioresour. Technol. **158**, 193–200 (2014)
87. Q. Dong, M.M. Niu, D.M. Bi, W.Y. Liu, X.X. Gu, C. Lu, Microwave-assisted catalytic pyrolysis of moso bamboo for high syngas production. Bioresour. Technol. **256**, 145–151 (2018)
88. B.A. Mohamed, C.S. Kim, N. Ellis, X.T. Bi, Microwave-assisted catalytic pyrolysis of switchgrass for improving bio-oil and biochar properties. Bioresour. Technol. **201**, 121–132 (2016)
89. D. Chung, Electromagnetic interference shielding effectiveness of carbon materials. Carbon **39**(2), 279–285 (2001)

90. A.N. Yusoff, M. Abdullah, S. Ahmad, S. Jusoh, A. Mansor, S. Hamid, Electromagnetic and absorption properties of some microwave absorbers. J. Appl. Phys. **92**(2), 876–882 (2002)
91. G. Sun, B. Dong, M. Cao, B. Wei, C. Hu, Hierarchical dendrite-like magnetic materials of Fe_3O_4, γ-Fe_2O_3, and Fe with high performance of microwave absorption. Chem. Mater. **23**(6), 1587–1593 (2011)
92. Q. Liu, Q. Cao, H. Bi, C. Liang, K. Yuan, W. She, Y. Yang, R. Che, CoNi@SiO_2@TiO_2 and CoNi@Air@TiO_2 microspheres with strong wideband microwave absorption. Adv. Mater. **28**(3), 486–490 (2016)
93. X. Bai, Y. Zhai, Y. Zhang, Green approach to prepare graphene-based composites with high microwave absorption capacity. J. Phys. Chem. C **115**(23), 11673–11677 (2011)
94. R.C. Che, L.M. Peng, X.F. Duan, Q. Chen, X. Liang, Microwave absorption enhancement and complex permittivity and permeability of Fe encapsulated within carbon nanotubes. Adv. Mater. **16**(5), 401–405 (2004)
95. M.-S. Cao, J. Yang, W.-L. Song, D.-Q. Zhang, B. Wen, H.-B. Jin, Z.-L. Hou, J. Yuan, Ferroferric oxide/multiwalled carbon nanotube vs polyaniline/ferroferric oxide/multiwalled carbon nanotube multiheterostructures for highly effective microwave absorption. ACS Appl. Mater. Interfaces **4**(12), 6949–6956 (2012)
96. R. Che, C. Zhi, C. Liang, X. Zhou, Fabrication and microwave absorption of carbon nanotubes/$CoFe_2O_4$ spinel nanocomposite. Appl. Phys. Lett. **88**(3), 033105 (2006)
97. M. Zhou, X. Zhang, J. Wei, S. Zhao, L. Wang, B. Feng, Morphology-controlled synthesis and novel microwave absorption properties of hollow urchinlike α-MnO_2 nanostructures. J. Phys. Chem. C **115**(5), 1398–1402 (2010)
98. X. Guo, Y. Deng, D. Gu, R. Che, D. Zhao, Synthesis and microwave absorption of uniform hematite nanoparticles and their core-shell mesoporous silica nanocomposites. J. Mater. Chem. **19**(37), 6706–6712 (2009)
99. F. Razmjooei, K.P. Singh, M.Y. Song, J.-S. Yu, Enhanced electrocatalytic activity due to additional phosphorous doping in nitrogen and sulfur-doped graphene: a comprehensive study. Carbon **78**, 257–267 (2014)
100. H.-W. Liang, W. Wei, Z.-S. Wu, X. Feng, K. Müllen, Mesoporous metal–nitrogen-doped carbon electrocatalysts for highly efficient oxygen reduction reaction. J. Am. Chem. Soc. **135**(43), 16002–16005 (2013)
101. J. Song, M.L. Gordin, T. Xu, S. Chen, Z. Yu, H. Sohn, J. Lu, Y. Ren, Y. Duan, D. Wang, Strong lithium polysulfide chemisorption on electroactive sites of nitrogen-doped carbon composites for high-performance lithium–sulfur battery cathodes. Angew. Chem. **127**(14), 4399–4403 (2015)
102. S. Liu, J. Tian, L. Wang, Y. Zhang, X. Qin, Y. Luo, A.M. Asiri, A.O. Al-Youbi, X. Sun, Hydrothermal treatment of grass: a low-cost, green route to nitrogen-doped, carbon-rich, photoluminescent polymer nanodots as an effective fluorescent sensing platform for label-free detection of Cu (II) ions. Adv. Mater. **24**(15), 2037–2041 (2012)
103. T. Ji, Z. Li, C. Liu, X. Lu, L. Li, J. Zhu, Niobium-doped TiO_2 solid acid catalysts: strengthened interfacial polarization, amplified microwave heating and enhanced energy efficiency of hydroxymethylfurfural production. Appl. Catal. B **243**, 741–749 (2019)
104. Y. Zhukova, Y. Pustov, A. Konopatsky, S. Dubinskiy, M. Filonov, V. Brailovski, Corrosion fatigue and electrochemical behavior of superelastic Ti-Nb-Ta alloy for medical implants under cyclic load conditions. Mater. Today Proc. **2**, S991–S994 (2015)
105. J.-P. Niemelä, Y. Hirose, T. Hasegawa, M. Karppinen, Transition in electron scattering mechanism in atomic layer deposited Nb: TiO_2 thin films. Appl. Phys. Lett. **106**(4), 042101 (2015)
106. P. De Wild, W. Huijgen, H. Heeres, Pyrolysis of wheat straw-derived organosolv lignin. J. Anal. Appl. Pyrol. **93**, 95–103 (2012)

Chapter 3
Self-responsive Nanomaterials for Flexible Supercapacitors

Daolan Liu, Yueyu Tong, Lei Wen and Ji Liang

Abstract The worldwide demand for green and renewable energy resources as well as the development of electronic devices has greatly boosted the improvement of energy storage systems. As one of the major types of energy storage devices, supercapacitors, with ultrahigh power densities, long-term cycling lives, and rapid charge and discharge capabilities, have been extensively investigated at the current stage, especially for those flexible or wearable electronic devices, which could be integrated into a smart system. In this chapter, the basic structures, the energy storage mechanisms, the categorization, and the characteristics of supercapacitors are comprehensively discussed. This chapter mainly focuses on different major components of flexible supercapacitors, ranging from the flexible electrode structure, the flexible substrates, and the improved electrolyte, to the construction of self-responsive flexible devices. Meanwhile, the emerging flexible integrated systems with these devices have also been illustrated, such as the energy sensor integrated systems and the energy collection-storage-sensing systems. Furthermore, the future trend of flexible supercapacitors based on future demands will be lastly discussed, focusing on the feasible and efficient strategies for designing novel and high-performance supercapacitors in future research.

D. Liu · Y. Tong · J. Liang (✉)
Institute for Superconducting and Electronic Materials, University of Wollongong, Wollongong, NSW 2522, Australia
e-mail: liangj@uow.edu.au

L. Wen
Shenyang National Laboratory for Materials Science, Institute of Metal Research, Chinese Academy of Sciences, Shenyang 110016, Liaoning, China

J. Liang
Key Laboratory for Advanced Ceramics and Machining Technology of Ministry of Education, School of Materials Science and Engineering, Tianjin University, Tianjin 300350, China

Z. Sun and T. Liao (eds.), *Responsive Nanomaterials for Sustainable Applications*, Springer Series in Materials Science 297,
https://doi.org/10.1007/978-3-030-39994-8_3

3.1 Introduction

In recent decades, the extensive worldwide exploitation of fossil fuels has made people face increasingly serious environmental issues and resource shortages. And this pushes us to develop new, efficient, and sustainable methods for utilizing and converting energy with high environmental friendliness and low carbon emissions [1]. At present, although various types of renewable energy resources, such as wind energy, nuclear energy, solar energy, and biomass energy, have already replaced traditional petroleum and natural gas resources partially, the seasonal and regional distribution of those green energy resources makes them hard to be directly used for practical applications [1–3]. If these energy resources can be properly converted into electricity and stored in energy storage devices, then it will greatly facilitate the delivery and convenient usage of energy, especially during peak hours. Moreover, the rapid development of portable electronic devices and electric vehicles has also posed higher requirements for energy storage systems on their energy and power densities, as well as safety. Currently, fuel cells, lead–acid batteries, lithium ion batteries, and capacitors are the four major types of energy storage devices [1, 4–8]. Although the current battery technology can successfully satisfy most of those requirements, the relatively low coulombic efficiency at high rates makes it difficult to be rapidly charged and discharged. Supercapacitors (SCs), with ultrahigh power densities, rapid charge and discharge capabilities, and long-term cycle lives (>500,000 times), are expected to be a potential solution for the above-mentioned issues. Consequently, an increasing amount of attention has been paid to further enhance the performance of supercapacitors, which has already become one of the most promising energy storage devices [4, 5].

More importantly, there has been much recent research on the development of flexible or wearable electronic devices that can be integrated into clothes, glasses, watches, and even skin, and the bottleneck in this technology appears to be the power supply. Thus, one of the most significant issues impeding the success of these newly developed portable gadgets is the development of smart electrochemical energy storage devices, which can convert chemical energy and electricity, reversibly. Currently, the most suitable candidates for this purpose are flexible and smart SCs. Such SCs for mobile electronics have some important features [9, 10]. (1) Flexibility: Smart SCs are devices that can work in a range of elastic states and retain electrochemical performance during various deformation processes, such as stretching and bending. (2) Self-response properties: In addition to being directly used as the power supply for flexible devices, future flexible SCs need to be combined with other flexible devices, such as solar cells, nanogenerators, and various sensors to develop integrated equipment. Future flexible batteries should have the characteristic of self-responding, which is the fundamental and essential concept of the Smart Planet and the Internet of Things. In this chapter, we first give an introduction about SCs, and then, review the progress of flexible SCs and self-responsive flexible integrated system of SCs.

3.2 Introduction of Supercapacitors

3.2.1 *The Structure of Supercapacitors*

The supercapacitor, also known as an electrochemical capacitor, is an energy storage device similar to a battery, but to some extent has differences. It is an electrochemical device between an ordinary battery and a conventional capacitor. Energy storage in a supercapacitor is achieved by adsorption and desorption of ions or rapid redox reactions on the surface of the electrodes, which is highly reversible. Thus, a supercapacitor can charge and discharge repeatedly, without significantly affecting its capacitance. Usually, supercapacitors (Fig. 3.1) are composed of a positive electrode, a negative electrode, electrolyte, separators, and current collectors, etc. [11]. The voltage of a supercapacitor during the charge and discharge processes is determined by the used electrolyte, and the breakdown voltage of the electrolyte cannot be exceeded by the working potential. The separators of supercapacitors need to have high electronic resistance but high ionic permeability, to prevent the internal short circuit and maintain a smooth ion/charge transfer. Generally, the power density of a supercapacitor is 1–2 orders of magnitude higher than that of secondary batteries (Fig. 3.2). So, supercapacitors are good choices for driving consumer electronics, electric vehicles, military equipment, etc. [6].

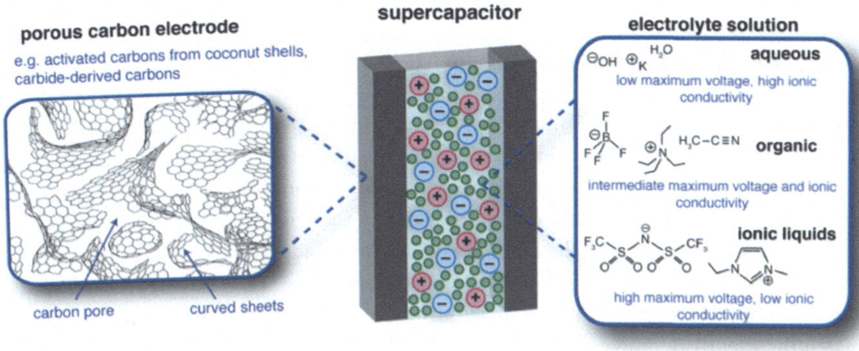

Fig. 3.1 Schematic view of a supercapacitor. Porous carbon materials with disordered structures are used as the electrodes, and the cell is soaked with an electrolyte that may be organic, aqueous, or ionic liquid-based, with some typical electrolytes shown. Note, for simplicity the separator (which prevents short circuit), the binder that holds the electrode materials together and the current collectors are not shown [11]

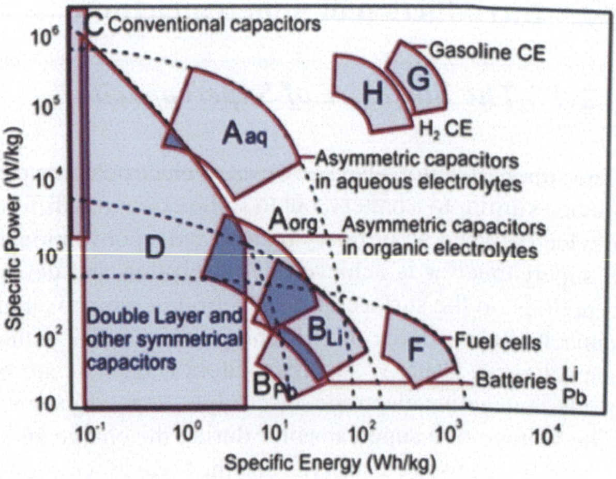

Fig. 3.2 Ragone plot with specific energy and power for different energy storage devices [6]

3.2.2 *The Energy Storage Mechanisms of Supercapacitors*

There are two types of energy storage mechanisms for supercapacitors: the double-layer energy storage mechanism and the pseudocapacitive energy storage mechanism.

3.2.2.1 Double-Layer Energy Storage Mechanism

The electric double-layer mechanism is the dominated energy storage mechanism for traditional supercapacitors. Among the various electrode materials, carbon-based materials with large specific surface areas and high porosities are mostly studied. They store energy via oppositely charged layers on the surface of electrode materials (Fig. 3.3a). The electrode and the electrolyte are in contact with each other in a capacitor. When the functional groups on the electrode surface are charged by adsorbing the ions in the electrolyte, a charge distribution layer will be formed on the interface between the electrode and electrolyte in a short time with the same amount of opposite charge, which is called the electric double layer. Thus, the corresponding energy storage devices are named as electric double-layer capacitors [12]. The charge and discharge processes of an electric double-layer supercapacitor can be described as followed:

Charging process: When an electric field is applied on two electrodes, the anions and cations in the electrolyte move to the positive and negative electrodes, respectively, thereby forming an electric double layer, which remains stable after the removal of the outer electric field [13].

Discharging process: When the electrodes are connected with the external circuit, electrons will directionally migrate to form an external current due to the potential differences. In the meantime, the absorbed anions and cations on the surface of the

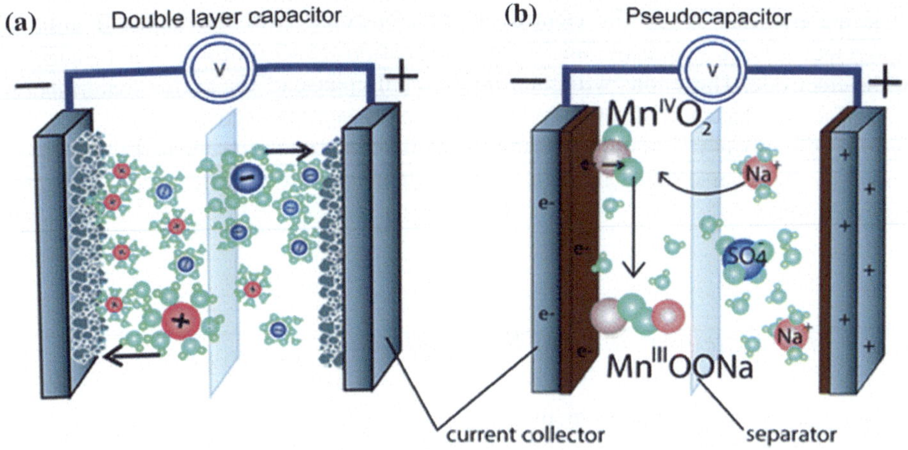

Fig. 3.3 Basic schematics for **a** an all carbon electric double-layer capacitors (left), **b** a pseudocapacitor (MnO_2-depicted center). All devices have an active material (e.g., carbon, MnO_2, $LiCoO_2$), a current collector, a separating membrane, and electrolyte (e.g., Na_2SO_4, or $LiPF_6$ solutions) [12]

electrode will return to the electrolyte, accompanying with the disintegration of the electric double layer.

Because the electric double-layer capacitor is storing energy through the high-diffusion speed charge layer, extremely fast charge and discharge processes and intrinsically high power densities can be guaranteed [14]. However, the physically adsorbed charges in the electric double layer are very limited in number, which makes it hard to achieve a high energy density for such capacitors. In general, conductivity, specific surface area, porosity, pore size distribution, and ion sizes of electrolyte involved in reactions are the key factors in determining the energy storage characteristics of the electric double-layer capacitors [15].

3.2.2.2 Pseudocapacitive Energy Storage Mechanism

In this type of supercapacitor, electrochemically active materials, such as transition metal oxides and sulfides, are loaded on the surface of the electrode, which are capable of reversible energy storage by redox reactions (Fig. 3.3b) [11]. Since the reaction is carried out throughout the bulk phase, rather than merely on the surface of the electrode, the capacitance that can be achieved with this system is relatively large. The pseudocapacitance-dominated energy storage process may also accompany by an electric double-layer energy storage process. Therefore, the energy storage capability of pseudocapacitances is simultaneously determined by the active materials, specific surface area, and redox reaction of the electrode materials [6]. In this system, the charge–discharge process can be described as followed:

Charging process: When the electrodes are connected with the external voltage input/electric field, a large amount of anion and cation are accumulated on the surface of the electrode. These ions will enter into the bulk phase of the active materials on the surface of electrodes through redox reactions to store charges.
Discharging process: The ions entering the active material then returned to the electrolyte by the reverse redox reactions while releasing stored electricity through the external circuit.

3.2.3 *The Categorization of Supercapacitors*

3.2.3.1 By Working Mechanisms

Supercapacitors can be classified into double-layer supercapacitors, pseudocapacitors, and hybrid supercapacitors according to their different energy storage mechanisms. Generally, activated carbons and carbon aerogels are used as electrode active materials for double-layer supercapacitors. For pseudocapacitors, metal oxides/sulfides and polymers are usually used as the active materials for electrodes. Comparing with the formal two kinds, hybrid supercapacitors can successfully overcome the low power density of pseudocapacitors and the low energy density of normal supercapacitors, and in the meantime, meet the high operating voltage requirements.

3.2.3.2 By Electrode Materials

Supercapacitors can be classified into carbon electrode capacitors, conducting polymer capacitors, and noble metal oxides electrode capacitors according to different electrode materials.

3.2.3.3 By Electrolytes

Electrolytes are electrically conductive compounds after being dissolved in an aqueous solution. The electrolytes for supercapacitors are usually water-based electrolytes or organic electrolytes. The aqueous electrolytes can be mainly divided into acidic electrolytes (e.g., 36% sulfuric acid aqueous solution), alkaline electrolytes (e.g., KOH, NaOH aqueous solution), and neutral electrolytes. The organic electrolytes generally contain lithium salts mainly composed of $LiClO_4$ and other quaternary ammonium salts composed of ammonium tetraethylammonium tetrafluoroborate, which are dissolved in the corresponding solvents according to specific requirements.

3.2.3.4 By the Reaction and the Structure of Electrodes

Supercapacitors can also be classified into symmetric supercapacitors (Fig. 3.4a) and asymmetric ones (Fig. 3.4b) according to the composition of the electrodes. A capacitor in which the two electrodes have the same composition is called asymmetrical capacitor [5]. On a symmetrical capacitor, electrochemical reactions proceed in opposite directions on the two electrodes. In contrast, an asymmetric capacitor contains different active materials on their two electrodes, with different types of reactions occurring on them, which is able to combine the high energy and power density features to better meet the demands for high power applications [5].

3.2.4 *Characteristics of Supercapacitors*

Comparing with other energy storage devices, supercapacitors possess the following advantages [17]:

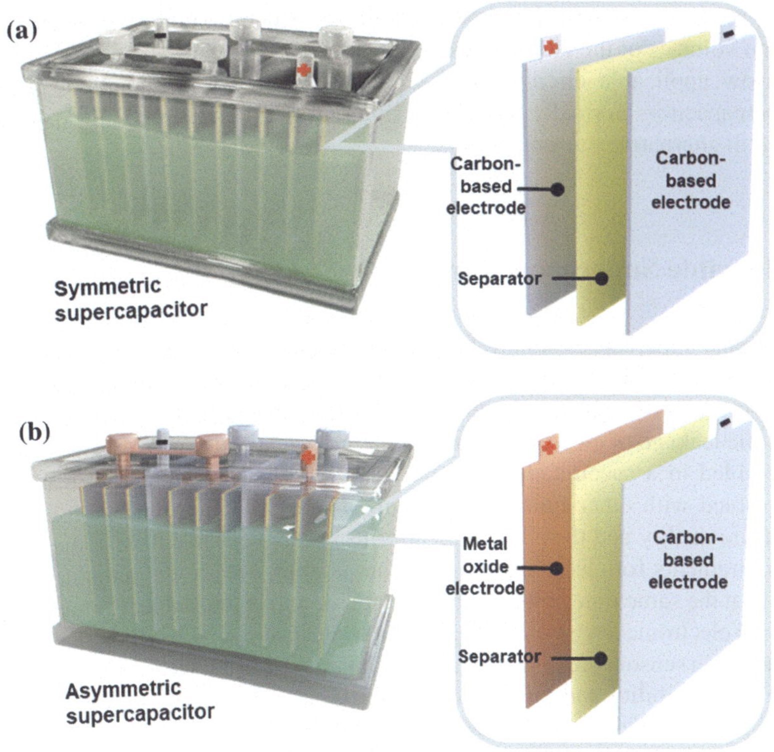

Fig. 3.4 Schematic view of **a** a symmetric supercapacitor, and **b** an asymmetric supercapacitor [16]

① High power densities: The power density of supercapacitors is 10–100 times that of ordinary batteries.
② Broad operating temperature ranges: During the charge and discharge processes of supercapacitors, the transfer of charges is mainly carried out on the surface or near the surface region of the electrode active materials. With the change of temperature, its capacity does not change greatly, enabling them to operate normally in the environment of −40 to 70 °C.
③ High reliability: Ordinary batteries will have suffered a voltage drop when leaving unused for a long time, which is call self-discharge. Unlike ordinary them, the capacity performance of supercapacitors will not be affected by this.
④ Environmental-friendly with little pollution: Without using heavy metals and other toxic substances during the production of supercapacitors makes supercapacitors an environmental-friendly power source with long life spans. Moreover, the circuit is simple, and no special forming process is required. And the dump energy can be directly read and easy to be detected.

However, supercapacitors also have some deficiencies [18].

① Low energy densities: The energy density of supercapacitors is much lower than that of ordinary batteries.
② Low pressure resistance: limited cell voltage, which is subject to the decomposition voltages of the electrolytes within supercapacitors.
③ Narrow applicable circuit range: Compared with conventionally capacitors, supercapacitors normally have larger internal resistance, impeding them to be used in alternating current circuits.

3.3 Flexible Supercapacitors

Conventional supercapacitors are bulky and cumbersome; thus, they have significant drawbacks for their applications in wearable devices. For example, toxic liquid electrolytes require high-security packaging materials and techniques to effectively prevent the leakage of electrolytes during use. Supercapacitor components can only be assembled in a specific shape, such as spiral cylinders, which are difficult to be assembled with other functional systems. In the meantime, with the improvement of life quality and the advancement of technology, people have come up with new requirements for electronic devices; that is, they need to be miniaturized and wearable at the same time, while being humanized and comfortable. Compared with traditional electronic devices, flexible/wearable electronic devices are widely used in electronic skin sensors, flexible displays, health monitoring devices, etc., due to their lightweight, flexibility, and ease of wear, which greatly changed the way people work and live [19–22]. Key technologies for the development of flexible/wearable electronic devices include flexible energy storage technology, elastic conductor technology, chip technology, and integration technology, while flexible energy storage technology is mainly controlled by the development of flexible supercapacitors/batteries.

The flexible supercapacitor is an energy storage device with a certain flexibility, which can be bent or stretched. It overcomes many shortcomings of traditional supercapacitors and has been widely used in wearable electronics [23–25]. Except for its high power density, flexible supercapacitors also have other advantages, like fast charge and discharge processes, long lifetime, and broadly applicable temperature range, which can better serve wearable electronic or portable electronic devices. Flexible integrated photodetection devices and flexible sensing devices driven by flexible/stretchable supercapacitors have been developed. It is believed that in the near future, it will become one of the indispensable devices in human daily life or polar exploration.

3.3.1 *Electrode Materials*

The performance of flexible supercapacitors largely depends on the electrode materials and electrolytes. The electrode materials for electric double-layer supercapacitors are mainly carbon materials with high specific surface areas, for example, nanocarbon materials, activated carbons [26, 27], carbon nanotubes [28–30], graphene [31, 32], carbon onions [33], etc. The electrode materials for pseudocapacitors mainly include transition metal oxides (e.g., RuO_2, MnO_2) [34, 35], transition metal hydroxides (e.g., $Ni(OH)_2$, $Co(OH)_2$) [36, 37], transition metal sulfides (e.g., $NiCo_2S_4$, CoS) [38, 39], transition metal nitrides (e.g., TiN) [40], transition metal carbides (e.g., TiC) [41], and conducting polymers (e.g., polypyrrole, polyaniline) [42, 43], etc. Beyond that, some newly emerged electrode materials, such as MXenes, and black phosphor (BP) have gradually become the research hot spot in recent years.

3.3.1.1 Carbon-Based Electrode Materials

Carbon materials have many outstanding properties, including excellent electrical conductivity, flexibility, low cost, lightweight, etc., and are ideal electrode materials. Currently, activated carbon, graphene, and carbon nanotubes are primary carbon-based supercapacitor electrodes.

Activated carbon is a kind of carbon material with plenty of pore structures, leading to a specific surface area up to 3000 $m^2\ g^{-1}$. It is an ideal double-layer electrode material [45]. Currently, commercial mass production has been realized with high capacitance properties and low price. To fabricate activated carbons, carbon-containing organic precursors are carbonized and then activated to form pores by physical or chemical methods to obtain the desired activated carbon materials. The resulted activated carbons can be divided into activated carbon powders and activated carbon cloths. The activated carbon in the powder form (Fig. 3.5) is hard to handle as an electrode material, which always needs to be mixed with conductive agents (e.g., carbon nanotubes) and binders. But activated carbon clothes are good electrode materials, especially for flexible supercapacitors. The electrochemical performance

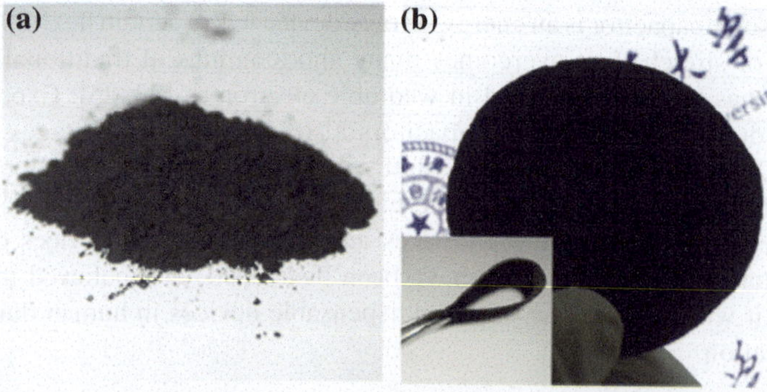

Fig. 3.5 Macroscopic photographs of **a** activated carbon powder and **b** activated carbon–carbon nanotube 5% paper. The insert image shows the bending of the paper using a pair of tweezers [44]

of activated carbon materials is mainly determined by their specific surface area, pore size, and the conductivity of materials. However, since it is difficult to control the distribution and size of pores during the fabrication process, a part of the surface area cannot be effectively utilized in practical applications, and thus, the actual specific capacity of the activated carbon is much smaller than the theoretical specific capacity.

Hexagonal carbon networks, which are interconnected by carbon atoms, are the primary units for graphene (Fig. 3.6), the two-dimensional layered carbon material. The high conductivity, large theoretical specific surface area (2675 m^2 g^{-1}), and the special two-dimensional layer structure all make graphene rich in active sites and ideal for manufacturing flexible double-layer supercapacitor electrodes [47]. The main preparation methods of graphene are mechanical stripping, chemical

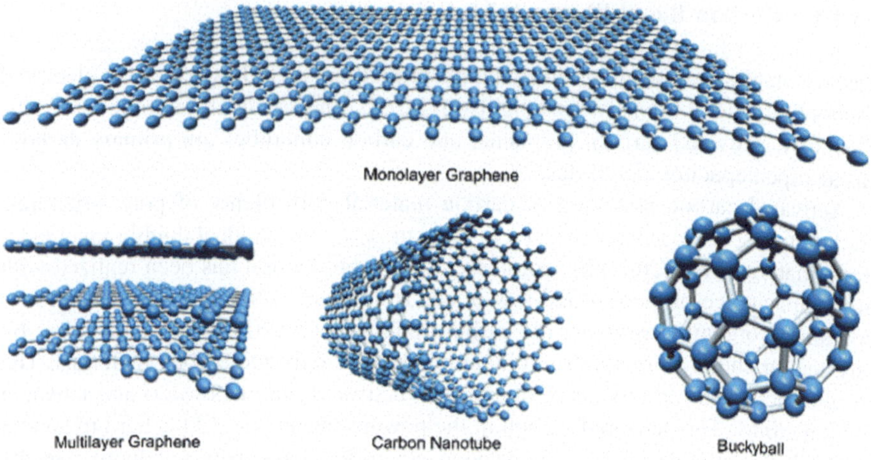

Fig. 3.6 Structure of monolayer and multilayer graphene, carbon nanotubes, and buckyballs [46]

vapor deposition epitaxial growth, and chemical stripping/exfoliation. Comparing with other preparation methods, the chemical stripping/exfoliation method is a more cost-effective method, which is also the dominating method for graphene preparation nowadays. However, graphene that is obtained from this method is in an oxidized form, which needs to be further reduced for supercapacitor electrode fabrication. During this process, the surface functional groups of graphene are destroyed, resulting in graphene agglomeration, electrolyte permeability reduction, and electrochemical performance degradation [48]. The specific surface area, active sites, and conductivity of graphene can be improved by chemically activation pore-making and element doping (N, P, etc.) for higher specific capacitance [49–51]. The manufacture of large-area graphene should be further investigated to meet the requirements of the current market.

Carbon nanotubes (Fig. 3.6) are closed, hollow tubular carbon materials formed by the curling of hexagonal mesh graphite sheets [52]. Depending on the number of carbon layers on the surface, carbon nanotubes can be divided into single-walled carbon nanotubes (one layer of carbon atoms) and multi-walled carbon nanotubes (more than one layer of carbon atoms). The diameters of carbon nanotubes range from smaller than 1 nm to tens of nanometers, and their lengths range from tens of nanometers to micrometers. The large aspect ratios give carbon nanotubes large specific surface areas (100–400 m^2 g^{-1}), which are especially suitable for double-layer electrode materials. However, comparing with activated carbons, the specific capacitance of carbon nanotubes is relatively lower, which is mainly resulted from the relatively smaller micropore volumes in carbon nanotubes, resulting in a reduction in the utilization of the active area and, thus, a reduced specific capacity. Therefore, many types of research combine carbon nanotubes with other pseudocapacitive electrode materials to acquire better performances, using their unique structure and high conductivity to compensate for the poor specific capacity of carbon nanotubes and low conductivity of pseudocapacitive materials [53–56]. Carbon nanotubes are the most widely used double-layer electrode materials in the field of flexible supercapacitors for the preparation of flexible electrodes.

3.3.1.2 Metal Oxides/Nitrides/Sulfides-Based Electrode Materials

Generally, the electrostatic charge storage mechanism of carbon-based materials gives them a lower specific capacitance than that of pseudocapacitive electrode materials. The charge storage of pseudocapacitive electrode materials is realized by a fast, reversible surface redox reaction, thus exhibiting a considerable capacitance value (300–2000 F g^{-1}) [58]. Pseudocapacitive electrode materials can be classified into transition metal oxides/hydroxides/sulfides/nitrides [59] and conducting polymers (e.g., polyaniline, polypyrrole, polyethylene dioxythiophene) [60]. However, not all transition metal oxides can be used as pseudocapacitive electrode materials. Only those metallic elements with multiple valence states can manage the valence state change during energy storage processes. In addition, the crystal structures of those transition metal oxides should be stable during the valence state change. Among various transition metal oxides, RuO_2 is the first and most widely studied electrode material for its wide operating voltage window (1.2 V), highly reversible redox reaction with three oxidation states, a high proton conductivity, a high mass ratio capacitance, and a long cycle life [61]. However, the high cost, toxicity, requirement of acidic electrolytes, and low natural abundance of RuO_2 greatly hinder its potential applications in real markets.

Therefore, many efforts have been directly exerted to alternative and inexpensive metal oxides. Manganese-based oxides have been widely studied as a promising RuO_2 replacement for its low-cost and high theoretical specific capacitance (1400 F g^{-1}) [62]. However, the relatively low conductivity (10^{-5} to 10^{-6} S cm^{-1}) and chemical stability of MnO_2 have already limited its electrochemical properties. Therefore, an effective strategy for improving the electrical conductivity and stability of MnO_2 is to combine MnO_2 with other highly conductive materials, such as carbon-based materials or conductive polymers [63]. Vanadium-based oxides are also considered excellent candidates for replacing RuO_2 due to their high mass-specific capacitance [64]. Nickel-based and cobalt-based divalent or trivalent materials have received extensive attention in the application of supercapacitors because they have a relatively high theoretical specific capacitance through the Faraday reaction for charge storage [65]. Metal nitrides, such as titanium nitride, are considered as a new class of high-performance electrode materials due to their excellent electrical conductivity (e.g., TiN 4000–55,500 S cm^{-1}) [66]. Similarly, metal sulfides [67] and metal phosphorous [68] are also being widely explored as electrode materials for supercapacitors. However, metal nitrides are usually easy to be oxidized in aqueous solutions. To solve this problem, these materials need to cooperate with more stable materials through different processes, such as, encapsulated by carbon nanotubes [69] or wrapped by graphene [70] (Fig. 3.7).

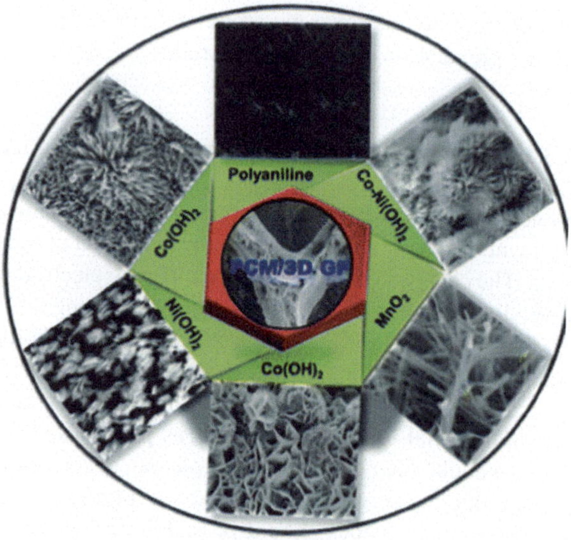

Fig. 3.7 Various pseudocapacitive materials, viz. metal oxides/hydroxides, polymers grown on 3D graphene frameworks as supercapacitor electrodes [57]

3.3.1.3 Conducting Polymer Electrode Materials

Conducting polymers, such as PANI, PPy, and PTh (Fig. 3.8), are another kind of pseudocapacitive electrode materials. They have the following advantages: low preparation cost, easy preparation on any substrate, excellent conductivity, reversible redox reaction, environmental friendliness. Moreover, their redox activity can be adjusted by chemical modification; thus, they are considered as a kind of ideal pseudocapacitive electrode material [72] and are widely used in flexible supercapacitors. High specific capacitance, excellent stability, and considerable energy storage performance are always available for pseudocapacitors with conducting polymer-based electrodes either under positive or negative potentials. Energy storage is mainly achieved by a redox reaction for conducting polymers. When an oxidation process occurs, ions in the electrolyte will enter into the frameworks of conducting polymers, i.e., the doping process. According to the electrical properties of the inserted ions, the oxidation process and reduction process can be divided into p-type doping and n-type doping, respectively. Conducting polymers will lose electrons when anions enter the frameworks of conducting polymers, which is p-type doping (Fig. 3.9a). Conversely, when cations enter, the frameworks of conducting polymers will obtain electrons, which is n-type doping (Fig. 3.9b). The doping type of conducting polymers is mainly determined by the conjugated π bond of the main chain, and generally, the p-type

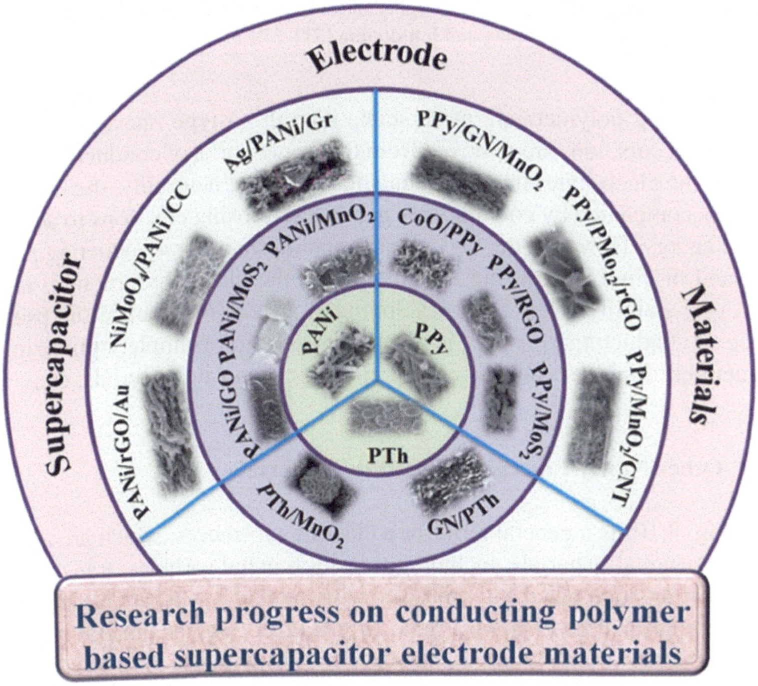

Fig. 3.8 Conducting polymer-based supercapacitor electrode materials [42]

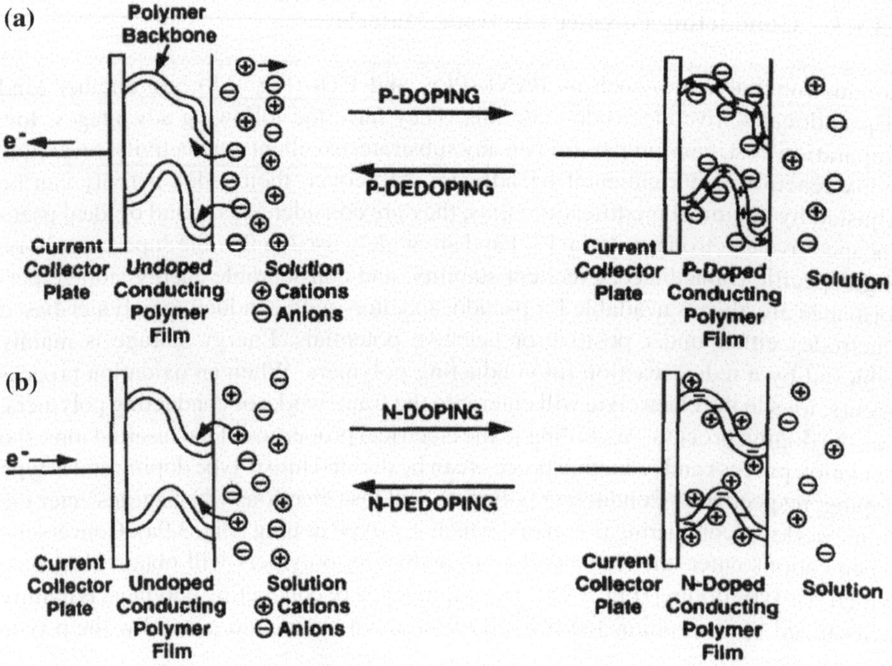

Fig. 3.9 Schematic representation of the charging and discharging processes at conducting polymer electrodes associated with **a** p-doping and **b** n-doping [71]

doped conducting polymers are more stable than the n-type ones. When a reduction reaction occurs, ions are removed from the frameworks of conducting polymers and reenter the electrolyte. In order to maintain electric neutrality, the entire redox process is accompanied by continuously gaining and losing electrons to achieve the storage of energy. However, during charging and discharging, conducting polymers may expand or contract, causing the collapse of the electrode structure, resulting in poor cycle stability. Therefore, developing strategies to enhance the pseudocapacitance of conducting polymers is greatly in need, for example, optimizing their nanostructures, doping surfactants, and preparing composite materials, etc.

3.3.1.4 Other Newly Emerging Electrode Materials

MXene (Fig. 3.10) is a general term for a class of substances, which are a member of two-dimensional materials, including transition metal carbides, transition metal nitrides, and transition metal carbonitrides, winning widespread attention since they were discovered in 2011 [74]. $M_{n+1}X_nT_x$ (n = 1, 2 or 3) is the general formula for those compounds, where M represents early transition metals (e.g., Ti, V, Nb, Cr, Mo); X represents C, N, or CN; and T_x represents surface terminal groups, including hydroxyl, oxygen, and fluorine [75]. Their high conductivity, good mechanical

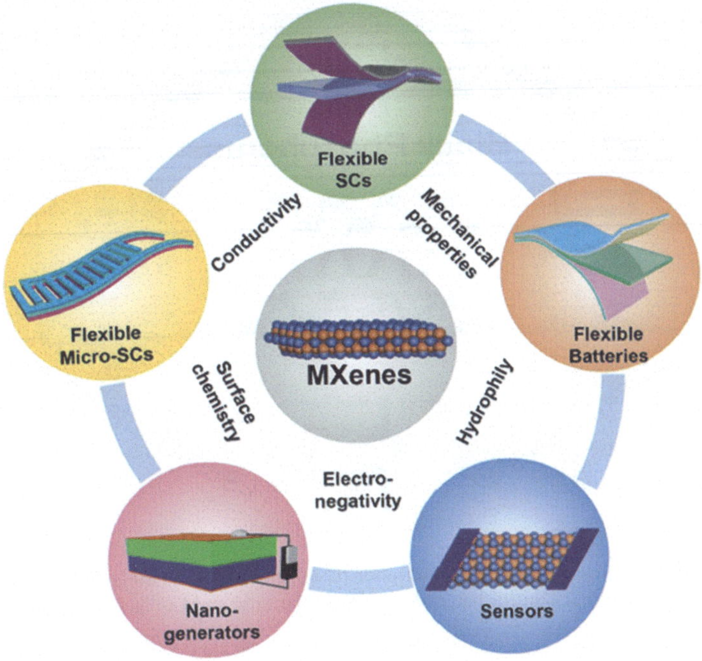

Fig. 3.10 Schematic illustration of MXene-based nanomaterials for flexible energy storage devices, including flexible SCs, Micro-SCs, batteries, and other flexible electronic devices such as nanogenerators and sensors [73]

properties, and hydrophilicity make them outstanding candidates for energy storage applications. For instance, $Ti_3C_2T_x$ that synthesized by various methods embraces a wide scope of conductivity: from 1000 to 6500 S cm^{-1} [76].

Black phosphor (Fig. 3.11), a new member of two-dimensional materials family, is the most stable allotrope of phosphor. It is a layered structure accumulated by the folded planes that are made of phosphorus atoms, which are interacted by van der Waals forces [68]. Similar to other layered structures, single or multiple layers of black phosphorus nanosheets can be obtained by mechanical stripping and liquid stripping techniques [70]. Multilayer black phosphorus is a layered p-type direct gap semiconductor. In addition, the interlayer spacing between adjacent layers of black phosphorus is 5.3 Å, which is larger than the layer spacing of graphene (3.6 Å) and comparable to that of 1 T MoS_2 (6.15 Å), making it an ideal material for flexible energy storage.

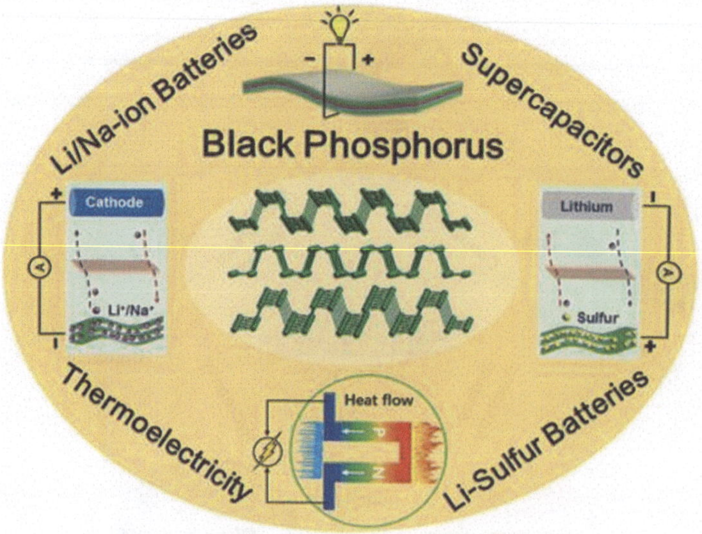

Fig. 3.11 Schematic illustration of black phosphor-based nanomaterials for energy storage devices, including supercapacitors, Li-sulfur batteries, thermoelectricity, and Li/Na-ion batteries [77]

3.3.2 Flexible Substrates

The flexible substrate is a flexible, stretchable, or compressible substrate that holds electrode active materials. Thus, the mechanical properties and structures of electrodes can be directly determined by the selection of substrates. Comparing with those self-supporting electrodes, the electrodes with flexible substrates not only possess good flexibility, but also outstanding tensile and compression properties, making them capable of maintaining stable conductivity and electrochemical activity under stretching and compressing. During the stretching and compressing processes, flexible substrates are the main bearing carriers for stress, which can greatly reduce the damage on electrode materials and maintain stable performances for electrode materials. Currently, growth, coating, and winding are primary methods reported for flexible substrate electrodes preparation. One or more preparation methods can be used to prepare flexible substrate electrodes depending on the type of substrates and electrode materials, such as printing, adsorption, and laser etching. According to the conductivity of substrates, they can be divided into conducting substrates and insulating substrates.

3.3.2.1 Conducting Substrates

Metal foils/fibers/foams (Fig. 3.12) are ideal flexible substrates due to their outstanding conductivity, good flexibility, and the convenience of electrochemical deposition

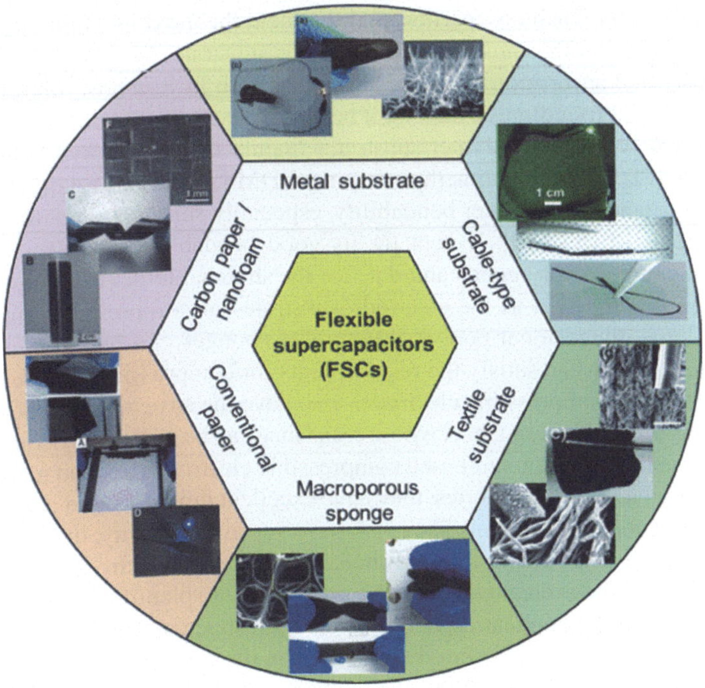

Fig. 3.12 An overview of support substrates, such as metal, carbon-based electrodes, cable-type electrodes, and porous materials such as conventional paper, textiles, and sponges to fabricate high-performance FSCs [78]

for various electrode materials [79]. Common substrates are nickel, copper, titanium, stainless steel, carbon cloth, etc. Carbon clothes, except for their outstanding conductivity, are also easy to process, corrosion-resisting, and lightweight, making them suitable as flexible substrates as well. In the meantime, the large specific surface areas and multiparous structures of carbon can greatly enhance the contact between electrodes and electrolytes, further facilitating the transfer of electrons and ions. Although both metal- and carbon-based flexible substrates possess good conductivity and flexibility, they also have some drawbacks. For the metallic substrates, they normally cannot be fully restored to their original state after being stretched or bent to a high degree. The carbon-based flexible substrates usually have relatively low tensile strength, so they break easily.

3.3.2.2 Insulating Substrates

Paper, polymers, sponge, and fabrics are the main insulating substrates (Fig. 3.12). Compared with the conducting substrates, these insulating substrates are lower in cost and have better tensile properties, making them the most available flexible substrates at present. As the most common flexible substrate, the paper has a unique pore structure and adsorption properties, together with its low cost, high flexibility, and environmental friendliness, making it be widely used as a flexible electrode substrate in recent years. The polymer substrates mainly included polyethylene glycol terephthalate (PET) and polydimethylsiloxane (PDMS). PET has relatively stronger corrosion resistance and a better bendability, especially suitable to be used as bendable flexible substrates [80]. Except for its good bendability, PDMS can also be stretched, making it a very good candidate for the stretchable flexible substrate [81]. Sponges are mainly used as the electrode substrates of compressible supercapacitors. The three-dimensional porous structure of sponges can be compressed to a certain extent, which can satisfy the requirements for compressible supercapacitors. The sponge-based compressible electrodes also have the structural characteristic of compression adjustable resistance, whose resistance can be reduced by compression. This feature endows the sponge-based compressible electrodes a broad market in the supercapacitors field. As for fabrics, they have excellent flexibility and can stand various stretching and deformation. They also have versatile porosity, lightweight, and low price, making them one of the best choices as a flexible electrode substrate. The machinability of fabrics enables them to be processed into plentiful sizes and shapes, greatly expanding the application ranges of flexible supercapacitors [78]. Two primary methods for preparing electrode materials on the surface of fabrics have been developed by far. One is to prepare a conductive layer on the surface of fabrics and then grow the electrode materials on it; the other is to in situ prepare an electrode material with good conductivity on the surface of fabrics. The latter can greatly curtail the preparation processes and promote the development of fabrics in the flexible supercapacitors field. The high conductivities of carbon materials and conducting polymers make them good candidate to be in situ grown on the surface of fabrics as flexible electrodes. However, the intrinsic low specific capacity of carbon-based materials makes it necessary for them to combine with other electrode materials (e.g., conducting polymers, transition metal oxides, etc.), which is a cumbersome process. In comparison, conducting polymers not only possess a better conductivity, but also have a higher specific capacity and an easier processing process. In addition, the in situ growth of conducting polymers on the surface of the fabrics will not damage the mechanical properties of fabrics, making them the most ideal electrode materials in matching with the fabric substrates.

3.3.3 Electrolytes

The electrolytes of supercapacitors can mainly be divided into water-based electrolytes, organic liquids, ionic liquids, solid/quasi-solid electrolytes, and redox electrolytes. Water-based electrolytes are aqueous solutions of certain substances, which can be divided into alkaline electrolytes (e.g., KOH solution), acidic solutions (e.g.,

H_2SO_4 solution), and neutral solutions (e.g., Na_2SO_4 solution), etc. A major drawback of the water-based electrolytes is that their operating voltage is limited to the decomposition voltage of water. Theoretically, water will decompose into oxygen and hydrogen when the applied voltage exceeds 1.23 V. The decomposition of water will affect the performances of supercapacitors and even destroy their structure. However, the water-based electrolytes can achieve a high ion concentration and thus high ionic conductivity, leading to higher output powers than other types of electrolytes. Thus, current research on water-based electrolytes is mainly aimed to increase the operating voltage and expand the potential window.

Organic electrolytes have a high share in the commercial market due to their high operating voltage (2.3–2.5 V) and mass productivity. Moreover, when organic electrolytes are used, low-density current collectors, such as Al foil, can be used, further increasing their energy density. The commonly used solvents for organic electrolytes are acetonitrile, polycarbonate, etc., while the electrolyte salts are tetraethylammonium tetrafluoroborate, triethyl methyl ammonium tetrafluoroborate, etc. The ionic radii of these salts are relatively large, and the ionic resistance of the organic electrolytes is relatively larger than that of the water-based electrolytes. Therefore, the conductivity improvement of organic electrolytes has been the research focus at the moment. Moreover, some organic solvents, such as acetonitrile, are toxic, so the environmental-friendly problem is another factor that must be considered.

Ionic liquids are a kind of electrolyte with considerable research value and application prospects [82]. Usually, this liquid, a molten salt at room temperature, is composed of an asymmetric organic cation and an organic/inorganic anion. The ionic concentration and working voltage of ionic liquids are very high. The latter can reach up to 6 V, so, obtaining a high potential window and a high energy density can be expected by using such electrolytes. However, the viscosity of ionic liquids is usually very high, greatly restricting the movement of ions, resulting in low conductivity and the corresponding reduced power density. Thus, enhancing the compatibility between ionic liquids and electrode materials, increasing the infiltration between ionic liquids and electrode materials, and reducing the migration resistance of anions and cations in ionic liquids are the key points for improving its performance.

Solid/quasi-solid polymer electrolytes not only can be used as ion-conductive media, but also as separators. They are also controllable in shape, leak-free, and reliable, thus, becoming the first choice for flexible supercapacitors and providing a way for the miniaturization of capacitors. Solid/quasi-solid polymer electrolytes can be classified into the following types: all-solid-state polymer electrolytes, gel polymer electrolytes, and composite polymer electrolytes.

The functional groups of some polymers will form a polymer cationic complex by complexation processes without the addition of solvents. This formed complex is called an all-solid polymer electrolyte. The ionic conductivity of all solid polymer electrolytes at room temperature is 10^{-8} to 10^{-7} S cm^{-1}. For most polymer electrolytes, their relatively low ionic conductivities at room temperature, poor consistencies of the interfaces between electrodes and electrolytes, and the low solubility for electrolyte salts lead to the crystallization of electrolytes near the electrode.

Thus, the resulted extremely low ionic conductivity of all-solid polymer electrolytes in return largely limits its application in supercapacitors.

Compared with all solid polymer electrolytes, the gel polymer electrolytes possess much better ionic conductivities at room temperature (10–10^{-3} S cm^{-1}), certain mechanical properties, and high cycling efficiencies, thus, being considered as one of the most ideal polymer electrolytes. Gel polymer electrolytes consist of polymer skeletons, organic/aqueous solutions serving as dispersion media, and supporting electrolytes (salt/acid/alkali) working as conductive ions. Generally, polymers contain the following types: polyethylene oxide, polyvinylpyrrolidone, polylactic acid, polyvinyl alcohol, polyacrylic acid, polymethacrylic acid, polyacrylamide, polyvinylidene fluoride hexafluoropropylene, polyethylene glycol, or biopolymers, such as alginic acid, carrageenan, hyaluronic acid, and carboxymethyl cellulose.

For gel polymer electrolytes, water can be used instead of organic solvents in the electrolytes, which not only can reduce environmental pollution, but also can significantly cut down equipment costs. This aqueous gel polymer electrolyte is known as a hydrogel polymer electrolyte. Hydrogel polymer electrolytes are widely used in flexible supercapacitors due to their simple preparation, good hydrophilicity, good film formation, non-toxicity, and low cost.

The hosts of composite polymer electrolytes are mainly inorganics that blend, plasticize, and compound with various polymers, which possess similar structures. The inorganic additives can be divided into two types: One is inert without ion-transporting capacity and the other is a type of active additives, which can transport ions. Those electrolyte ions can migrate in both inorganic filler phases and polymer phases, significantly improving their ionic conductivity. For example, additives, such as SiO_2, TiO_2, Sb_2O_3, graphene oxide, and hydroxyethyl cellulose, are used to improve the performances of polyvinyl alcohol-based hydrogels. Two-dimensional graphene and graphene oxides have also been introduced into gel polymer electrolytes. These two-dimensional nano-fillers can help to reduce the crystallization of gel polymer electrolytes and form an interconnected high-speed channel for continuous ion motion to improve the ionic conductivity of gel polymer electrolytes.

In addition, adding the electrochemically active substance into a conventional electrolyte can form a redox electrolyte. On the one hand, the electrolyte can form an electric double-layer capacitance on the surface of the electrode; on the other, the electrolyte can undergo redox reactions with electrochemically active substances. Thus, this strategy will endow capacitors with "battery" characteristics to enhance their energy storage performances.

3.4 Strategies for Flexible Supercapacitors Construction

Since the discovery of transistors by Bardeen and Brattain, electronics have revolutionized our life. Especially in recent years, portable electronic devices have penetrated every aspect of our daily life. A more recent direction for electronic devices is the development of modern flexible devices, such as touch screens, wearable communication devices, and surgical and diagnostic implements. These applications should naturally integrate with the human body to provide unique performances and require lightweight, rugged construction in thin, conformal formats [83].

Flexible integrated devices will be used in smart and self-powered sensory, wearable, and portable electronics. Therefore, in a sense, integration or the action of incorporating multifunctions into one device to adapt itself to a changing environment is an essential trend in the evolution of flexible SCs. In the following section, we first review the construction of flexible SCs and then summarize the recent progress of the self-responsive SCs system.

3.4.1 1D Wire Supercapacitors

So far, most of the designed and developed flexible supercapacitors have been fabricated into the flat "sandwiched" architecture similar to the traditional supercapacitors, and a few pieces of research have focused on the wire-type supercapacitors. Compared with the membrane-like structure, the wire structure has the special advantage of weavable shape, which could meet the requirement of miniaturization and integration for the future development of mobile electronic devices. Lots of factors could influence the performance of wire supercapacitors, such as the fiber electrode materials, electrolyte, encapsulation pattern, and so on. Among these factors, seeking suitable wire electrode materials is the key to improving its intrinsic properties. By now, the widely applied electrode materials for wire supercapacitor include carbon fiber, polymer fiber, and metal fiber, and their modification or composition with the second and even the third components. Recently, the development of novel materials with wire structure, such as graphene fibers and carbon nanotubes fibers, provides new research directions for wire supercapacitors. On the basis of structure, there are two types of architectures for 1D wire supercapacitors, spirally twisted wire supercapacitors and coaxial wire supercapacitors, the fabrication of which are different from each other as well. In the following section, we will give an introduction of two kinds of wire supercapacitors.

(1) Fabrication of spirally twisted wire supercapacitors

Figure 3.13 shows a typical spirally twisted wire supercapacitor. As shown in Fig. 3.13, the flexible wire substrate is firstly required for fabrication of such a wire supercapacitor. The applied substrate is carbon fibers, on which $CoNiO_2$ nanowires and active carbon were grown via a hydrothermal process as the cathode and anode electrode materials, respectively. With KOH-PVA as the electrolyte, the cathode fiber and anode fiber were spirally twisted together around a polymethylmethacrylate (PMMA) backbone and encapsulated with a layer of PDMS as the shell [84]. The wire supercapacitors could be fabricated with a length of 1.2 m, which could be further increased. Meanwhile, the electrochemical performance is also excellent

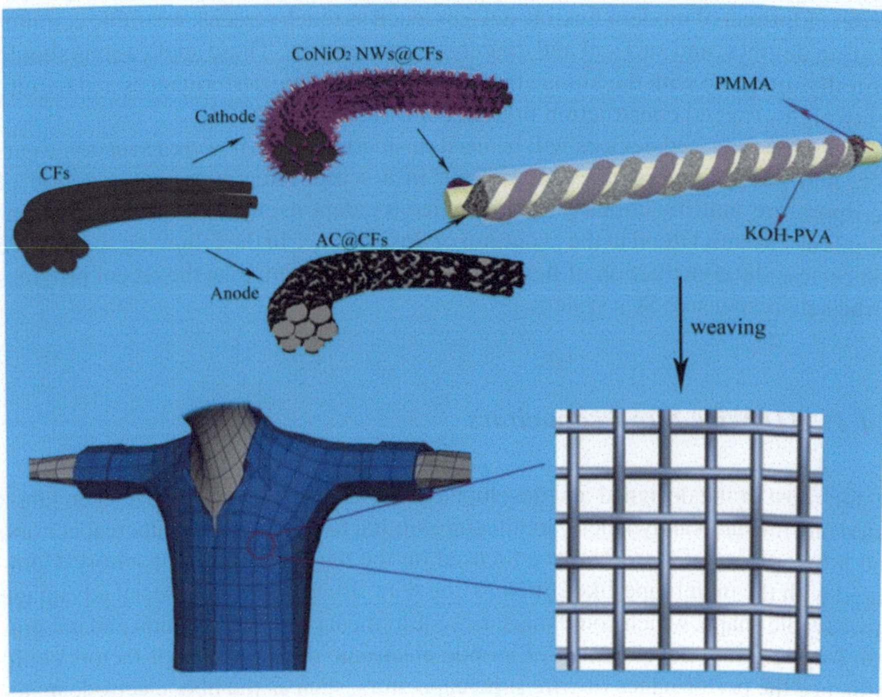

Fig. 3.13 Fabrication process illustration for the spirally twisted wire supercapacitors (AWSs) [84]

with the potential window reaching 1.8 V, a high capacity of 1.68 mF cm^{-1} at the current of 0.05 mA cm^{-1}, and a high energy density of 0.95 mWh cm^{-3}. In addition, this unique structure endowed the supercapacitor with good flexibility, toughness, weaveability, and skin-affinity, making it possible to be weaved as a Chinese knot, watchband, belt, clothes textile, and even as wearable energy storage sources to power portable electronics.

Liu et al. [85] also applied carbon fibers as flexible substrate to construct a novel type of flexible all-solid-state planar integrated fiber supercapacitors by planar integrated assembly of hierarchical $ZnCo_2O_4$ nanowire arrays/carbon fibers electrodes, as shown in Fig. 3.14. More intriguing, it exhibited an enhanced distributed-capacitance effect while integrated 2, 6, 10, 14, 20, and 30 composite fibers on flexible carbon fiber substrate, making it possible to obtain the relative maximum functionality and minimized size at the same time.

In addition, carbon fiber, metal wire, and plastic fiber coated with Au film were also used as substrates to fabricate novel and efficient flexible fiber supercapacitors, which consist of two fiber electrodes, a helical spacer wire, and an electrolyte, as shown in Fig. 3.15. In such a supercapacitor, the active material is commercial pen ink, which was added into the gel electrolyte, and a spacer wire plays the role of

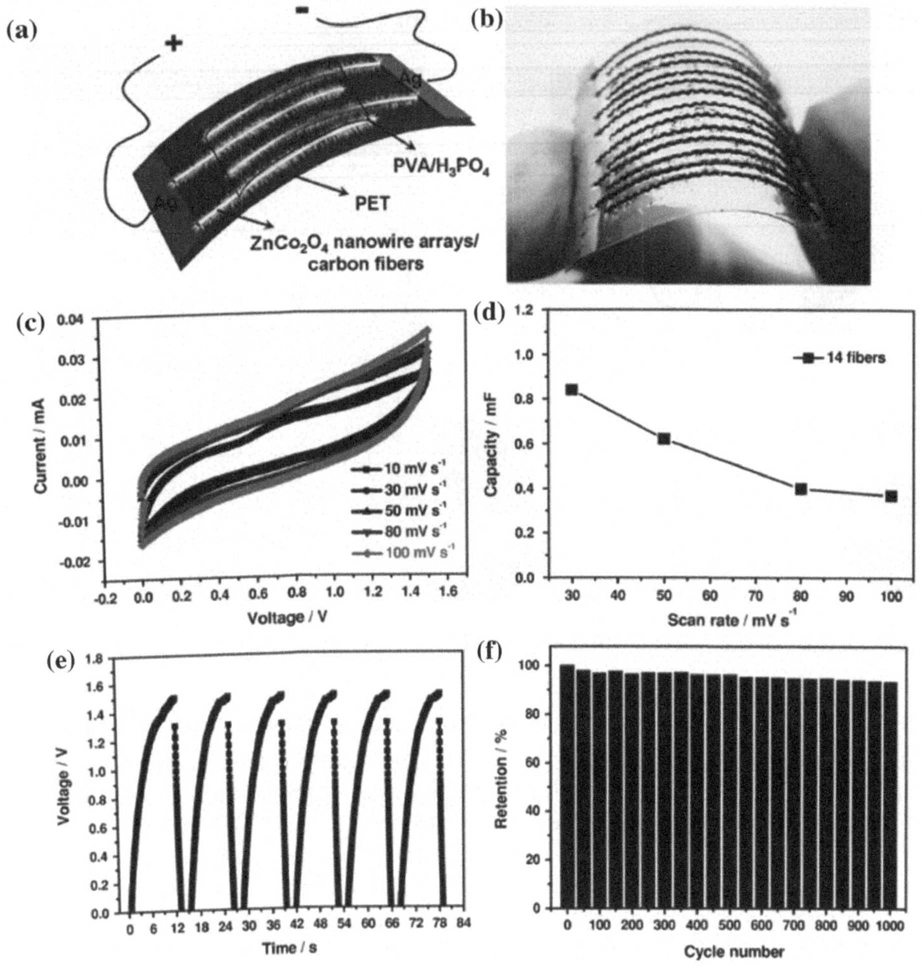

Fig. 3.14 Performance of the flexible planar integrated fiber supercapacitors with hierarchical $ZnCo_2O_4$ nanowire arrays/carbon fibers electrodes [85]

separating the two fiber electrodes. With the unique structure, the Au-coated plastic fiber-based supercapacitor could obtain the area-specific capacitance of approximately 5.5 mF cm^{-2}, while the carbon fiber-based supercapacitor about 26.4 mF cm^{-2} [86].

As a two-dimensional material with atomic thickness, graphene exhibits good electrical conductivity deriving from its 2D structure. Moreover, the relatively high specific surface area enables graphene a promising electrode candidate for double-layer electrochemical capacitors. Furthermore, it could be easily composited with

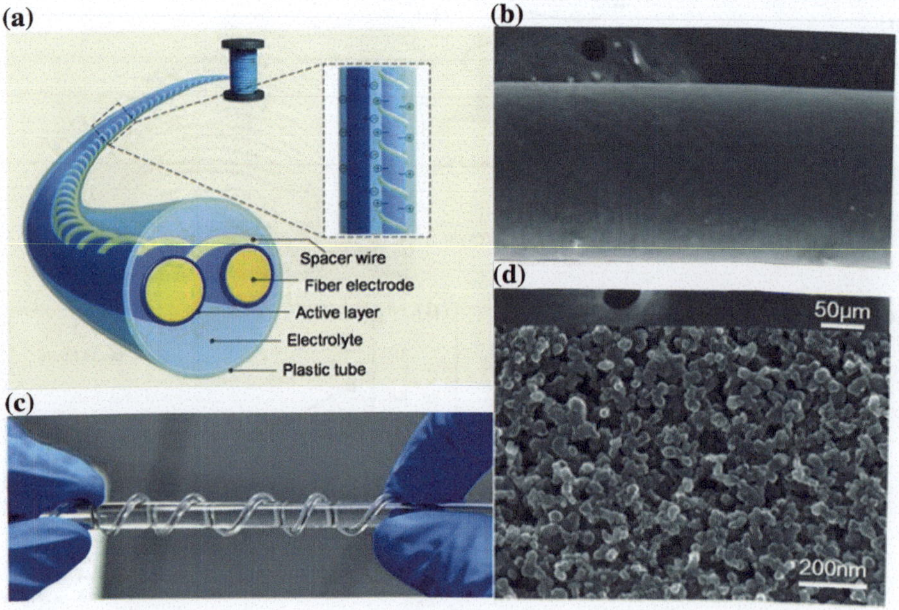

Fig. 3.15 Wire supercapacitor with commercial pen ink as the active material [86]

other materials and used as the electrode of pseudocapacitance by introducing various functional groups. Meanwhile, graphene could be fabricated into flexible, self-supporting films and fibers for flexible, portable supercapacitor electrodes by means of filtration, wet-spinning, and other methods [87–91]. Qu et al. fabricated a unique all graphene core-sheath fiber by applying a graphene fiber core and a sheath of 3D porous network-like graphene framework and further assembled it into spirally twisted wire supercapacitors with good flexibility and portability, just as shown in Fig. 3.16 [91].

Conductive polymers, such as polyaniline, possessing high conductivity and high pseudocapacitance, have been widely employed as electrode active materials for electrochemical supercapacitors [92]. However, the polyaniline presents obvious volume change during the repeated charging and discharging process, resulting in low stability for the capacitor. In this regard, the carbon nanotube is often added into polyaniline to improve the performance of supercapacitors due to its high surface area and good mechanical, electrical, and thermal stability. But this strategy has limited effects, often attributed to the irregular accumulation of carbon nanotube in the composite electrode material, in which the charge has to cross the numerous boundaries in the random carbon nanotube networks, leading to low efficiency. Therefore, it is very likely to solve this problem by simply regulating the distribution of those carbon nanotubes. On this basis, Cai et al. [93] designed a new class of polyaniline composite fibers employing aligned multi-walled carbon nanotubes. Novel wire

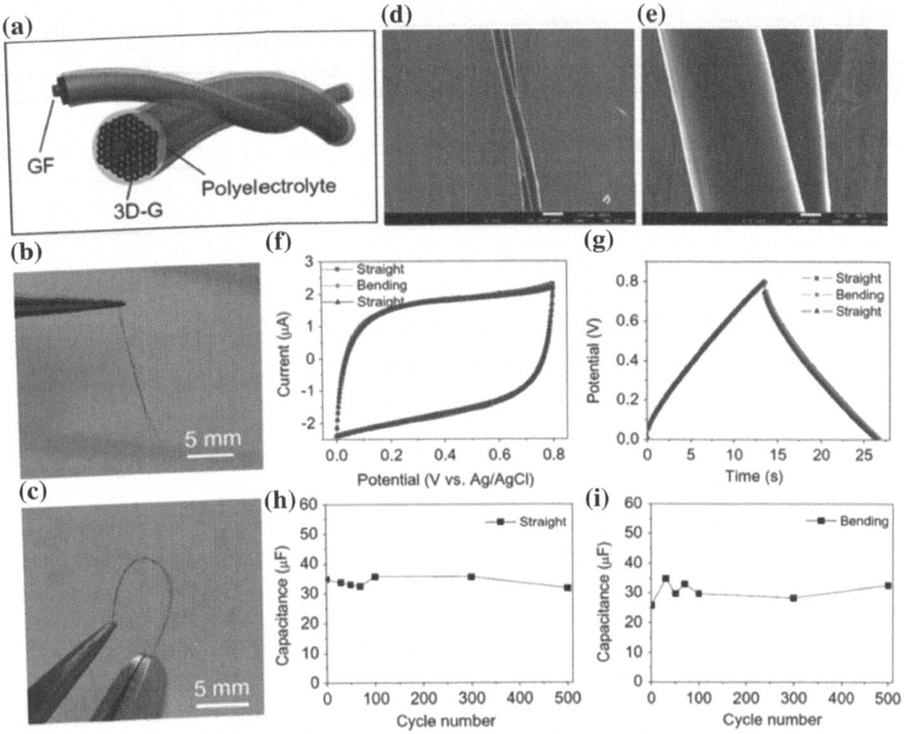

Fig. 3.16 **a** Schematic illustration of an all graphene core-sheath fiber wire supercapacitor. **b**, **c** Photographs of the supercapacitor in the free and bending state. **d**, **e** SEM images. **f**, **g** CV curves at 50 mV/s and charge–discharge curves at 2 μA in straight and bending states. **h**, **i** The capacitance stability in bending and straight state undergoing 500 straight-bending cycles [91]

microsupercapacitors were fabricated by twisting such two composite fibers, obtaining a high specific capacitance of 274 F g^{-1} or 263 mF cm^{-1}. Moreover, these wire microsupercapacitors with high flexibility and unique knittability could be produced by conventional textile technology on a large scale.

Baughman et al. [94] firstly obtained single-wall carbon nanotube (SWCNT) wires with high toughness via a wet-spinning process, which could even be made into body armor. The specific fabrication procedure is as follows: (1) prepare carbon nanotube gel fiber by injecting surfactant-dispersed SWCNT into a rotating bath of aqueous polyvinyl alcohol (PVA); (2) wash and desorb the polyvinyl alcohol and surfactant; and (3) spin solid nanotube fibers with lengths of 100 m in a continuous process at a rate of more than 70 cm min^{-1}, then incorporate the SWCNT fiber into PVA solution to produce SWCNT/PVA composite gel fiber, winding on a mandrel. The resulting composite fibers are about 50 mm in diameter, containing 60 wt% of SWCNT. Then, it was further put into PVA/phosphoric acid aqueous solution for surface coating treatment and woven into textiles. After the second coating treatment, it could be used as the electrode of twisted wire supercapacitors. This fiber supercapacitor with

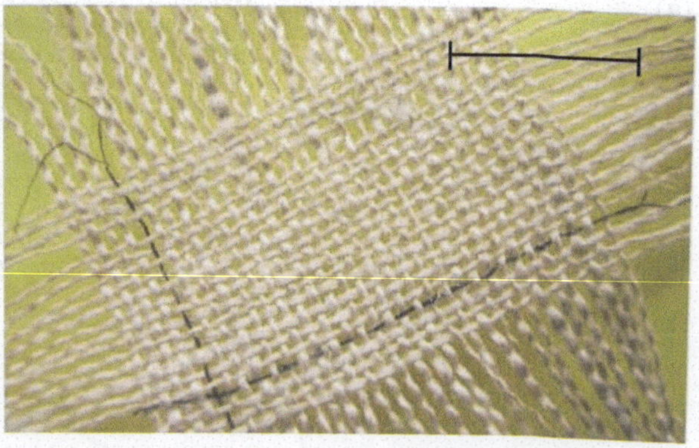

Fig. 3.17 Photographs of a type of textile with two nanotube fiber supercapacitors woven in orthogonal directions [94]

100 mm in diameter could provide a capacitance of 5 F g^{-1} and an energy storage density of 0.6 Wh kg^{-1} at 1 V, as shown in Fig. 3.17.

Chou et al. [95] fabricated an all-solid stretchable wire supercapacitor by assembling two CNT fiber electrodes, spandex fiber substrate, and H_2SO_4-PVA gel electrolyte via a prestraining-then-releasing approach. This supercapacitor presented high areal specific capacitance of 4.63–4.99 mF cm^{-2}, and good durability with 108% of the capacitance retention after 10,000 charge–discharge cycles and 20 mechanical stretching–releasing cycles with a tensile strain of 100%. In addition, Ren et al. also developed wire microsupercapacitors by twisting two aligned MWCNT fibers. It exhibited a mass-specific capacitance of 13.31 F g^{-1}, areal specific capacitance of 3.01 mF cm^{-2}, or length specific capacitance of 0.015 mF cm^{-1} [96].

Recently, a variety of small devices for directly weaving spirally twisted wire supercapacitors have been developed, which enhances the efficiency of the braided devices and reduces the cost of manufacturing linear supercapacitors. For example, Peng et al. have designed a rotation-translation method that could take advantage of the elasticity of polymer and the excellent electrical and mechanical properties of carbon nanotubes to fabricate stretchable wire supercapacitors. The capacitor can be bent, folded, and stretched, and retains 100% of the properties of the capacitor at a stretch of 75%. Such wire capacitors can be further woven into textiles with different shapes and integrated into various microelectronic devices to meet the future micro-energy demand [97].

(2) Fabrication of coaxial wire supercapacitors

As to coaxial wire supercapacitors, the fabrication process is illustrated in Fig. 3.18. Firstly, the elastic fiber substrate was coated with a layer of gel electrolyte composed of H_3PO_4-poly(vinyl alcohol) (PVA), followed by wrapping with the inner CNT sheets electrode, separating with the second layer of electrolyte, wrapping with the outer CNT sheets electrode, which should possess the same thickness as the inner

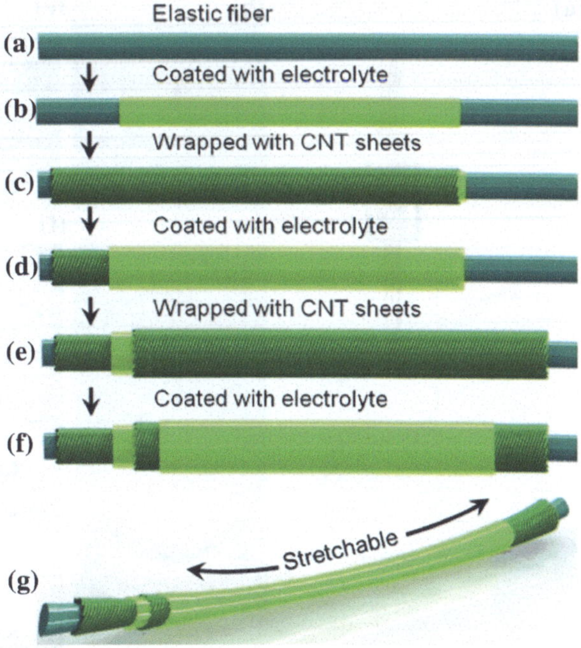

Fig. 3.18 Fabrication procedure of a highly stretchable wire supercapacitor with a coaxial structure [98]

electrode, and finally coating with the third layer of electrolyte, achieving the whole assembling process for supercapacitors [98]. Compared with spirally twisted wire supercapacitors, the materials used for coaxial supercapacitors require higher standards, and this type of capacitor is not a universal device fabrication method. However, coaxial capacitors have the advantages of shorter spacing between the positive and the negative electrodes, leading to a higher utilization rate of electrode materials. Moreover, it has been evidenced experimentally that the specific capacitance of coaxial capacitors is much higher than that of spirally twisted wire supercapacitors with the same electrode materials. Therefore, further improvement and optimization of such coaxial supercapacitors are still of significance.

Kou et al. [99] proposed a coaxial wet-spinning method to fabricate polyelectrolyte-wrapped graphene fibers. In this system, the core solution and the sheath solution could be spun onto the fiber substrate simultaneously through the inner and the outer channels, respectively, which is an evident feature for the spinner, just shown as in Fig. 3.19. The core solution is composed of graphene oxide liquid crystals (GO-LC), while the sheath material is sodium carboxymethyl cellulose (CMC), which exhibits the property of ionic conductivity and electric insulativity, avoiding the occurrence of short circuit. This supercapacitor presented high capacitances of 269 and 177 mF cm^{-2} and energy densities of 5.91 and 3.84 mWh cm^{-2} with liquid and solid electrolytes, respectively.

Besides, Chen et al. [100] designed a novel class of coaxial electric double-layer capacitor (EDLC) fibers, consisting of aligned CNT fiber and CNT sheet acting

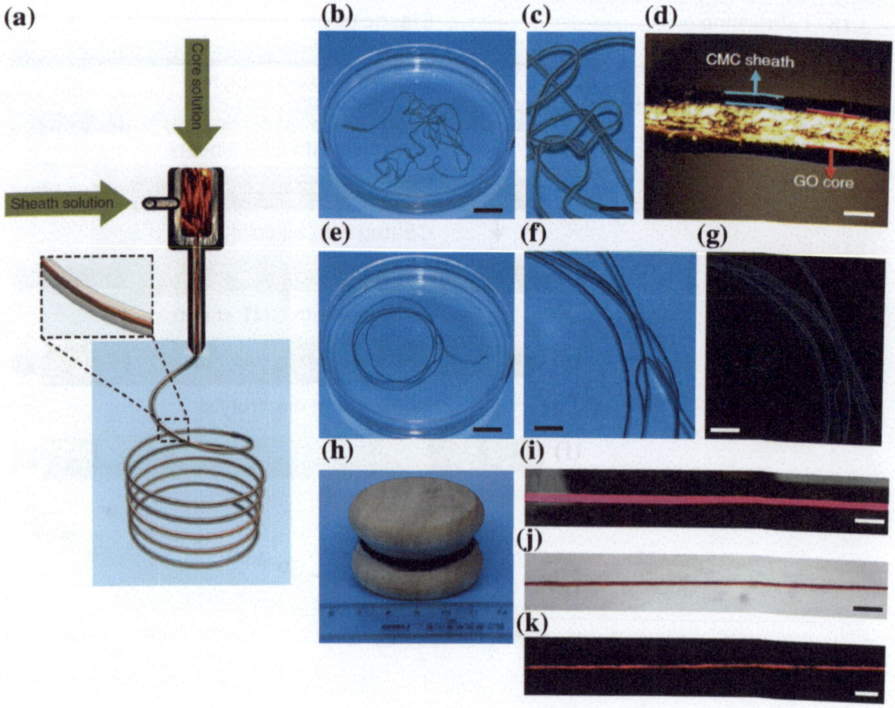

Fig. 3.19 Coaxial wet-spinning assembly procedure and the as-synthesized core-sheath fibers [99]

as two electrodes, separated by a polymer electrolyte gel (Fig. 3.20). It was found that the maximum discharge capacitance could reach 59 F g^{-1} (32.09 F cm^{-3} or 29 μF cm^{-1} or 8.66 mF cm^{-2}), much higher than 4.5 F g^{-1} of the EDLC fabricated by twisting two CNT fibers together, which was attributed to the decreased contact resistance between the two electrodes derived from the unique coaxial structure.

3.4.2 2D Flexible Planar Supercapacitors

The assembly strategies for planar supercapacitors mainly include photolithography, printing, laser etching, and template. This section will focus on the application of these four methods to fabricate 2D planar supercapacitors.

(1) Photolithography method

Photolithography refers to the process of transferring images from a mask to a substrate with the assistance of a photoresist under the light. The main operating process is as follows: First of all, evenly coat a layer of photoresist on the surface of the substrate. After completely dried, the photoresist is covered with the mask, followed by irradiating the substrate through the mask with ultraviolet light. Then, the

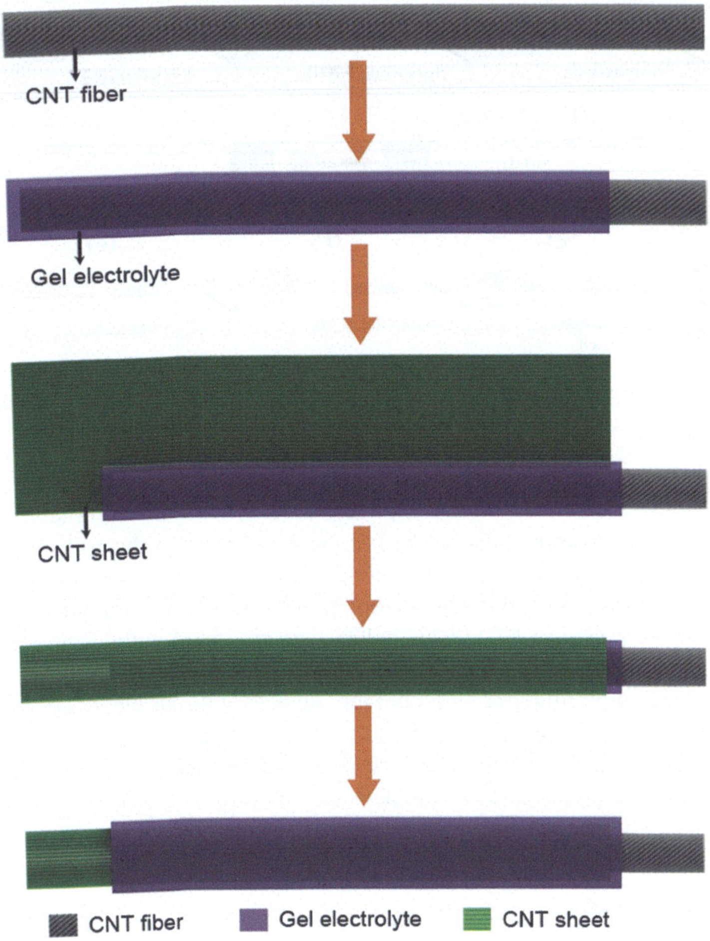

Fig. 3.20 Fabrication process of the coaxial EDLC fiber [100]

photoresist, both on the exposed and unexposed area (the former is called positive photoresist, while the later is negative photoresist), is dissolved and removed via imaging techniques, leaving the image on the substrate surface.

Figure 3.21 is a typical process of fabricating a planar fork-finger capacitor via a photolithography method. Polyethylene terephthalate (PET) as a substrate was firstly cleaned and covered with a 40-nm-thick Ni film via sputtering using a vacuum thermal coating technique at a vacuum below 8×10^{-4} mbar (Fig. 3.21a and b). Subsequently, $Ni(OH)_2$ nanoplates were spin-coated on the PET/Ni substrate, and then another layer of the Ni foam (40 nm) was sputtering deposited on it (Fig. 3.21c and d). After spinning a layer of photoresist onto its surface, put the finger-like mask on the substrate and perform photolithography to pattern and expose the surface of the Ni-coated PET with photoresist (Fig. 3.21e). Afterward, the Ni-coated PET

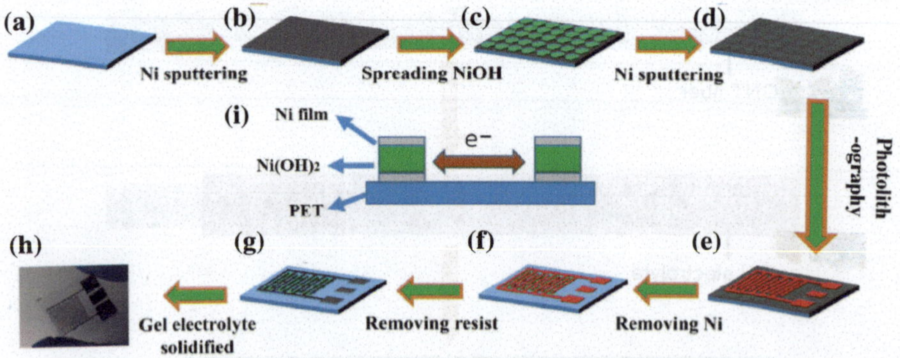

Fig. 3.21 Schematic process of fabricating planar capacitors via photolithography technique [101]

was soaked in the developer to remove the photoresist on the exposure area and then remove the nickel film on the exposure area with the sulfuric acid solution. The unexposed photoresist shielded by the mask can protect the fork-finger metal nickel electrode from sulfuric acid corrosion. After cleaning with water, the residual photoresist was also removed using ethanol or acetone. Finally, the prepared device was coated with a layer of PVA/KOH gel and encapsulated with a transparent soft film, achieving the capacitor assembling (Fig. 3.21f–h) [101].

By using the photolithography technique, devices with an electrode spacing of several microns or even smaller can be fabricated, and the structure of the devices is diversified, such as round and wavy devices. Furthermore, it could also achieve the fabrication of integrated electronic devices with different functions on the same substrate, which paves the way to integrated multifunctional wearable electronics [102].

(2) Patterning method

Patterning method refers to the process of fabricating devices with a metal or plastic recycled template with a certain electrode shape achieved by 3D printing, metal cutting, and other methods. There are two main strategies: First, the template is placed on the surface of the flexible substrate, and the active material liquid with a certain concentration is left on the surface of the flexible substrate via the template by means of filtration. After removing the template, electrode material with a specific shape can be left on the substrate. Second, the substrate is firstly coated with a uniform layer of electrode materials, followed by being covered with the template. The acid or alkali or corrosive ionic liquid is dropped on the corresponding anti-corrosion template, and the corrosive liquid flows on the surface of the material along the template channel to etch the material, leaving the materials covered by the template unchanged. When the etch process is finished, the electrode material with the template shape could be left on the substrate after the removal of the template.

Figure 3.22 illustrates the fabrication process by using the second patterning method [103]. By using recycled templates, this method is relatively simple to perform and applicable to a variety of materials; however, its accuracy is not high and the edge of the electrode is usually rough; therefore, it is limited to be employed in the fabrication of precise, miniaturized, and large-scale flexible supercapacitor devices.

(3) Printing method

Printing electronics is an arisen novel technology with high efficiency and good effect. It features high printing accuracy, universal applicability to materials, and no mask, avoiding the corrosion of materials from acid or base photoresist. Inkjet printing is the most widely applied technology in printing electronics. The key to inkjet printing is the formula proportion of ink materials, which could be greatly influenced by a variety of parameters, such as viscosity and dispersion. Therefore, the proportion of different ink materials is not the same due to the different physical properties of various materials. When the ink is qualified, the spray head of the inkjet printer will absorb the ink reagent through the microporous plate and transfer it to the treated substrate, spraying the droplets onto the surface of the substrate in a preset pattern with the power of a thermal or acoustic injector, leaving the device with specific patterns of electrodes on the substrate. By using more nozzles and spraying the same area repeatedly, most inkjet printers can produce high-resolution images.

Figure 3.23 shows the inkjet printing of custom-designed microsupercapacitor arrays from a "home computer and printer" using pristine exfoliated graphene ink [104]. Therefore, an ordinary household printer could even print capacitors, which perform excellent electrochemical properties as well, and the capacitor can be connected in series and parallel on a piece of paper, which greatly reduces the pollution to the environment and the waste of materials.

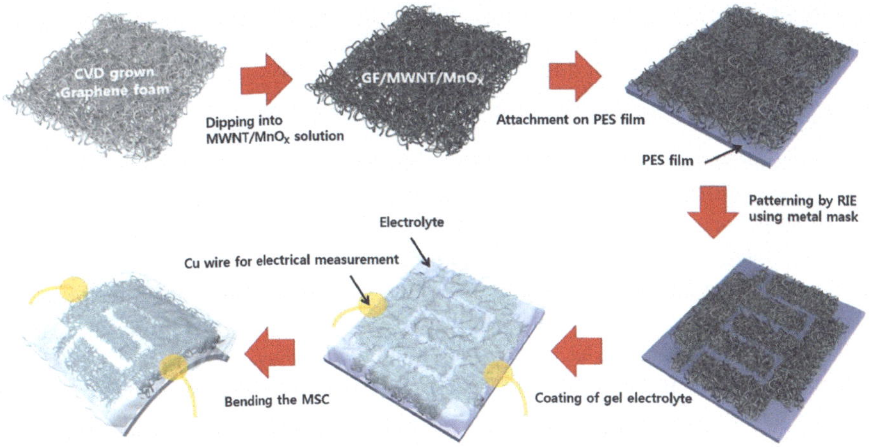

Fig. 3.22 Illustration of the fabrication process with the patterning method [103]

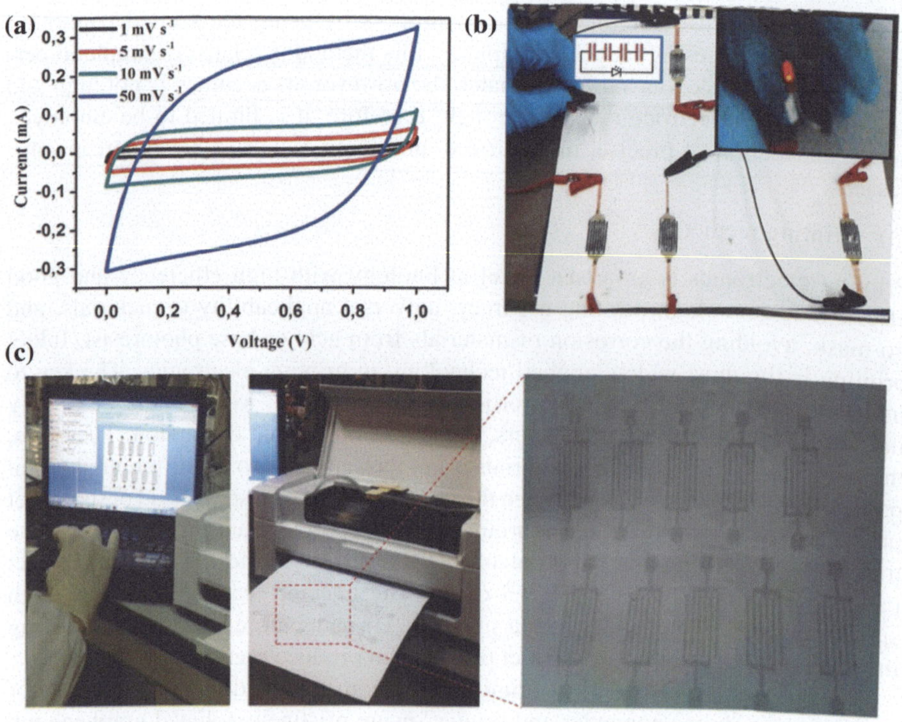

Fig. 3.23 Inkjet printing of custom-designed microsupercapacitor arrays from a "home computer and printer" using pristine-exfoliated graphene ink [104]

Inkjet printing has played an irreplaceable role in flexible electronics due to its low cost, high efficiency, and high accuracy. The flexible electronic circuit printed by this technology has been put into practical application, which improves the mechanical stability of electronic devices, reduces the production cost of circuit boards, and brings new experience for human beings.

(4) Laser etching method

Laser etching technology is developed on the basis of the traditional technology of etching lines with a chemical reagent [105]. In this system, the high energy laser beam directly focused on the substrate coated with the thin material film through adjusting the focus position, making the material film layer instantaneously gasified, achieving the etching process. Moreover, this method has little influence on the substrate materials due to the adjustable nature of laser energy at the same time. Compared with chemical reagent etching, this method is smaller in size, easier to operate, lower in cost, and higher in yield. Specifically, by simply controlling computer software, expected laser-etched patterns could be obtained by importing electrode structure drawings. After feeding, adjusting parameters, capacitors can be processed and carried on the test. By saving setting parameters, it could realize batch production, greatly improving the efficiency and yield.

Figure 3.24 shows the laser etching process for preparing planar capacitors. Firstly, spray a layer of conductive materials on the glass substrate as the current collector, followed by spraying a layer of supercapacitor electrode materials. After that, the fork structure electrode could be achieved after the laser etching treatment. Finally, the capacitor assembly will be completed after coated with electrolyte material [106].

El-Kady et al. designed graphene-based and all-solid-state electrochemical capacitors on a standard LightScribe DVD optical drive to achieve the direct laser reduction of graphite oxide films to graphene (Fig. 3.25) [107]. The produced films could be directly applied as supercapacitor electrodes without using any additional binders or current collectors, as is the case for conventional supercapacitors. Moreover, the capacity of the all-solid-state device using gel electrolyte is equivalent to that of the supercapacitor using a water-containing electrolyte, and the preparation method is relatively simple. In addition, the capacity of the supercapacitor remained almost unchanged after 1000 bending cycles from different angles. The flexible and optically transparent all-solid supercapacitor can maintain a capacity of more than 96.5% of its initial capacitance after bending for 10,000 times at 45°; therefore, the performance of the supercapacitor is relatively stable.

With advanced technology, laser etching has become a mature technology that has been put into production and application, and its precision has reached several microns in industrial application, which provides a powerful technical guarantee for the development of flexible electronics.

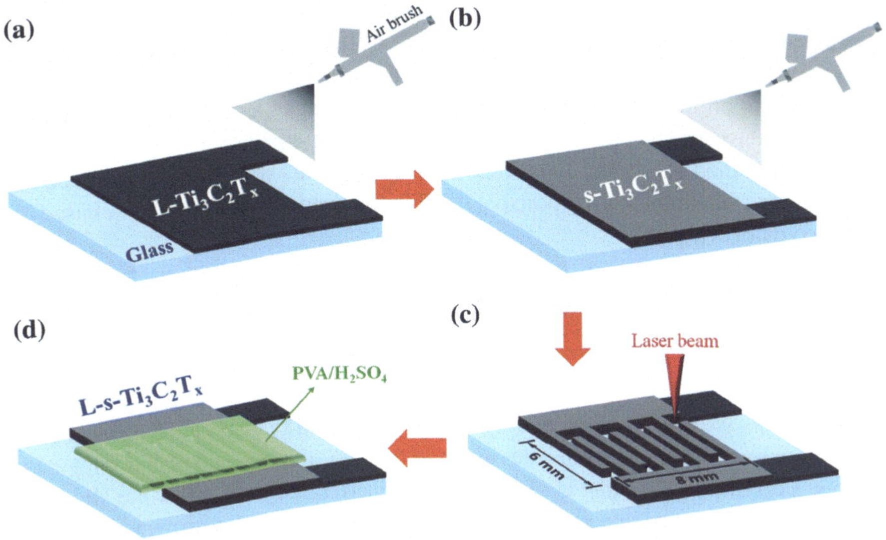

Fig. 3.24 Illustration of the laser etching process for preparing planar capacitors [106]

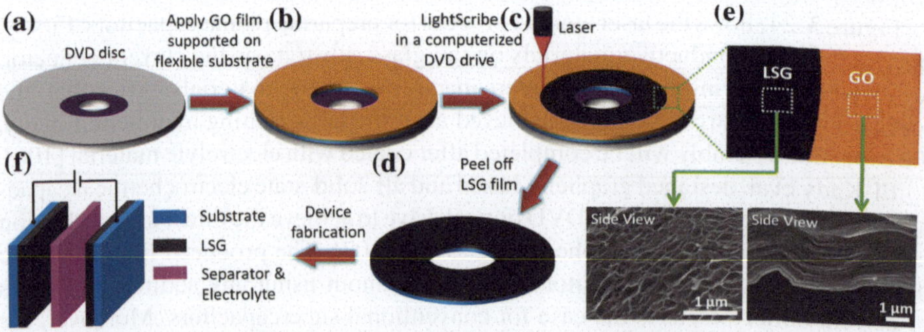

Fig. 3.25 Fabrication process of the all-solid-state laser-scribed graphene LSG-EC electrochemical capacitors [107]

3.5 Self-responsive Flexible Integrated System

Flexible/stretchable integrated systems, such as self-actuated photoelectric detection systems and self-actuated sensor systems, have developed rapidly in recent years due to their features of high integration, multifunction, comfort, and contact with human skin [108, 109]. Supercapacitors have evident advantages as energy suppliers and storage units. The existing integrated systems can be divided into two categories based on the roles that supercapacitors play. The first is the energy sensor integrated system, in which the supercapacitors are used as the power supply support function units to make sure its normal movement [110, 111]. In the second type, which refers to the energy collection-storage-sensing system, the supercapacitors bridge the energy collection unit and the functional unit. Specifically, the supercapacitors firstly save the energy converted by energy collectors, such as nanogenerators and solar energy collectors and then power the functional units [112]. This section will introduce the development status of a self-responsive flexible integration system, mainly based on these two categories.

3.5.1 Flexible Capacitor-Sensor Integrated System

In this section, we introduce the energy sensor integrated system on the aspect of the one-dimensional linear integrated devices and the two-dimensional planar integrated devices. Early in 2014, Wang et al. [113] developed a fiber-based flexible all-solid-state asymmetric supercapacitor, in which the Co_3O_4 nanowires growing on nickel fibers act as the positive electrode while graphene on carbon fibers as both the negative electrode and light-sensitive material. Therefore, a flexible linear integrated system was obtained, which could simultaneously achieve the function of energy storage and optoelectronic detection (Fig. 3.26a). Photograph in Fig. 3.26b directly demonstrates the fiber-based all-solid-state flexible asymmetric supercapacitor with

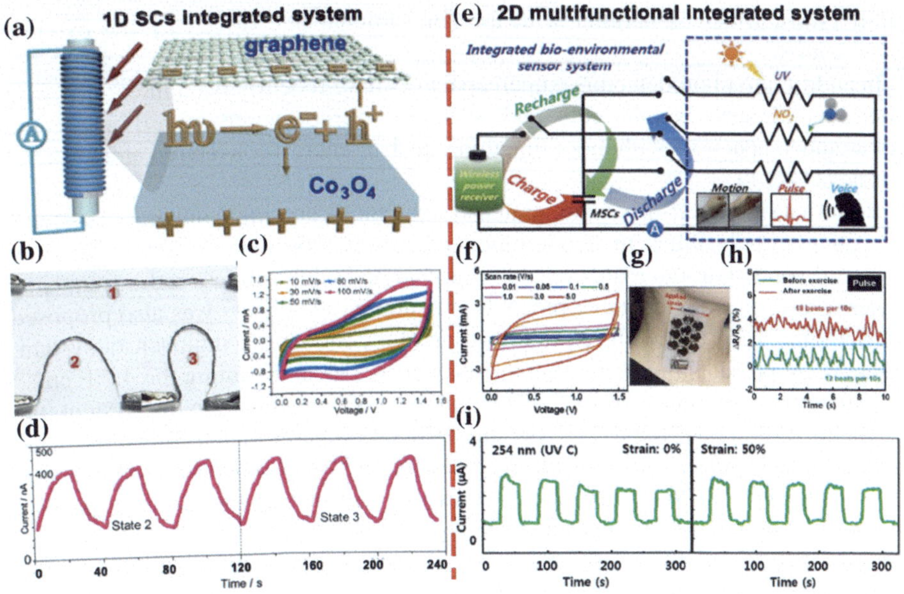

Fig. 3.26 Flexible capacitor-sensor integrated systems [113, 114]

high mechanical stability at different bending states. Figure 3.26c presents the CV curves of this supercapacitor at different scan rates ranging from 10 to 100 mV s^{-1}. It performed a good rate capability both at a scan rate as high as 100 mV s^{-1} and a maximum cell voltage of 1.5 V. And the volumetric capacitance could reach 2.1 F cm^{-3}, the maximum energy density is 0.62 mWh cm^{-3}, and the power density is 1.47 W cm^{-3} at a current density of 20 mA cm^{-3}. Moreover, it also exhibited good cycling stability with approximately 84% of its initial capacitance preserved. This excellent electrochemical performance of this supercapacitor can be attributed to the improvement of the voltage window, the high specific capacitance provided by Co_3O_4 nanowires with pseudocapacitance as well as electric double-layer capacitor properties, and the excellent cycling stability resulting from graphene-based carbon materials. Figure 3.26d displays the response of the integrated system to white light at different bending states under a light intensity of 40 mW cm^{-2}. The test started with charging the supercapacitor to 1.5 V from an external power supply, then placing the system on the probe table and periodically turning the white light on and off. The integrated system can reliably detect white light switches as the light response curve presented, demonstrating the feasibility of integrating supercapacitors and photodetectors on a single fiber.

Afterward, a series of planar supercapacitors integrated with a self-driven photodetector system has been developed as well. Xu and Shen [115] assembled a reduced graphene oxide (rGO)-based planar supercapacitor with 8.01 F cm^{-3} of volumetric capacitance and 6.204 Wh cm^{-3} of energy density, integrated with a CdS nanowire-based photodetector. One single such a microdevice could reach 34.50 of current on/off ratios, which is parallel with the conventional one driven by the external power source.

In addition to photodetectors, supercapacitors have recently been integrated with other devices, such as solar cells and detectors, to improve energy efficiency and endow supercapacitors with more functions, such as energy conversion and artificial muscles [116, 117].

Besides a single self-driven integrated system, multifunctional integrated systems that can monitor health or the environment are also proposed. For example, an array planar all-solid-state microsupercapacitor integrated with wireless radio frequency power receivers and stretchable bioenvironmental multi-sensors was also proposed [114]. Figure 3.26e shows the circuit diagram of the stretchable multisensor system. Figure 3.26f presents the cyclic voltammetry curves, demonstrating the 4.7 F cm^{-3} of volumetric capacitance and 1.5 mWh cm^{-3} of energy density. In this system, the fragmentized graphene foam (FGF) strain sensor takes charge of detecting biosignals, such as a neck pulse, saliva swallowing, voice, and body movement (Fig. 3.26g, h). And the MWNT/SnO_2 nanowire hybrid film sensor bioenvironmental sensors response to the signal upon exposure to NO_2 gas and UV light (Fig. 3.26i). Therefore, this multifunctional system paves the way to the next-generation body-attachable health/environmental monitoring devices.

3.5.2 Flexible Capacitor-Energy-Collection-Storage-Sensing System

Although the above-integrated systems have successfully realized monitoring the signals from the biological environment, such as temperature, sound, light, gas, and pressure, they generally lack energy collection units, requiring large charging facilities, leading to independently working mode. Therefore, it is of great significance to employ energy collection and wireless charging units to achieve an integrated system including energy collection, storage, and application [118–120].

Gao et al. [121] developed a flexible integrated system comprising a wireless charging coil, an all-solid-state asymmetric microsupercapacitor (MSC), and a photoconductive-type photodetector of perovskite nanowires (NWs), as shown in Fig. 3.27a. In this system, MSC as the power source saves the energy via the wireless charging coil received from the wireless power transmitter and then drives the photoconductive perovskite NWs detector in sequence. The MSC presented a stable CV curve and capacitance retention under different bending angles, which is illustrated in Fig. 3.27b. Besides, the MSC also displayed good cycling stability, preserving 92% of capacitance after 7000 cycles. Figure 3.27c shows the self-discharge curve of the MSC. When the capacitor is charged to 1.6 V, it starts to drive the perovskite nanowires-based photodetector. And the photocurrent response curve in Fig. 3.27d proves the stability of the integrated system.

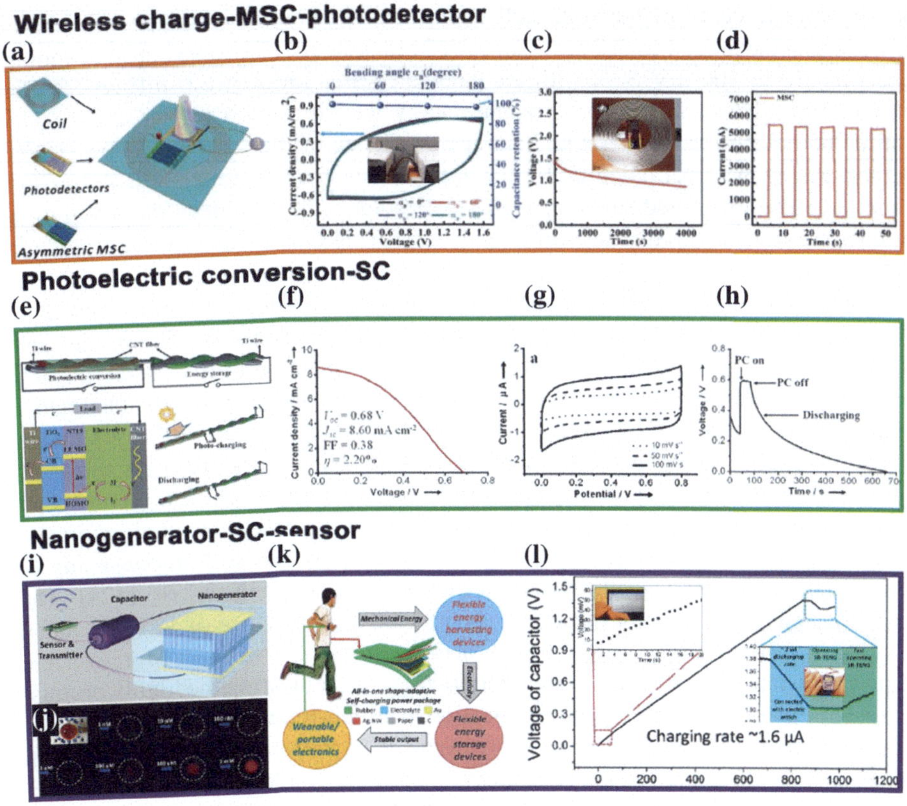

Fig. 3.27 Energy-collection-storage-sensing system [121–124]

Solar energy is considered one of the most promising renewable and clean energy sources, and energy conversion patterns stored in capacitors by harvesting solar energy and converting it into electricity are also received a lot of attention [125–130]. In 2013, Xu et al. [131] reported an integrated solar energy-supercapacitor system based on bipolar TiO_2 nanotube arrays. In this system, the dye-sensitized solar cells act as energy collectors and the supercapacitors as energy storage units. The areal capacitance of the supercapacitor is as high as 1.1 mF cm^{-2}, and the power conversion efficiency of the whole integrated circuit reaches 1.64%. Peng et al. [123] designed an integrated energy wire system, comprised of a photoelectric conversion section and an energy storage part, as shown in Fig. 3.27e–h. The schematically illustration of the working mechanism for this system is shown in Fig. 3.27e. Specifically, the transportation of electrons to the CNT fiber electrode via an external circuit starts with the incorporation of the released electrons from dye molecules to the conduction band of titania nanotubes. In the internal circuit, the electrons from the CNT fiber electrode will be captured by the I_3^- ions to produce I^- ions in order to make the activated dye molecules reduced to the ground state to achieve the cycle. The current density/voltage curve of the solar cell in Fig. 3.27f demonstrating the conversion efficiency is calculated as 2.2%. Figure 3.27g depicts the CV curve at a scanning rate from 10 to 100 mV S^{-1}, with the calculated area ratio of 2.0 mF cm^{-2}. The photocharging–discharging curve of the energy wire is shown in Fig. 3.27h. The entire photoelectric conversion and storage efficiency of the system reaches 1.5%, demonstrating the great potential for the applications in electronic textiles. Zou and Peng have also integrated supercapacitors with solar cells to achieve the integration of energy conversion, storage, and output, improving energy utilization efficiency [132, 133].

In addition to solar energy, human motion, such as walking or running, may generate significant amounts of mechanical energy. Nanogenerators can capture and harness the energy produced via human movement. Since the first prototype of a nanogenerator was proposed by Wang, a variety of one- and two-dimensional nanogenerators have been developed [134–146]. In 2012, Wang came up with an integrated system including nanogenerators, supercapacitors, and sensors [122], as shown in Fig. 3.27i. Figure 3.27j presents the optical images of the brightness changing of an LED lamp over time, powered by a nanogenerator. In 2016, Guo et al. improved the nanogenerator by assembling an all-in-one self-powered system consisting of nanogenerators, supercapacitors, and an electronic watch (Fig. 3.27k) [124]. The as-fabricated supercapacitor displays a specific mass capacitance of 12 F g^{-1} and a specific surface capacitance of 1 mF cm^{-2} at the scan rate of 10 mV s^{-1}. Moreover, the capacitance has no evident decrease after 5000 cycles, indicating the highly promising reliability and stability for the application in a high-frequency charge/discharge system. Figure 3.27l shows the *V*–*t* curve of the integrated system. The energy from the nanogenerator stored in the capacitor can drive the stable operation of the electronic watch, demonstrating its feasibility and reliability and paving way to the new research directions of the self-powered wearable electronic products in the future.

3.6 Future Trends

On the basis of the current developing trends, the future demand for flexible/stretchable integrated electronic devices requires the energy storage devices with high ability of miniaturization, flexibility, and integration. Coincidentally, the emergence of flexible microsupercapacitor highly satisfied these requirements. In this paper, different types of flexible/stretchable supercapacitors are introduced based on the flexible/stretchable substrates, the electrode materials, the electrode structure, the improved electrolyte, and the constitution of the devices. Meanwhile, the application of these devices in the emerging field of integrated systems is also simply illustrated, such as the integration of micro-planar supercapacitors with pressure sensors, linear supercapacitors with photoelectric detectors, and supercapacitors with gas sensors and the corresponding in situ signal acquisition and display system, etc. By now, a variety of different types of flexible/stretchable supercapacitors and self-driven multifunctional wearable devices have been developed and even put into practice, greatly enlarging the application form of wearable electronic devices and improving comfort experience for human beings. But there are still many issues requiring further exploration. For example, a single microsupercapacitor with small voltage window and low energy density is generally unable to power the sensor unit; therefore, the integrated system with several supercapacitors in series is always the simple operation, leading to volume expansion of the integrated system, which, however, obviously contradicts the basic requirements for comfortable experience. In order to improve the electrochemical performance of flexible/stretchable supercapacitors, more attention could be paid to the following aspects in future research:

(1) A flexible/stretchable substrate is the most basic component for flexible electronic devices. Till now, a variety of cheap and excellent substrate materials have been widely employed in the fabrication of supercapacitors; however, such issues as poor hydrophilicity seriously limit their further development. PET substrate is a commonly used flexible substrate, but this material is sharp, easy to scratch the skin when used in wearable devices; PDMS is very soft and high comfort, but it requires high standards for the combined materials due to its poor hydrophilicity. Therefore, more attention should be paid to those flexible/stretchable substrates with excellent performance and universal applicability.
(2) Optimizing all-solid electrolyte is also an effective strategy to improve the properties of capacitors, to which less attention has been paid.
(3) Improving electrode materials with high performance is the most direct way to enhance the electrochemical performance of supercapacitors. It is still of great significance to improve the utilization rate of the materials by fabricating novel nanomaterials with high performance under the guidance in theory.
(4) Stretchable/self-healing supercapacitors will be a promising device in the near future. Although various types of stretchable/self-healing supercapacitors have been successfully assembled using stretchable/self-healing substrates or wavy electrode structures, other new assembling methods enabling large-scale production for stretchable/self-healing supercapacitors should also be considered.
(5) As for the integrated system, it will become the main research direction to realize multi-intelligentization for supercapacitors via an integrated self-driven system including energy collection unit (nanogenerator, photovoltaic cell)/wireless charging ring, energy storage unit (supercapacitor), and functional unit (various sensors and detectors). Generally, the energy required by the functional unit is very low as the low working current and fast responding; however, the stable energy output of the energy device is still challengeable. Therefore, how to stabilize the output voltage of the microcapacitor is one of the main issues that need to be figured out as soon as possible.

References

1. J. Nowotny, J. Dodson, S. Fiechter, T.M. Gür, B. Kennedy, W. Macyk, T. Bak, W. Sigmund, M. Yamawaki, K.A. Rahman, Towards global sustainability: education on environmentally clean energy technologies. Renew. Sustain. Energy Rev. **81**, 2541–2551 (2018)
2. X. Hong, J. Mei, L. Wen, Y. Tong, A.J. Vasileff, L. Wang, J. Liang, Z. Sun, S.X. Dou, Nonlithium metal-sulfur batteries: steps toward a leap. Adv. Mater. **31**(5), 1802822 (2019)
3. D. Liu, Y. Tong, X. Yan, J. Liang, S.X. Dou, Recent advances in carbon-based bifunctional oxygen catalysts for zinc-air batteries. Batter. Supercaps. **2**(9), 743–765 (2019)
4. S.-L. Li, Q. Xu, Metal-organic frameworks as platforms for clean energy. Energy Environ. Sci. **6**(6), 1656–1683 (2013)
5. K. Chen, D. Xue, Materials chemistry toward electrochemical energy storage. J. Mater. Chem. A **4**(20), 7522–7537 (2016)
6. P. Simon, Y. Gogotsi, Materials for electrochemical capacitors, in *Nanoscience and Technology*, pp. 320–329
7. L. Wang, Z. Zhou, X. Yan, F. Hou, L. Wen, W. Luo, J. Liang, S.X. Dou, Engineering of lithium-metal anodes towards a safe and stable battery. Energy Storage Mater. **14**, 22–48 (2018)
8. T. Yang, J. Liang, I. Sultana, M.M. Rahman, M.J. Monteiro, Y. Chen, Z. Shao, S.R.P. Silva, J. Liu, Formation of hollow MoS_2/carbon microspheres for high capacity and high rate reversible alkali-ion storage. J. Mater. Chem. A **6**(18), 8280–8288 (2018)
9. L. Wen, J. Chen, J. Liang, L. Feng, H.M. Cheng, Flexible batteries ahead. Natl. Sci. Rev. **4**(1), 20–23 (2016)
10. L. Wen, J. Liang, J. Chen, Z.-Y. Chu, H.-M. Cheng, F. Li, Smart materials and design toward safe and durable lithium ion batteries. Small Methods **3**(11), 1900323 (2019)
11. A.C. Forse, C. Merlet, J.M. Griffin, C.P. Grey, New perspectives on the charging mechanisms of supercapacitors. J. Am. Chem. Soc. **138**(18), 5731–5744 (2016)
12. K. Jost, G. Dion, Y. Gogotsi, Textile energy storage in perspective. J. Mater. Chem. A **2**(28), 10776–10787 (2014)
13. P. Sharma, T.S. Bhatti, A review on electrochemical double-layer capacitors. Energy Convers. Manag. **51**(12), 2901–2912 (2010)
14. O. Barbieri, M. Hahn, A. Herzog, R. Kötz, Capacitance limits of high surface area activated carbons for double layer capacitors. Carbon **43**(6), 1303–1310 (2005)
15. A. García-Gómez, G. Moreno-Fernández, B. Lobato, T.A. Centeno, Constant capacitance in nanopores of carbon monoliths. Phys. Chem. Chem. Phys. **17**(24), 15687–15690 (2015)
16. Y. Tong, J. Liang, H.K. Liu, S.X. Dou, Energy storage in Oceania. Energy Storage Mater. **20**, 176–187 (2019)
17. X. Peng, L. Peng, C. Wu, Y. Xie, Two dimensional nanomaterials for flexible supercapacitors. Chem. Soc. Rev. **43**(10), 3303–3323 (2014)
18. V.D. Nithya, N. Sabari Arul, Progress and development of Fe_3O_4 electrodes for supercapacitors. J. Mater. Chem. A **4**(28), 10767–10778 (2016)
19. Z. Song, W. Li, Y. Bao, W. Wang, Z. Liu, F. Han, D. Han, L. Niu, Bioinspired microstructured pressure sensor based on a janus graphene film for monitoring vital signs and cardiovascular assessment. Adv. Electron. Mater. **4**(11), 1800252 (2018)
20. Y. Sun, J.A. Rogers, Inorganic semiconductors for flexible electronics. Adv. Mater. **19**(15), 1897–1916 (2007)
21. X. Wang, L. Dong, H. Zhang, R. Yu, C. Pan, Z.L. Wang, Recent progress in electronic skin. Adv. Sci. **2**(10), 1500169 (2015)
22. W. Guo, X. Yan, F. Hou, L. Wen, Y. Dai, D. Yang, X. Jiang, J. Liu, J. Liang, S.X. Dou, Flexible and free-standing SiO_x/CNT composite films for high capacity and durable lithium ion batteries. Carbon **152**, 888–897 (2019)
23. J. Xu, K. Wang, S.-Z. Zu, B.-H. Han, Z. Wei, Hierarchical nanocomposites of polyaniline nanowire arrays on graphene oxide sheets with synergistic effect for energy storage. ACS Nano **4**(9), 5019–5026 (2010)

24. J. Zhao, Y. Su, Z. Yang, L. Wei, Y. Wang, Y. Zhang, Arc synthesis of double-walled carbon nanotubes in low pressure air and their superior field emission properties. Carbon **58**, 92–98 (2013)
25. Y. Qiao, C.M. Li, S.-J. Bao, Q.-L. Bao, Carbon nanotube/polyaniline composite as anode material for microbial fuel cells. J. Power Sources **170**(1), 79–84 (2007)
26. H. Jiang, J. Ma, C. Li, Mesoporous carbon incorporated metal oxide nanomaterials as supercapacitor electrodes. Adv. Mater. **24**(30), 4197–4202 (2012)
27. J. Gamby, P.L. Taberna, P. Simon, J.F. Fauvarque, M. Chesneau, Studies and characterisations of various activated carbons used for carbon/carbon supercapacitors. J. Power Sources **101**(1), 109–116 (2001)
28. K.H. An, W.S. Kim, Y.S. Park, J.-M. Moon, D.J. Bae, S.C. Lim, Y.S. Lee, Y.H. Lee, Electrochemical properties of high-power supercapacitors using single-walled carbon nanotube electrodes. Adv. Funct. Mater. **11**(5), 387–392 (2001)
29. K. Jurewicz, K. Babeł, R. Pietrzak, S. Delpeux, H. Wachowska, Capacitance properties of multi-walled carbon nanotubes modified by activation and ammoxidation. Carbon **44**(12), 2368–2375 (2006)
30. T. Zhang, S. Han, W. Guo, F. Hou, J. Liu, X. Yan, S. Chen, J. Liang, Continuous carbon nanotube composite fibers for flexible aqueous lithium-ion batteries. SM&T **20**, e00096 (2019)
31. C. Liu, Z. Yu, D. Neff, A. Zhamu, B.Z. Jang, Graphene-based supercapacitor with an ultrahigh energy density. Nano Lett. **10**(12), 4863–4868 (2010)
32. J. Feng, L. Dong, X. Li, D. Li, P. Lu, F. Hou, J. Liang, S.X. Dou, Hierarchically stacked reduced graphene oxide/carbon nanotubes for as high performance anode for sodium-ion batteries. Electrochim. Acta **302**, 65–70 (2019)
33. C. Zhang, J. Li, E. Liu, C. He, C. Shi, X. Du, R.H. Hauge, N. Zhao, Synthesis of hollow carbon nano-onions and their use for electrochemical hydrogen storage. Carbon **50**(10), 3513–3521 (2012)
34. Z.-H. Huang, Y. Song, D.-Y. Feng, Z. Sun, X. Sun, X.-X. Liu, High mass loading MnO_2 with hierarchical nanostructures for supercapacitors. ACS Nano **12**(4), 3557–3567 (2018)
35. T. Liu, W.G. Pell, B.E. Conway, Self-discharge and potential recovery phenomena at thermally and electrochemically prepared RuO_2 supercapacitor electrodes. Electrochim. Acta **42**(23), 3541–3552 (1997)
36. L. Cao, F. Xu, Y.-Y. Liang, H.-L. Li, Preparation of the novel nanocomposite $Co(OH)_2$/ultra-stable Y zeolite and its application as a supercapacitor with high energy density. Adv. Mater. **16**(20), 1853–1857 (2004)
37. Y. Fu, J. Song, Y. Zhu, C. Cao, High-performance supercapacitor electrode based on amorphous mesoporous $Ni(OH)_2$ nanoboxes. J. Power Sources **262**, 344–348 (2014)
38. Z. Yang, C.-Y. Chen, H.-T. Chang, Supercapacitors incorporating hollow cobalt sulfide hexagonal nanosheets. J. Power Sources **196**(18), 7874–7877 (2011)
39. J. Xiao, L. Wan, S. Yang, F. Xiao, S. Wang, Design hierarchical electrodes with highly conductive $NiCo_2S_4$ nanotube arrays grown on carbon fiber paper for high-performance pseudocapacitors. Nano Lett. **14**(2), 831–838 (2014)
40. C. Zhu, P. Yang, D. Chao, X. Wang, X. Zhang, S. Chen, B.K. Tay, H. Huang, H. Zhang, W. Mai, H.J. Fan, All metal nitrides solid-state asymmetric supercapacitors. Adv. Mater. **27**(31), 4566–4571 (2015)
41. X. Xia, Y. Zhang, D. Chao, Q. Xiong, Z. Fan, X. Tong, J. Tu, H. Zhang, H.J. Fan, Tubular TiC fibre nanostructures as supercapacitor electrode materials with stable cycling life and wide-temperature performance. Energy Environ. Sci. **8**(5), 1559–1568 (2015)
42. Q. Meng, K. Cai, Y. Chen, L. Chen, Research progress on conducting polymer based supercapacitor electrode materials. Nano Energy **36**, 268–285 (2017)
43. X. Hong, R. Wang, Y. Liu, J. Fu, J. Liang, S. Dou, Recent advances in chemical adsorption and catalytic conversion materials for Li-S batteries. J. Energy Chem. **42**, 144–168 (2020)
44. G. Xu, C. Zheng, Q. Zhang, J. Huang, M. Zhao, J. Nie, X. Wang, F. Wei, Binder-free activated carbon/carbon nanotube paper electrodes for use in supercapacitors. Nano Res. **4**(9), 870–881 (2011)

45. M. Endo, T. Maeda, T. Takeda, Y.J. Kim, K. Koshiba, H. Hara, M.S. Dresselhaus, Capacitance and pore-size distribution in aqueous and nonaqueous electrolytes using various activated carbon electrodes. J. Electrochem. Soc. **148**(8), A910–A914 (2001)
46. J. Lloyd-Hughes, T.-I. Jeon, A review of the terahertz conductivity of bulk and nano-materials. J. Infrared Millim. Terahertz Waves **33**(9), 871–925 (2012)
47. B.Z. Jang, A. Zhamu, Processing of nanographene platelets (NGPs) and NGP nanocomposites: a review. J. Mater. Sci. **43**(15), 5092–5101 (2008)
48. I.K. Moon, J. Lee, R.S. Ruoff, H. Lee, Reduced graphene oxide by chemical graphitization. Nat. Commun. **1**(1), 73 (2010)
49. Z. Wen, X. Wang, S. Mao, Z. Bo, H. Kim, S. Cui, G. Lu, X. Feng, J. Chen, Crumpled nitrogen-doped graphene nanosheets with ultrahigh pore volume for high-performance supercapacitor. Adv. Mater. **24**(41), 5610–5616 (2012)
50. Y. Wen, B. Wang, C. Huang, L. Wang, D. Hulicova-Jurcakova, Synthesis of phosphorus-doped graphene and its wide potential window in aqueous supercapacitors. Chem. Eur. J. **21**(1), 80–85 (2015)
51. S. Wu, G. Chen, N.Y. Kim, K. Ni, W. Zeng, Y. Zhao, Z. Tao, H. Ji, Z. Lee, Y. Zhu, Creating pores on graphene platelets by low-temperature KOH activation for enhanced electrochemical performance. Small **12**(17), 2376–2384 (2016)
52. H. Pan, J. Li, Y. Feng, Carbon nanotubes for supercapacitor. Nanoscale Res. Lett. **5**(3), 654 (2010)
53. Z. Tang, C.-H. Tang, H. Gong, A high energy density asymmetric supercapacitor from nano-architectured $Ni(OH)_2$/carbon nanotube electrodes. Adv. Funct. Mater. **22**(6), 1272–1278 (2012)
54. Q. Xiao, X. Zhou, The study of multiwalled carbon nanotube deposited with conducting polymer for supercapacitor. Electrochim. Acta **48**(5), 575–580 (2003)
55. J.H. Park, J.M. Ko, O. Ok Park, Carbon nanotube/RuO_2 nanocomposite electrodes for supercapacitors. J. Electrochem. Soc. **150**(7), A864–A867 (2003)
56. X.-W. Wang, H.-P. Guo, J. Liang, J.-F. Zhang, B. Zhang, J.-Z. Wang, W.-B. Luo, H.-K. Liu, S.-X. Dou, An integrated free-standing flexible electrode with holey-structured 2D bimetallic phosphide nanosheets for sodium-ion batteries. Adv. Funct. Mater. **28**(26), 1801016 (2018)
57. U. Patil, S.C. Lee, S. Kulkarni, J.S. Sohn, M.S. Nam, S. Han, S.C. Jun, Nanostructured pseudocapacitive materials decorated 3D graphene foam electrodes for next generation supercapacitors. Nanoscale **7**(16), 6999–7021 (2015)
58. J. Sun, C. Wu, X. Sun, H. Hu, C. Zhi, L. Hou, C. Yuan, Recent progresses in high-energy-density all pseudocapacitive-electrode-materials-based asymmetric supercapacitors. J. Mater. Chem. A **5**(20), 9443–9464 (2017)
59. X. Rui, H. Tan, Q. Yan, Nanostructured metal sulfides for energy storage. Nanoscale **6**(17), 9889–9924 (2014)
60. Y. Shi, L. Peng, Y. Ding, Y. Zhao, G. Yu, Nanostructured conductive polymers for advanced energy storage. Chem. Soc. Rev. **44**(19), 6684–6696 (2015)
61. C. Zhang, T.M. Higgins, S.-H. Park, S.E. O'Brien, D. Long, J.N. Coleman, V. Nicolosi, Highly flexible and transparent solid-state supercapacitors based on RuO_2/PEDOT:PSS conductive ultrathin films. Nano Energy **28**, 495–505 (2016)
62. N.R. Chodankar, D.P. Dubal, G.S. Gund, C.D. Lokhande, A symmetric MnO_2/MnO_2 flexible solid state supercapacitor operating at 1.6 V with aqueous gel electrolyte. J. Energy Chem. **25**(3), 463–471 (2016)
63. P. Shi, L. Li, L. Hua, Q. Qian, P. Wang, J. Zhou, G. Sun, W. Huang, Design of amorphous manganese oxide@multiwalled carbon nanotube fiber for robust solid-state supercapacitor. ACS Nano **11**(1), 444–452 (2017)
64. B. Pandit, D.P. Dubal, B.R. Sankapal, Large scale flexible solid state symmetric supercapacitor through inexpensive solution processed V_2O_5 complex surface architecture. Electrochim. Acta **242**, 382–389 (2017)
65. Y. Qian, R. Liu, Q. Wang, J. Xu, D. Chen, G. Shen, Efficient synthesis of hierarchical NiO nanosheets for high-performance flexible all-solid-state supercapacitors. J. Mater. Chem. A **2**(28), 10917–10922 (2014)

66. P. Pande, P.G. Rasmussen, L.T. Thompson, Charge storage on nanostructured early transition metal nitrides and carbides. J. Power Sources **207**, 212–215 (2012)
67. M.S. Javed, S. Dai, M. Wang, Y. Xi, Q. Lang, D. Guo, C. Hu, Faradic redox active material of Cu_7S_4 nanowires with a high conductance for flexible solid state supercapacitors. Nanoscale **7**(32), 13610–13618 (2015)
68. X. Li, A.M. Elshahawy, C. Guan, J. Wang, Metal phosphides and phosphates-based electrodes for electrochemical supercapacitors. Small **13**(39), 1701530 (2017)
69. A. Achour, J.B. Ducros, R.L. Porto, M. Boujtita, E. Gautron, L. Le Brizoual, M.A. Djouadi, T. Brousse, Hierarchical nanocomposite electrodes based on titanium nitride and carbon nanotubes for micro-supercapacitors. Nano Energy **7**, 104–113 (2014)
70. Y. Yue, P. Han, X. He, K. Zhang, Z. Liu, C. Zhang, S. Dong, L. Gu, G. Cui, In situ synthesis of a graphene/titanium nitride hybrid material with highly improved performance for lithium storage. J. Mater. Chem. **22**(11), 4938–4943 (2012)
71. A. Rudge, J. Davey, I. Raistrick, S. Gottesfeld, J.P. Ferraris, Conducting polymers as active materials in electrochemical capacitors. J. Power Sources **47**(1), 89–107 (1994)
72. M. Kalaji, P.J. Murphy, G.O. Williams, The study of conducting polymers for use as redox supercapacitors. Synth. Met. **102**(1), 1360–1361 (1999)
73. Q. Yang, Y. Wang, X. Li, H. Li, Z. Wang, Z. Tang, L. Ma, F. Mo, C. Zhi, Recent progress of MXene-based nanomaterials in flexible energy storage and electronic devices. Energy Environ. Mater. **1**(4), 183–195 (2018)
74. T.M. Ng, M.T. Weller, G.P. Kissling, L.M. Peter, P. Dale, F. Babbe, J. de Wild, B. Wenger, H.J. Snaith, D. Lane, Optoelectronic and spectroscopic characterization of vapour-transport grown Cu_2ZnSnS_4 single crystals. J. Mater. Chem. A **5**(3), 1192–1200 (2017)
75. B. Anasori, Y. Xie, M. Beidaghi, J. Lu, B.C. Hosler, L. Hultman, P.R.C. Kent, Y. Gogotsi, M.W. Barsoum, Two-dimensional, ordered, double transition metals carbides (MXenes). ACS Nano **9**(10), 9507–9516 (2015)
76. F. Shahzad, M. Alhabeb, C.B. Hatter, B. Anasori, S. Man Hong, C.M. Koo, Y. Gogotsi, Electromagnetic interference shielding with 2D transition metal carbides (MXenes). Science **353**(6304), 1137–1140 (2016)
77. Y. Zhang, Y. Zheng, K. Rui, H.H. Hng, K. Hippalgaonkar, J. Xu, W. Sun, J. Zhu, Q. Yan, W. Huang, 2D black phosphorus for energy storage and thermoelectric applications. Small **13**(28), 1700661 (2017)
78. Z. Liu, F. Mo, H. Li, M. Zhu, Z. Wang, G. Liang, C. Zhi, Advances in flexible and wearable energy-storage textiles. Small Methods **2**(11), 1800124 (2018)
79. P. Lu, X. Wang, L. Wen, X. Jiang, W. Guo, L. Wang, X. Yan, F. Hou, J. Liang, H.-M. Cheng, S.X. Dou, Silica-mediated formation of nickel sulfide nanosheets on CNT films for versatile energy storage. Small **15**(15), 1805064 (2019)
80. M. Kaempgen, C.K. Chan, J. Ma, Y. Cui, G. Gruner, Printable thin film supercapacitors using single-walled carbon nanotubes. Nano Lett. **9**(5), 1872–1876 (2009)
81. A. Lamberti, F. Clerici, M. Fontana, L. Scaltrito, A highly stretchable supercapacitor using laser-induced graphene electrodes onto elastomeric substrate. Adv. Energy Mater. **6**(10), 1600050 (2016)
82. H. Yang, L. Yin, J. Liang, Z. Sun, Y. Wang, H. Li, K. He, L. Ma, Z. Peng, S. Qiu, C. Sun, H.-M. Cheng, F. Li, An aluminum-sulfur battery with a fast kinetic response. Angew. Chem. Int. Ed. **57**(7), 1898–1902 (2018)
83. L. Wen, F. Li, H.M. Cheng, Carbon nanotubes and graphene for flexible electrochemical energy storage: from materials to devices. Adv. Mater. **28**(22), 4306–4337 (2016)
84. Y. Ai, Z. Lou, L. Li, S. Chen, H.S. Park, Z.M. Wang, G. Shen, Meters-long flexible $CoNiO_2$-nanowires@ carbon-fibers based wire-supercapacitors for wearable electronics. Adv. Mater. Technol. **1**(8), 1600142 (2016)
85. B. Liu, D. Tan, X. Wang, D. Chen, G. Shen, Flexible, planar-integrated, all-solid-state fiber supercapacitors with an enhanced distributed-capacitance effect. Small **9**(11), 1998–2004 (2013)

86. Y. Fu, X. Cai, H. Wu, Z. Lv, S. Hou, M. Peng, X. Yu, D. Zou, Fiber supercapacitors utilizing pen ink for flexible/wearable energy storage. Adv. Mater. **24**(42), 5713–5718 (2012)
87. S. Shi, C. Xu, C. Yang, Y. Chen, J. Liu, F. Kang, Flexible asymmetric supercapacitors based on ultrathin two-dimensional nanosheets with outstanding electrochemical performance and aesthetic property. Sci. Rep. **3**, 2598 (2013)
88. T. Huang, B. Zheng, L. Kou, K. Gopalsamy, Z. Xu, C. Gao, Y. Meng, Z. Wei, Flexible high performance wet-spun graphene fiber supercapacitors. RSC Adv. **3**(46), 23957–23962 (2013)
89. Y. Huang, J. Liang, Y. Chen, An overview of the applications of graphene-based materials in supercapacitors. Small **8**(12), 1805–1834 (2012)
90. H. Gao, F. Xiao, C.B. Ching, H. Duan, Flexible all-solid-state asymmetric supercapacitors based on free-standing carbon nanotube/graphene and Mn_3O_4 nanoparticle/graphene paper electrodes. ACS Appl. Mater. Interfaces **4**(12), 7020–7026 (2012)
91. Y. Meng, Y. Zhao, C. Hu, H. Cheng, Y. Hu, Z. Zhang, G. Shi, L. Qu, All-graphene core-sheath microfibers for all-solid-state, stretchable fibriform supercapacitors and wearable electronic textiles. Adv. Mater. **25**(16), 2326–2331 (2013)
92. X. Hong, Y. Lu, S. Li, X. Wang, X. Wang, J. Liang, Carbon foam@reduced graphene oxide scaffold grown with polyaniline nanofibers for high performance symmetric supercapacitor. Electrochim. Acta **294**, 376–382 (2019)
93. Z. Cai, L. Li, J. Ren, L. Qiu, H. Lin, H. Peng, Flexible, weavable and efficient microsupercapacitor wires based on polyaniline composite fibers incorporated with aligned carbon nanotubes. J. Mater. Chem. A **1**(2), 258–261 (2013)
94. A.B. Dalton, S. Collins, E. Munoz, J.M. Razal, V.H. Ebron, J.P. Ferraris, J.N. Coleman, B.G. Kim, R.H. Baughman, Super-tough carbon-nanotube fibres. Nature **423**(6941), 703 (2003)
95. P. Xu, T. Gu, Z. Cao, B. Wei, J. Yu, F. Li, J.H. Byun, W. Lu, Q. Li, T.W. Chou, Carbon nanotube fiber based stretchable wire-shaped supercapacitors. Adv. Energy Mater. **4**(3), 1300759 (2014)
96. J. Ren, L. Li, C. Chen, X. Chen, Z. Cai, L. Qiu, Y. Wang, X. Zhu, H. Peng, Twisting carbon nanotube fibers for both wire-shaped micro-supercapacitor and micro-battery. Adv. Mater. **25**(8), 1155–1159 (2013)
97. X. Chen, H. Lin, J. Deng, Y. Zhang, X. Sun, P. Chen, X. Fang, Z. Zhang, G. Guan, H. Peng, Electrochromic fiber-shaped supercapacitors. Adv. Mater. **26**(48), 8126–8132 (2014)
98. Z. Yang, J. Deng, X. Chen, J. Ren, H. Peng, A highly stretchable, fiber-shaped supercapacitor. Angew. Chem. Int. Ed. **52**(50), 13453–13457 (2013)
99. L. Kou, T. Huang, B. Zheng, Y. Han, X. Zhao, K. Gopalsamy, H. Sun, C. Gao, Coaxial wet-spun yarn supercapacitors for high-energy density and safe wearable electronics. Nat. Commun. **5**, 3754 (2014)
100. X. Chen, L. Qiu, J. Ren, G. Guan, H. Lin, Z. Zhang, P. Chen, Y. Wang, H. Peng, Novel electric double-layer capacitor with a coaxial fiber structure. Adv. Mater. **25**(44), 6436–6441 (2013)
101. H. Wu, K. Jiang, S. Gu, H. Yang, Z. Lou, D. Chen, G. Shen, Two-dimensional $Ni(OH)_2$ nanoplates for flexible on-chip microsupercapacitors. Nano Res. **8**(11), 3544–3552 (2015)
102. M.F. El-Kady, R.B. Kaner, Scalable fabrication of high-power graphene micro-supercapacitors for flexible and on-chip energy storage. Nat. Commun. **4**, 1475 (2013)
103. Y.S. Moon, D. Kim, G. Lee, S.Y. Hong, K.K. Kim, S.M. Park, J.S. Ha, Fabrication of flexible micro-supercapacitor array with patterned graphene foam/MWNT-COOH/MnO_x electrodes and its application. Carbon **81**, 29–37 (2015)
104. Z. Liu, Z.S. Wu, S. Yang, R. Dong, X. Feng, K. Müllen, Ultraflexible in-plane micro-supercapacitors by direct printing of solution-processable electrochemically exfoliated graphene. Adv. Mater. **28**(11), 2217–2222 (2016)
105. J. Cai, C. Lv, A. Watanabe, Laser direct writing of high-performance flexible all-solid-state carbon micro-supercapacitors for an on-chip self-powered photodetection system. Nano Energy **30**, 790–800 (2016)
106. Y.-Y. Peng, B. Akuzum, N. Kurra, M.-Q. Zhao, M. Alhabeb, B. Anasori, E.C. Kumbur, H.N. Alshareef, M.-D. Ger, Y. Gogotsi, All-MXene (2D titanium carbide) solid-state microsupercapacitors for on-chip energy storage. Energy Environ. Sci. **9**(9), 2847–2854 (2016)

107. M.F. El-Kady, V. Strong, S. Dubin, R.B. Kaner, Laser scribing of high-performance and flexible graphene-based electrochemical capacitors. Science **335**(6074), 1326–1330 (2012)
108. X. Hou, B. Liu, X. Wang, Z. Wang, Q. Wang, D. Chen, G. Shen, SnO_2-microtube-assembled cloth for fully flexible self-powered photodetector nanosystems. Nanoscale **5**(17), 7831–7837 (2013)
109. X. Xiao, T. Li, P. Yang, Y. Gao, H. Jin, W. Ni, W. Zhan, X. Zhang, Y. Cao, J. Zhong, Fiber-based all-solid-state flexible supercapacitors for self-powered systems. ACS Nano **6**(10), 9200–9206 (2012)
110. Y. Ai, Z. Lou, S. Chen, D. Chen, Z.M. Wang, K. Jiang, G. Shen, All rGO-on-PVDF-nanofibers based self-powered electronic skins. Nano Energy **35**, 121–127 (2017)
111. C. Yan, P.S. Lee, Stretchable energy storage and conversion devices. Small **10**(17), 3443–3460 (2014)
112. X. Wang, S. Niu, F. Yi, Y. Yin, C. Hao, K. Dai, Y. Zhang, Z. You, Z.L. Wang, Harvesting ambient vibration energy over a wide frequency range for self-powered electronics. ACS Nano **11**(2), 1728–1735 (2017)
113. X. Wang, B. Liu, R. Liu, Q. Wang, X. Hou, D. Chen, R. Wang, G. Shen, Fiber-based flexible all-solid-state asymmetric supercapacitors for integrated photodetecting system. Angew. Chem. Int. Ed. **53**(7), 1849–1853 (2014)
114. D. Kim, D. Kim, H. Lee, Y.R. Jeong, S.J. Lee, G. Yang, H. Kim, G. Lee, S. Jeon, G. Zi, Body-attachable and stretchable multisensors integrated with wirelessly rechargeable energy storage devices. Adv. Mater. **28**(4), 748–756 (2016)
115. J. Xu, G. Shen, A flexible integrated photodetector system driven by on-chip microsupercapacitors. Nano Energy **13**, 131–139 (2015)
116. J. Benson, I. Kovalenko, S. Boukhalfa, D. Lashmore, M. Sanghadasa, G. Yushin, Multifunctional CNT-polymer composites for ultra-tough structural supercapacitors and desalination devices. Adv. Mater. **25**(45), 6625–6632 (2013)
117. S.R. Shin, C.K. Lee, I. So, J.-H. Jeon, T.M. Kang, C. Kee, S.I. Kim, G.M. Spinks, G.G. Wallace, S.J. Kim, DNA-wrapped single-walled carbon nanotube hybrid fibers for supercapacitors and artificial muscles. Adv. Mater. **20**(3), 466–470 (2008)
118. Y. Hu, H. Cheng, F. Zhao, N. Chen, L. Jiang, Z. Feng, L. Qu, All-in-one graphene fiber supercapacitor. Nanoscale **6**(12), 6448–6451 (2014)
119. Y. Huang, M. Zhu, Y. Huang, Z. Pei, H. Li, Z. Wang, Q. Xue, C. Zhi, Multifunctional energy storage and conversion devices. Adv. Mater. **28**(38), 8344–8364 (2016)
120. X. Pu, L. Li, M. Liu, C. Jiang, C. Du, Z. Zhao, W. Hu, Z.L. Wang, Wearable self-charging power textile based on flexible yarn supercapacitors and fabric nanogenerators. Adv. Mater. **28**(1), 98–105 (2016)
121. Y. Yue, Z. Yang, N. Liu, W. Liu, H. Zhang, Y. Ma, C. Yang, J. Su, L. Li, F. Long, Z. Zou, Y. Gao, A flexible integrated system containing a microsupercapacitor, a photodetector, and a wireless charging coil. ACS Nano **10**(12), 11249–11257 (2016)
122. Z.L. Wang, Self-powered nanosensors and nanosystems. Adv. Mater. **24**(2), 280–285 (2012)
123. T. Chen, L. Qiu, Z. Yang, Z. Cai, J. Ren, H. Li, H. Lin, X. Sun, H. Peng, An integrated "energy wire" for both photoelectric conversion and energy storage. Angew. Chem. Int. Ed. **51**(48), 11977–11980 (2012)
124. H. Guo, M.-H. Yeh, Y.-C. Lai, Y. Zi, C. Wu, Z. Wen, C. Hu, Z.L. Wang, All-in-one shape-adaptive self-charging power package for wearable electronics. ACS Nano **10**(11), 10580–10588 (2016)
125. H. Huang, L. Lu, J. Wang, J. Yang, S.-F. Leung, Y. Wang, D. Chen, X. Chen, G. Shen, D. Li, Performance enhancement of thin-film amorphous silicon solar cells with low cost nanodent plasmonic substrates. Energy Environ. Sci. **6**(10), 2965–2971 (2013)
126. S. Mathew, A. Yella, P. Gao, R. Humphry-Baker, B.F. Curchod, N. Ashari-Astani, I. Tavernelli, U. Rothlisberger, M.K. Nazeeruddin, M. Grätzel, Dye-sensitized solar cells with 13% efficiency achieved through the molecular engineering of porphyrin sensitizers. Nat. Chem. **6**(3), 242 (2014)

127. J. Xu, Z. Ku, Y. Zhang, D. Chao, H.J. Fan, Integrated photo-supercapacitor based on PEDOT modified printable perovskite solar cell. Adv. Mater. Technol. **1**(5), 1600074 (2016)
128. H. Yoon, S.M. Kang, J.-K. Lee, M. Choi, Hysteresis-free low-temperature-processed planar perovskite solar cells with 19.1% efficiency. Energy Environ. Sci. **9**(7), 2262–2266 (2016)
129. Z. Zhang, X. Chen, P. Chen, G. Guan, L. Qiu, H. Lin, Z. Yang, W. Bai, Y. Luo, H. Peng, Integrated polymer solar cell and electrochemical supercapacitor in a flexible and stable fiber format. Adv. Mater. **26**(3), 466–470 (2014)
130. M. Zhu, Y. Huang, Y. Huang, Z. Pei, Q. Xue, H. Li, H. Geng, C. Zhi, Capacitance enhancement in a semiconductor nanostructure-based supercapacitor by solar light and a self-powered supercapacitor-photodetector system. Adv. Funct. Mater. **26**(25), 4481–4490 (2016)
131. J. Xu, H. Wu, L. Lu, S.F. Leung, D. Chen, X. Chen, Z. Fan, G. Shen, D. Li, Integrated photo-supercapacitor based on bi-polar TiO_2 nanotube arrays with selective one-side plasma-assisted hydrogenation. Adv. Funct. Mater. **24**(13), 1840–1846 (2014)
132. Y. Fu, H. Wu, S. Ye, X. Cai, X. Yu, S. Hou, H. Kafafy, D. Zou, Integrated power fiber for energy conversion and storage. Energy Environ. Sci. **6**(3), 805–812 (2013)
133. Z. Yang, L. Li, Y. Luo, R. He, L. Qiu, H. Lin, H. Peng, An integrated device for both photo-electric conversion and energy storage based on free-standing and aligned carbon nanotube film. J. Mater. Chem. A **1**(3), 954–958 (2013)
134. Y. Zi, L. Lin, J. Wang, S. Wang, J. Chen, X. Fan, P.K. Yang, F. Yi, Z.L. Wang, Triboelectric-pyroelectric-piezoelectric hybrid cell for high-efficiency energy-harvesting and self-powered sensing. Adv. Mater. **27**(14), 2340–2347 (2015)
135. G. Zhu, Y.S. Zhou, P. Bai, X.S. Meng, Q. Jing, J. Chen, Z.L. Wang, A shape-adaptive thin-film-based approach for 50% high-efficiency energy generation through micro-grating sliding electrification. Adv. Mater. **26**(23), 3788–3796 (2014)
136. G. Zhu, B. Peng, J. Chen, Q. Jing, Z.L. Wang, Triboelectric nanogenerators as a new energy technology: from fundamentals, devices, to applications. Nano Energy **14**, 126–138 (2015)
137. Q. Zheng, Y. Zou, Y. Zhang, Z. Liu, B. Shi, X. Wang, Y. Jin, H. Ouyang, Z. Li, Z.L. Wang, Biodegradable triboelectric nanogenerator as a life-time designed implantable power source. Sci. Adv. **2**(3), e1501478 (2016)
138. C. Zhang, Z.L. Wang, Tribotronics—a new field by coupling triboelectricity and semiconductor. Nano Today **11**(4), 521–536 (2016)
139. Y. Xie, S. Wang, S. Niu, L. Lin, Q. Jing, J. Yang, Z. Wu, Z.L. Wang, Grating-structured freestanding triboelectric-layer nanogenerator for harvesting mechanical energy at 85% total conversion efficiency. Adv. Mater. **26**(38), 6599–6607 (2014)
140. W. Wu, L. Wang, Y. Li, F. Zhang, L. Lin, S. Niu, D. Chenet, X. Zhang, Y. Hao, T.F. Heinz, Piezoelectricity of single-atomic-layer MoS_2 for energy conversion and piezotronics. Nature **514**(7523), 470 (2014)
141. Z.L. Wang, J. Song, Piezoelectric nanogenerators based on zinc oxide nanowire arrays. Science **312**(5771), 242–246 (2006)
142. Z.L. Wang, J. Chen, L. Lin, Progress in triboelectric nanogenerators as a new energy technology and self-powered sensors. Energy Environ. Sci. **8**(8), 2250–2282 (2015)
143. S. Wang, Z.-H. Lin, S. Niu, L. Lin, Y. Xie, K.C. Pradel, Z.L. Wang, Motion charged battery as sustainable flexible-power-unit. ACS Nano **7**(12), 11263–11271 (2013)
144. J. Wang, X. Li, Y. Zi, S. Wang, Z. Li, L. Zheng, F. Yi, S. Li, Z.L. Wang, A flexible fiber-based supercapacitor-triboelectric-nanogenerator power system for wearable electronics. Adv. Mater. **27**(33), 4830–4836 (2015)
145. W. Tang, C.B. Han, C. Zhang, Z.L. Wang, Cover-sheet-based nanogenerator for charging mobile electronics using low-frequency body motion/vibration. Nano Energy **9**, 121–127 (2014)
146. X. Pu, M. Liu, X. Chen, J. Sun, C. Du, Y. Zhang, J. Zhai, W. Hu, Z.L. Wang, Ultrastretchable, transparent triboelectric nanogenerator as electronic skin for biomechanical energy harvesting and tactile sensing. Sci. Adv. **3**(5), e1700015 (2017)

Chapter 4
Magnetic Responsive MnO_2 Nanomaterials

Wenxian Li

Abstract The magnetisms of nanosized MnO_2 in different phases depend greatly on the morphology and the dimension of the nanoparticles and the interaction between the induced magnetism and the intrinsic magnetism. Different sharing modes of the basic structure in MnO_2 (i.e., MnO_6 octahedron) contribute to the low-temperature ground-state magnetism deviation from general antiferromagnetism. MnO_2 bulks possess an antiferromagnetic ordering between the corner-sharing MnO_6 octahedra and a ferromagnetic ordering between the edge-sharing MnO_6 octahedra. The essential reason is the change of interatomic distances, coordinate environment, and symmetry, resulting in the different surface states. For this reason, the effects of shapes, ions introducing, electronic structures, and exposed planes on magnetism are as important as sharing modes.

4.1 Introduction

4.1.1 Background

MnO_2 materials have attracted intensive research interest among medical scientists, biologists, and magnetism researchers [1, 2] due to their excellent magnetic performance in magnetic resonance imaging (MRI) enhancement, magnetic transport in biochemical engineering, and magnetic memory storage [3]. The typical non-toxic and environmental-friendly properties of MnO_2 precisely satisfy the requirements of medical, biological, and environmental materials. Liu et al. [4] prepared the MnO_2 nanoparticle encapsulated by NCP-shells constructed based on high-Z element hafnium (Hf) ions and c,c,t-(diamminedichlorodisuccinato)Pt(IV) (DSP) for tumor radiotherapy sensitizer. Song et al. [5] also synthesized the HA-coated and mannan-conjugated MnO_2 particles as an intelligent magnetic resonance molecular imaging

W. Li (✉)
Institute of Materials & Institute for Sustainable Energy, Shanghai University, Shanghai 200444, China
e-mail: shuliwx@t.shu.edu.cn

Z. Sun and T. Liao (eds.), *Responsive Nanomaterials for Sustainable Applications*, Springer Series in Materials Science 297,
https://doi.org/10.1007/978-3-030-39994-8_4

probe to positioning and diagnosing tumor. MnO_2 can become Mn^{2+} with magnetic resonance enhancement at specific sites, which significantly enhanced longitudinal relaxation and lateral relaxation properties of tumor sites. Therefore, it is critical to study the magnetic properties of manganese dioxide because it is beneficial to human health.

4.1.2 The Phase of MnO_2

The basic structure of MnO_2 is MnO_6 octahedron [7, 8]. In MnO_6 octahedron, oxygen atoms are distributed at the corners of octahedra, while Mn atoms are placed in the center, shown in Fig. 4.1a. One Mn atom is surrounded by six neighboring oxygen atoms. The structure of MnO_2 is quite flexible. There are a number of crystal forms, such as α, β-, γ-, λ-, and δ-MnO_2, which depends on the way of MnO_6 octahedral unit sharing [9–11]. The structures of different phases of MnO_2 can be characterized by the size of their tunnels and interlayer distance. Table 4.1 shows the information of structural characters of MnO_2.

For α-MnO_2, the double chains of corner-sharing MnO_6 octahedra form one-dimensional tunnel structure with T[2 × 2] and T[1 × 1] open tunnel with dimensions of 4.6 × 4.6 Å [18], shown in Fig. 4.1b. This tunnel structure has a large cross-sectional area, which can accommodate a large cation to enhance its stability of the structure, so it is suitable for the insertion/extraction of cations such as Li^+, Na^+, K^+, NH^{4+}, Ba^{2+}, or H_3O^+.

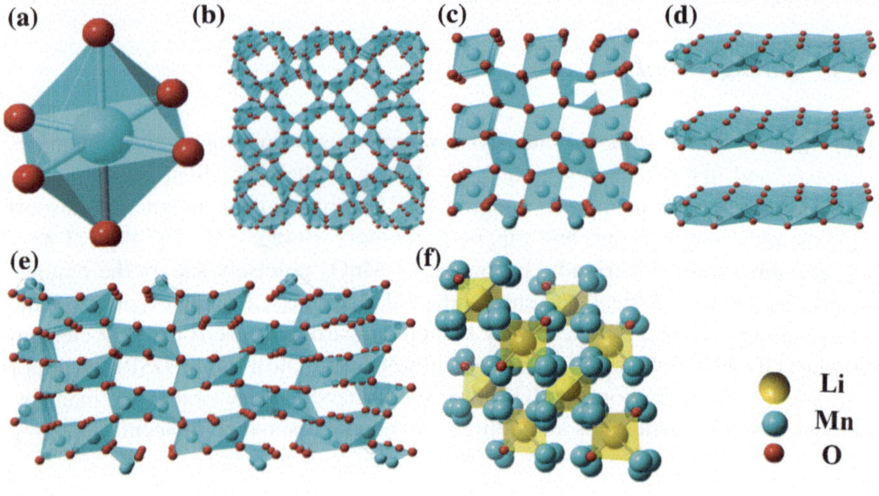

Fig. 4.1 **a** MnO_6 octahedral and the crystal structure of **b** α-MnO_2, **c** β-MnO_2, **d** δ-$MnO_{2\text{-}(001)}$, **e** γ-MnO_2 and **f** λ-MnO_2. Adapted with permission from Zhou et al. [6] © The Royal Society of Chemistry and the Centre National de la Recherche Scientifique, 2018

Table 4.1 Structural characters of MnO_2

Phases	Approximate chemical formula	Crystal systems	Structural feature
α-MnO_2 [13]	$(R)_2[Mn_s]O_{16}xH_2$ (R = K, Na, NH_4)	Monoclinic/quartet	T[2 × 2] T[1 × 1]
β-MnO_2 [14]	MnO_2	Quartet	T[1 × 1]
γ-MnO_2 [15]	$[Mn_{II}, Mn_{III}, Mn_{IV}](O, OH)_2$	Hexagonal	T[2 × 1] T[1 × 1]
δ-MnO_2 [16]	$K_{0.27}MnO_2 \cdot 0.54H_2O$ $Na_4Mn_{14}O_{27} \cdot 9H_2O$ $R_xMnO_2xH_2O$ (R = Li, K, Na, NH_4)	Orthogonal/hexagonal/diamond	Two-dimensional infinite sheet
λ-MnO_2 [17]	$(R1.0)[Mn_2]O_4$ (R = Li)	Cube	T[1 × 1] tunnel interconnection Three-dimensional

Reproduced with permission from Musil et al. [12] © the electrochemical society, 2015

β-MnO_2 owing a tetragonal rutile structure has T[1 × 1] tunnel [19, 20], formed by a single strand of corner-sharing MnO_6 octahedra, shown in Fig. 4.1c. The tunnel cross-sectional area of β-MnO_2 is so small that it does not provide ion channels to diffuse ions. β-MnO_2 is the most stable phase with relatively stable thermodynamic properties.

δ-MnO_2 is a two-dimensional layered birnessite structure with an interlayer separation of 7 Å [21], shown in Fig. 4.1d. This unique two-dimensional structure is characterized by the layer interspacing distance, which depends on the cations like H_2O.

γ-MnO_2 also has a complex one-dimensional tunnel structure formed by double-chain and single-stranded alternate, T[2 × 1] and T[1 × 1] tunnel structure, shown in Fig. 4.1e. Since the alternating arrangement of the double-stranded and single-stranded alternating structures is difficult to maintain long-range order, it is difficult for γ-MnO_2 to obtain single crystal.

λ-MnO_2 composes of coplanar tetrahedron and octahedron with a T[1 × 1] three-dimensional network tunnel structure, shown in Fig. 4.1f. Mn ions occupy 16d sites in Fd3m and form a three-dimensional array of corner-sharing tetrahedron. Ions are able to insert/extract freely in the three-dimensional space channel.

4.1.3 Electronic Distribution and d-Orbit of Mn

Electron distribution outside the nucleus of Mn is $1S_2$ $2S_2$ $2P_6$ $3S_2$ $3P_6$ $3d_5$ $4S_2$, in which the majority 3d-orbitals are completely filled with five electrons, while the minority 3d-orbitals are all left empty. This phenomenon would inevitably impact the magnetic coupling states of the Mn atoms in different valences [22]. Manganese has a variety of valence states ranging from +2 to +7 [15]. And manganese ion has a magnetic moment on its valence Mn^{2+}, Mn^{3+}, and Mn^{4+} [23], respectively, which are closely related to magnetism. The bond length of Mn–Mn is the other decisive factor in the magnetic couplings between Mn atoms [24, 25].

MnO_6 octahedron [28, 29], as the basic structure of MnO_2, is of great significance to study manganese dioxide. In the MnO_6 octahedral field [29], the d-orbit of Mn is repelled by the negative charge of the six ligands [30]. The spatial arrangement of the d-orbit is shown in Fig. 4.2a. d_{xy}, d_{xz}, and d_{yz} are between the gaps of the ligands, so the energy of three orbits increases slightly resulting in the small exclusion of the ligand. Yet the d_{z^2} and $d_{x^2} - d_{y^2}$ are head to head with the ligand, the orbital electrons are subject to a larger repulsive force, causing the energy of the two orbits increasing, which contributes to the division of d-orbit into a triple degenerate orbit t_{2g} and a

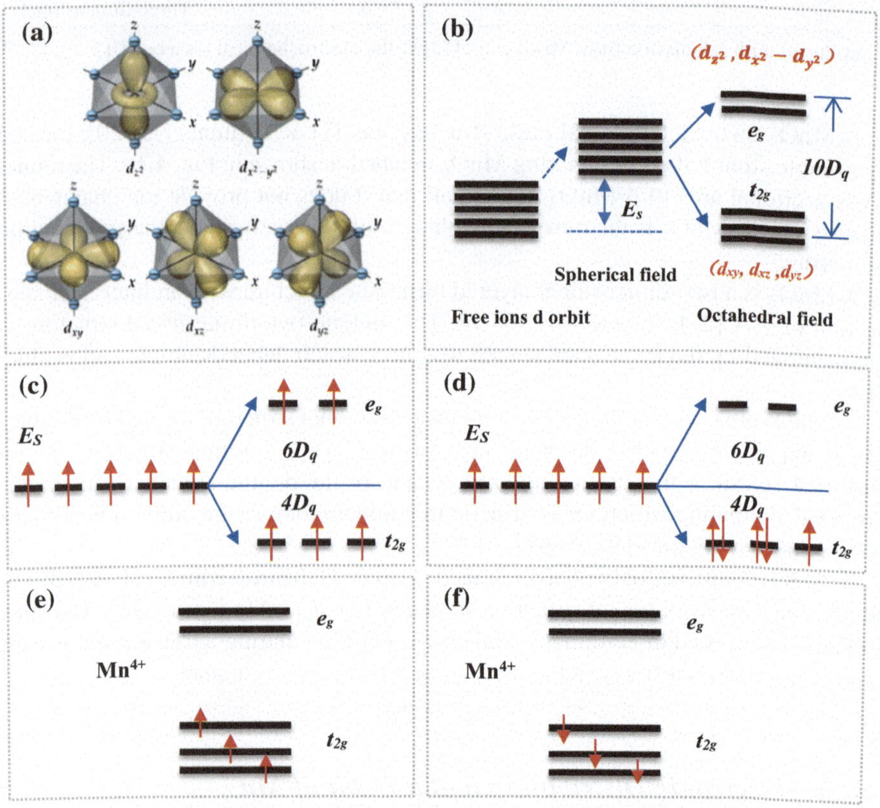

Fig. 4.2 **a** Octahedral complex d is the relative position of the orbital and ligand. Reproduced with permission from Li et al. [26] © Scientific Reports, 2015. **b** Energy splitting of the center atom d-orbit in the octahedral field. The distribution of Mn d5 electrons in the **c** weak and **d** strong octahedral field. The Mn^{4+} electrons parallel distribution **e** Spin-up and **f** Spin-Down. Reproduced with permission from Li et al. [27] © Scientific Report, 2014

double degenerate orbit e_g, shown in Fig. 4.2b. The major states of t_{2g} are thought to be occupied, whereas the minority states of t_{2g} and the whole e_g states keep empty [31]. According to the crystal field theory, the five d-orbitals of Mn are degenerated before d-orbital splitting into the free metal ions. Electrons are distributed in different orbits and spin parallel, satisfying the Hund rules, which is the only way of electrons arranging. Nevertheless, electrons appear with high-spin distribution and low-spin distribution after d-orbital splitting. Electrons occupy a high-spin state in the weak octahedral field, opposite in the strong octahedral field. The distribution of Mn d5 electrons in the weak and strong octahedral field is shown in Fig. 4.2c, d. Mn^{4+} electrons distribution is shown in Fig. 4.2e, f. The three electrons occupy the lowest energy orbit t_{2g} and are arranged in parallel.

In this review, detailed discussions about magnetism of MnO_2 with various phases are concluded from intrinsic magnetism, ions-induced magnetism, and the great influence of exposed plane and shape and size effect on the magnetism of MnO_2.

4.2 Magnetism

The magnetic mechanism of MnO_2 is quite complex to understand, which relates to the electronic structure in some way. Li et al. [32] plotted the spin-resolved density of states (DOS) of α-, β-, and δ-MnO_2 bulks to further reveal the differences in the electronic structure between the layered phase and two tunneled phases, shown in Fig. 4.3. They drew δ-MnO_2 has a distinct electronic structure compared to other two tunneled structures (α-MnO_2 and β-MnO_2). First, δ-MnO_2 is ferromagnetic (FM) while the others are in antiferromagnetic (AFM) ordering. Second, the conduction band (CB) and valence band (VB) of δ-MnO_2 are narrower compared to the others. They considered that the way of MnO_6 octahedral unit sharing plays an important role. All MnO_6 octahedra are edge-sharing in δ-MnO_2, but corner-sharing in β-MnO_2, while α-MnO_2 consists of 50% edge-sharing and 50% corner-sharing octahedra. The Mn–O–Mn in edge-sharing MnO_6 is 98°, which is smaller than that in the corner-sharing MnO_6 (129°). This causes much stronger superexchange antiferromagnetic [33] coupling between two corner-sharing MnO_6 than that in the edge-sharing MnO_6. Consequently, δ-MnO_2 bulks prefer FM ground state, while the spins of corner-sharing MnO_6 prefer to be antiparallel, leading to the AFM spin ordering [32]. Their views agree with the Duan's [34] opinion that the Mn–Mn coupling between corner-sharing MnO_6 is antiferromagnetic, while between edge-sharing MnO_6 is ferromagnetic, which contributes to the intrinsic magnetism of MnO_2 in different phases. Doping ions, exposed plane, and shape and size effect affect the magnetism of MnO_2 with different phases.

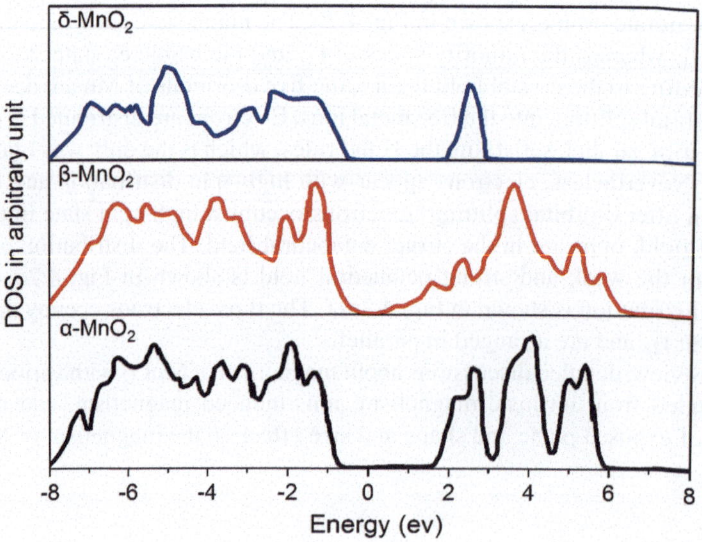

Fig. 4.3 Spin-resolved density of states (DOS) for α-, β- and δ-MnO_2 phases. Reproduced with permission from Li et al. [32] © American Chemical Society, 2016

4.2.1 Intrinsic Magnetism

4.2.1.1 Magnetism of α-MnO_2

The ground-state magnetic properties of α-MnO_2 bulks are generally considered as antiferromagnetic [35, 36] performance at low temperature due to the symmetric nature of Mn–O–Mn bonds, different from the nano-α-MnO_2. The Néel magnetic transition temperature T_N of α-MnO_2 bulks is about 24.5 K. Kuang et al. found that α-MnO_2 nanorods have a paramagnetic state at 300 K, while exhibit a weak ferromagnetic behavior at 5 K [37], shown in Fig. 4.4a, b. At 5 K, the coercivity field (H_c) of α-MnO_2 is 844.82 Oe. The corresponding remnant magnetization (M_r) is 4.48×10^{-4} emu/g, and the hysteresis loop shows a weak ferromagnetic behavior [38]. The M_r-H_c curves is nearly linear at 300 K. Yang et al. [39] pointed out that the *M-H* curve of α-MnO_2 microspheres shows a small hysteresis at 1.8 K and is also nearly linear at 300 K, shown in Fig. 4.4c. α-MnO_2 microspheres have the Néel magnetic transition temperature T_N ($T_N = 13$ K), shown in Fig. 4.4d. This temperature is lower than that of α-MnO_2 bulks due to the small grain size effect [40]. Below 13 K, α-MnO_2 microspheres exhibit antiferromagnetism with a parasitic ferromagnetic component. Antiferromagnetic α-MnO_2 showed the ferromagnetic nature at the lower temperature at the nano-size level. The magnetic hysteresis loop at 5.0 K, no magnetic hysteresis loop at 60 K, and a straight line of magnetic hysteresis loop at 120 K were also obtained by Zhu [41].

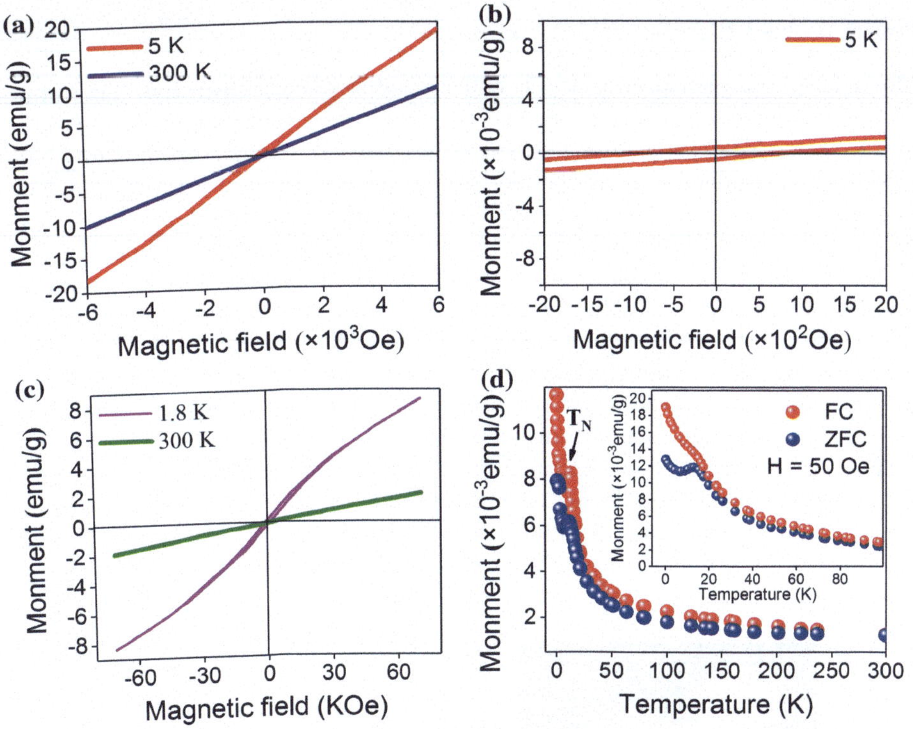

Fig. 4.4 **a** Hysteresis loops of α-MnO_2 nanorods at 5 and 300 K and **b** an enlarged view of 5 K. Reproduced with permission from Kuang et al. [37] © The Royal Society of Chemistry and the Centre National de la Recherche Scientifique, 2015. **c** The hysteresis loops of at 1.8 and 300 K and **d** the temperature dependence of magnetization curves under zero field cooling (ZFC) and field cooling (FC) of α-MnO_2 microspheres. Reproduced with permission from Yi et al. [42] © American Institute of Physics, 2004

It is supposed the ferromagnetic component at low temperature may result from the non-collinear molecular field that cants the two antiparallel sublattices due to the distortion of the crystal structure, especially arising from the oxygen vacancies. The mixture of Mn^{3+} and Mn^{4+} [43–45] valence state in α-MnO_2 has been considered as one of the reasons for the ferromagnetic ordering/spin glass behavior [46] at low temperature. Actually, the ferromagnetic-like ordering originates from the breaking of equivalence between two different Mn–O octahedrals in α-MnO_2 due to the filling of ions, especially because of neighboring tunnels which may be filled and unfilled lead to differences in the magnetic moment of Mn atoms in the apex site and plane site. Li et al. reported the effect of exposed crystal planes on the magnetism of α-MnO_2 nanowires [27]. Detailed discussion is at Sect. 4.2.3.

4.2.1.2 Magnetism of β-MnO_2

Ground-state magnetism of β-MnO_2 bulks is antiferromagnetism too. Tang et al. proposed that dandelion-like β-MnO_2 (Fig. 4.5a, b) has Néel temperature of about 100 K [47], which is about 6 K [45, 48] higher than that of the corresponding bulk β-MnO_2 crystals due to the special morphology and the effect of surface spins [49, 50], shown in Fig. 4.5c. The ZFC curve shows a small increase of the magnetization than the FC curve. This may be related to the large ratio of surface/volume, which inevitably leads to a much larger proportion of non-compensated surface spins on the antiferromagnetic core and thus reveals the difference between the ZFC curve and the FC curve [51]. No hysteresis loop at either 5 or 293 K is observed in Fig. 4.5d. The *M-H* curves were nearly linear above the Néel temperature, revealing that a paramagnetic state exists in this temperature. At 5 K, the *M-H* curve shows no hysteresis, indicating its antiferromagnetic nature below T_N.

However, Liang et al. considered that β-MnO_2 nanorods (shown in Fig. 4.6a, b) exhibits paramagnetic behavior and there was not an antiferromagnetic transition because of no hysteresis loop at either 5 or 300 K (shown in Fig. 4.6c) indicating that the magnetic state remains unchanged in the whole temperature range [37]. This is

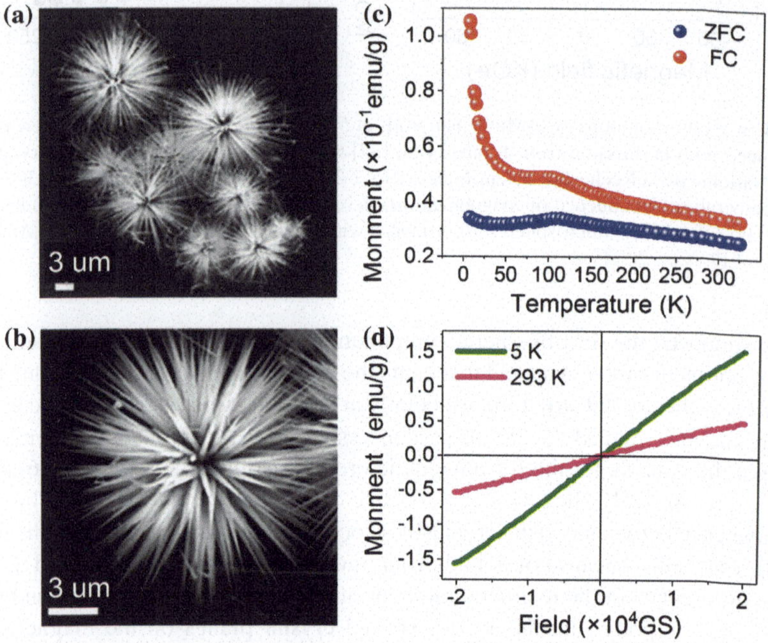

Fig. 4.5 SEM images **a** overall morphology and **b** detailed view of dandelion-like β-MnO_2. **c** ZFC and FC magnetization curves under an applied field of 1000 Oe and **d** field dependence of the magnetization at 5 K and 293 K of dandelion-like β-MnO_2. Reproduced with permission from Tang et al. [47] © IOP Publishing Ltd, 2006

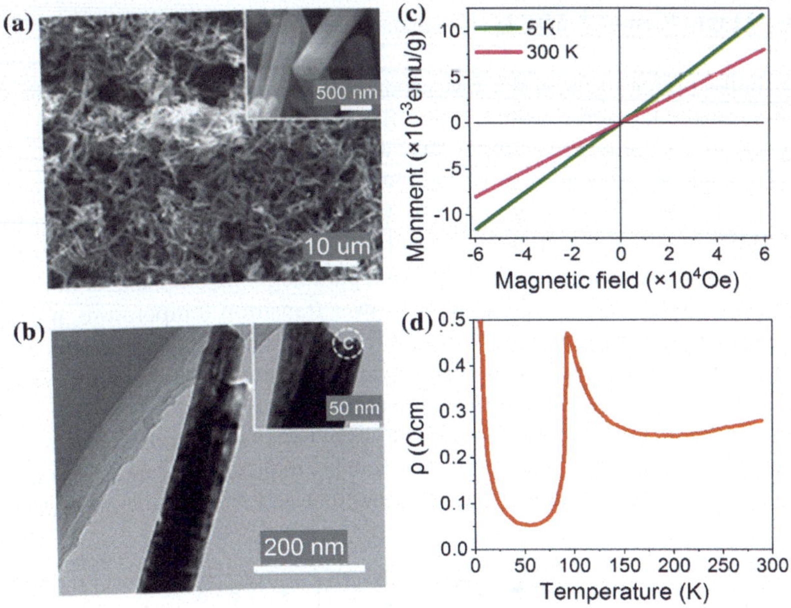

Fig. 4.6 **a** SEM image (inset is the higher magnification image), **b** TEM images and **c** the hysteresis loops at 5 and 300 K of β-MnO_2 nanorods. Reproduced with permission from Liang et al. [37] © 2015 The Royal Society of Chemistry and the Centre National de la Recherche Scientifique, 2006. **d** Resistivities of a single crystal of β-MnO_2 (The current was applied along the c axis). Reproduced with permission from Subhra et al. [55] © The American Physical Society, 2000

contrary to Tang's views. Subhra et al. put forward that the shape of β-MnO_2 makes difference in magnetism, which is discussed in the part of "The effect of sizes on magnetism of MnO_2." Then, the helimagnetism [14, 48, 52] of β-MnO_2 was confirmed by Sato et al. through the X-ray magnetic scattering characterization [53, 54] using synchrotron radiation. They gave a critical temperature point of 92 K, as is shown in Fig. 4.6d. Below around 200 K, it deviates from a metallic behavior and increases with decreasing temperature down to T_N (92 K), which coincides with the magnetic transition temperature into a helical state. Different from other manganates, β-MnO_2 shows neither ferromagnetism nor giant magnetoresistance at low temperature, but behaves anisotropic magnetoresistance and carrier localization. Sato et al. also explain the critical effect of the conduction e_g electrons and magnetic t_{2g} electrons in Mn d-orbitals on the β-MnO_2 magnetic properties. They considered that the t_{2g}-t_{2g} exchange interaction basically determines the magnetic structure in β-MnO_2, but double exchange interactions between e_g electrons have no fatal consequences [45].

4.2.1.3 Magnetism of δ-MnO_2

δ-MnO_2 bulks prefer to be in the FM ground state. Ge et al. [56] have been successfully prepared layered K_xMnO_2 [57, 58] ($x < 0.3$) nanoarchitectures and studied the magnetism. The obtained samples have a phase transition from ferromagnetic to paramagnetic phase at the corresponding low temperatures (T_c) of 32 K [59], shown in Fig. 4.7a. Figure 4.7b showed the field dependence of the magnetization for the layered MnO_2 nanowire bundles at 293 and 5 K. At 293 K, the magnetization does not show any hysteresis at all, revealing that a paramagnetic state exists in this temperature range. Whereas at 5 K, below the lower transition temperature, it exhibits a much larger hysteresis curve and shows a ferromagnetic behavior. Zhu et al. [60] reported birnessite-type MnO_2 nanowalls have a magnetic transition temperature of 9.2 K (shown in Fig. 4.7c, d) and show ferromagnetism below 9.2 K. While above 9.2 K, the paramagnetism dominates their physical behavior, whereas, the nanowalls exhibit prominent magnetic anisotropy at 5 K. This indicates that the easy magnetization direction of the δ-MnO_2 is out of plane, that is, the easy magnetic direction

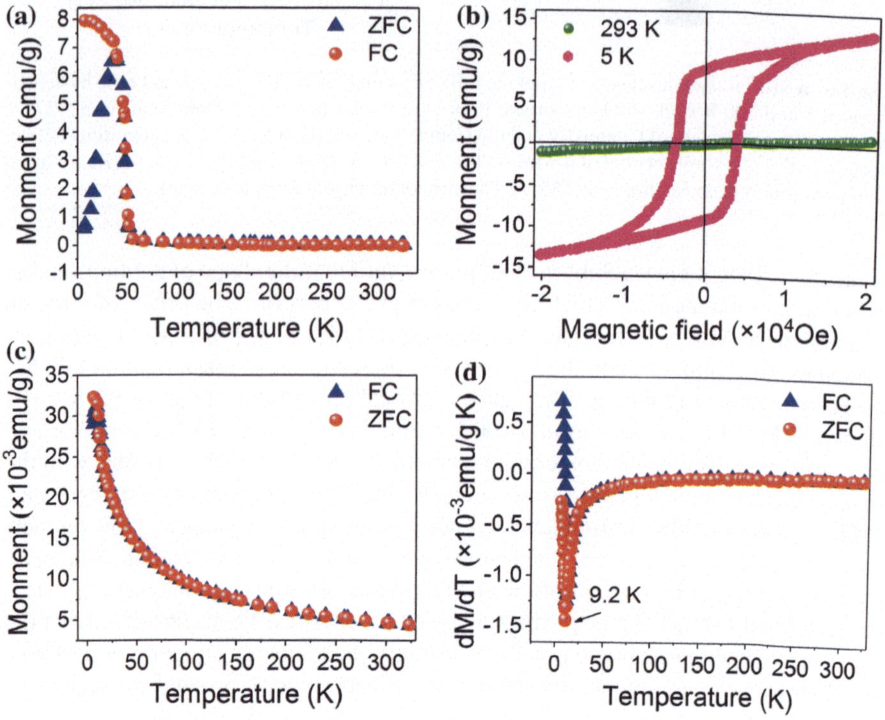

Fig. 4.7 **a** ZFC and FC magnetization curves under an applied field of 1000 Oe and **b** magnetization hysteresis loops of δ-MnO_2. Reproduced with permission from Ge et al. [56] © American Chemical Society, 2006. **c**, **d** M versus T of the birnessite-type MnO_2 powder in an applied field of 100 Oe. Reproduced with permission from Zhu et al. [60] © American Chemical Society, 2008

is in the *ab* plane. The magnetism of δ-MnO_2 is connected with the interlayer irons, which can be revealed by irons doped magnetism. Ions (K^+, Na^+ and H^+) have a critical influence on the magnetism of δ-MnO_2 (shown in Sect. 2.2).

4.2.1.4 Magnetism of γ-MnO_2 and λ-MnO_2

The ground magnetism of γ-MnO_2 and λ-MnO_2 bulks is similar to the ground magnetism of α-MnO_2 bulks. Zheng et al. [61] came up with that hexagonal nanosheets [62] of γ-MnO_2 possess a paramagnetism transition to antiferromagnetism temperature of 78 K, shown in Fig. 4.8a. There is a discontinuous change in the temperature range of 20–30 K on the zero field cooling curve, which may be caused by the surface electron spin unsaturation [63–65] due to the special morphology of the hexagonal nanosheets. This agrees with Zhou's [6] opinion, shown in Fig. 4.8b. The *M-H* curve is close to linear, shown in Fig. 4.8c, d, revealing the antiferromagnetic characteristics of the sample at 1.8 K. But the curve has a small hysteresis indicating that there are some ferromagnetic or ferrimagnetic structures formed in the compound,

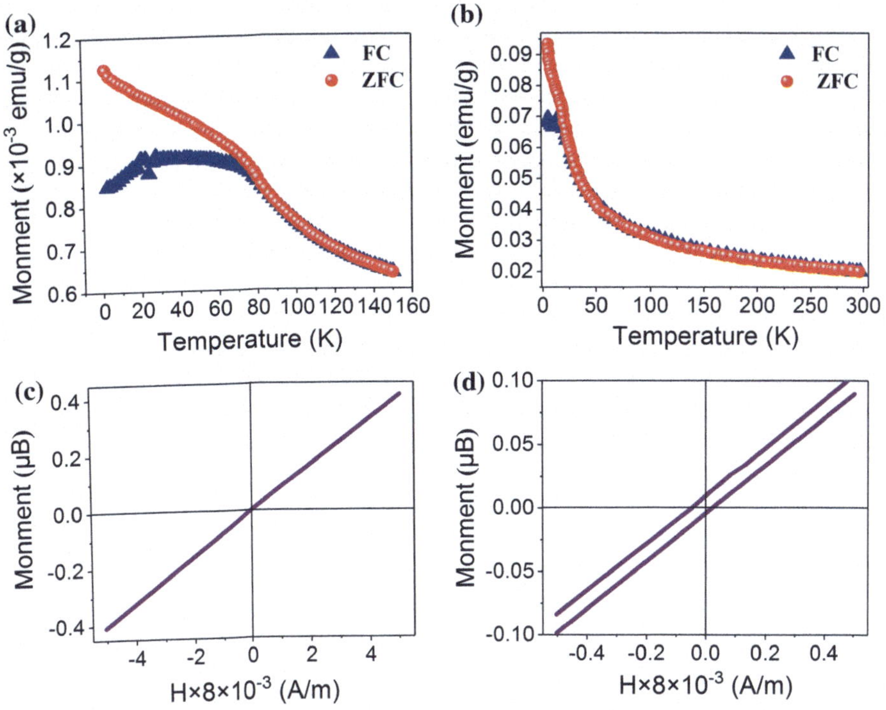

Fig. 4.8 **a** Relation of magnetic susceptibility with temperature, **b** the magnetization vs. applied magnetic field and **c** the ZFC and FC magnetization curves and **d** enlarged view of the magnetization vs. applied magnetic field of γ-MnO_2. Reproduced with permission from Zheng et al. [61] © The Royal Society of Chemistry and the Centre National de la Recherche Scientifique, 2018

which is related to small twist angles between sublattice magnetic moments of Mn atoms. It is discussed in the former part of α-MnO_2 microspheres [39]. λ-MnO_2, a metastable form of manganese dioxide, retains the cubic spinel structure. It shows antiferromagnetism found by Reedan [17] in 30–32 K.

Therefore, δ-MnO_2 of MnO_6 octahedras edge-sharing possesses FM ground state, while β-MnO_2 of MnO_6 octahedras corner-sharing and α-MnO_2 of MnO_6 octahedras edge-sharing and corner-sharing prefer AFM ground state.

4.2.2 *Ions-Induced Magnetism*

Magnetic properties of the materials are usually determined by its structural [66] and geometrical properties, such as the interatomic distances [67], coordination number and symmetry [68, 69]. Generally, the decrease of the coordination number causes the increase of the magnetic moment of per atom as a result of the decrease in the overlap of the nearby atomic orbitals. Compared to the coordination number, the effect of interatomic distances is remarkable. Magnetic properties of MnO_2 can be induced by the doping ions. They have a critical influence on the magnetism of MnO_2 because the introduction of the ions destroys the equilibrium of chemical bond valence resulting in the presence of Mn^{3+} and Mn^{4+} mixed valence state [70], which affects the magnetic exchange between Mn ions. The positional distribution of Mn^{3+} and Mn^{4+} in the crystal structure is closely linked to the position where the cation is inserted, which also results in a magnetic change of MnO_2 [46, 71]. Furthermore, the filling of alkaline ions in the tunnels leads to lattice distortion, playing an important role in the magnetism of MnO_2 [72].

Tseng et al. synthesized α-MnO_2 nanotubes with Li^+, Na^+, and K^+ ions using a hydrothermal technique, finding that when a doping concentration is lower than 12 at.%, the nanotubes performed ferromagnetic-like ordering [74] at low temperature (<50 K) [26], shown in Fig. 4.9a. Antiferromagnetic coupling emerged at the doping concentrations beyond 12 at.%, breaking the balance of Mn–O–Mn bonds. This results in the ferromagnetic-like ordering. Figure 4.9b shows the inclusion of K in the tunnels induces geometric frustration of the triangular lattices, which may destroy the balance of Mn–O–Mn antiferromagnetic ordering, leading to ferromagnetic-like ordering. 6 at.% Li or Na doping does not induce a large distortion in the first MnO_6 shell. This is due to the relatively lower dopant concentration of Li and Na. Luo et al. [75] investigated the effect of K^+ on the magnetic properties of α-MnO_2. The results showed that when the K^+ is less than half the space, the sample shows the behavior of the spin glass due to the triangular lattice arrangement of the magnetic Mn ions and the mixed valence state of the Mn^{3+} and Mn^{4+}. Their study shows that the magnetic properties of nanotubes can be adjusted by changing the K^+ concentration of α-K_xMnO_2. Duan et al. used first principles calculations to study the effect of Fe doping on the magnetic performance of α-MnO_2, shown in Fig. 4.9c. It turned out that the covalent bond length of M–O is reduced, while the degree of overlap of the M–O bond is increased after Fe doping, which is able to enhance

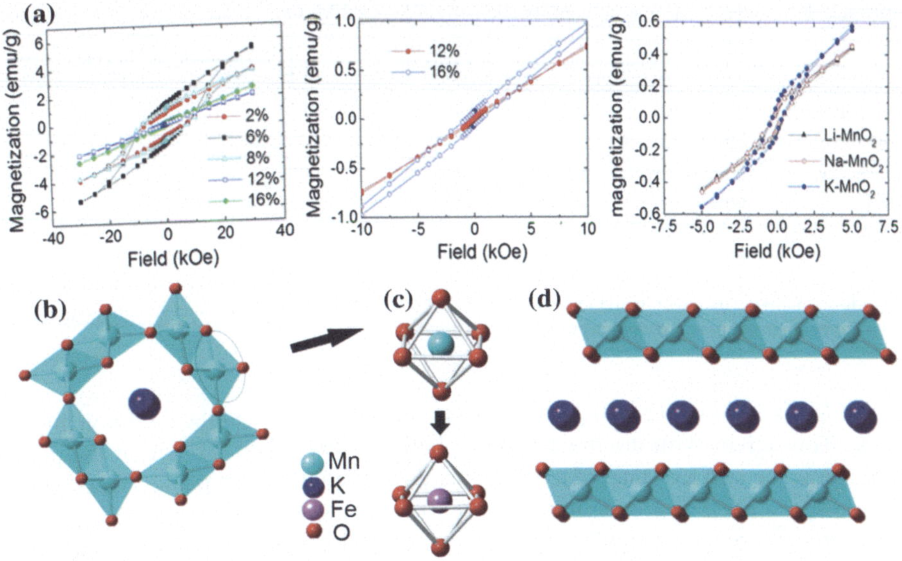

Fig. 4.9 **a** *M-H* loops of Li, Na, and K-doped MnO_2 nanotubes. Schematic diagram of the crystal structure for ion-containing **b** [2 × 2] tunnel structure α-MnO_2, **c** the basic MnO_6-octahedra and a possible substitution of Fe for Mn. Reproduced with permission from Jing et al. [73] © Published by Elsevier B.V. All rights reserved, 2012. **d** δ-MnO_2 with K^+ ions. Reproduced with permission from Yin et al. [18] © The Royal Society of Chemistry, 2015

bond strength. Consequently, Fe doping reinforces the spin polarization of MnO_2 and increases the total magnetic moment, leading to the magnetic enhancement, shown in Fig. 4.9b. Magnetic moments resulting from the orbiting and spinning of electrons around the nucleus determined the magnetic properties. The magnetism of δ-MnO_2 is typically induced by ions. Ge et al. [56] explored the relationship between different K^+ concentrations and the magnetic properties of samples, shown in Fig. 4.9d. The result shows that the samples with higher K^+ concentration would have larger coercivity (Hc), saturation magnetization (Ms), and remanent magnetization (Mr), shown in Table 4.2. Therefore, the magnetism is related to the concentrations of ions inclusion of K^+ or Na^+ in manganese oxides.

4.2.3 The Effect of Exposed Surfaces on Magnetism of MnO_2

The exposed surfaces of MnO_2 show significant influences on the magnetic performance [76]. Li et al. [26, 27] found α-MnO_2-(1 1 0) is ferromagnetic (FM) and α-MnO_2-(2 1 0) is mainly antiferromagnetic (AFM), shown in Fig. 4.10. α-MnO_2 (1 1 0) exhibits stronger magnetization and more stable magnetic coupling. The difference is caused by the surface structure that the (1 1 0) plane is formed by the coherent

Table 4.2 K^+ content and magnetic properties of 1D-layered MnO_2

Treatment	Morphology	K (wt%)	Ms (emu/g) at 5 K	Mr (emu/g) at 5 K	Hc (Oe) at 5 K
Washed with water	Nanowire bundles	3.2	14	9	4100
Washed with HCl	Nanowire bundles	0.27	4.4	2.5	3490
Without PEG 400	Nanorod bundles	3.45	5.2	1.8	3478

Reproduced with permission from Ge et al. [56] © 2006 American chemical society. Nanostructures by different treatments in the typical synthesis route

edge of MnO_6, and the (2 1 0) plane forms a stepped plane from the common angle MnO_6. This agrees with the magnetism of bulk α-MnO_2. The essence of the phenomenon is the surface state which depends greatly on different planes exposed to the free space [77]. The open tunnels of (1 1 0) plane form zigzag surfaces. The (2 1 0) plane forms an irregular step-type surface, which consists of steps and two adjacent grooves. There is a spin glass-like behavior [78] and with an exchange bias effect in α-MnO_2 because of the coupling of the surface spins with the antiferromagnetic matrix. The net magnetic moment during the coupling is contributing to the uncoupled spins in $3d^3$ orbitals of the Mn^{4+} ions in MnO_6 octahedra on the rough surface. The Mn 3d energy level can split into spin-up (↑) and spin-down (↓) states. Each spin state is further split by the octahedral ligand field splitting parameter 10 D_q into a low-energy triply degenerate orbital (t_{2g}) and a high-energy doubly degenerate orbital (e_g). In an octahedral ligand field, the three t_{2g} orbitals lie 4 Dq below the average energy and the two e_g orbitals lie 6 Dq above the average energy. Therefore, in the Mn $3d^3$ energy level, the three 3d electrons occupy t_{2g}(↑) orbitals as shown in Fig. 4.10e. Consequently, the Mn^{4+} shows very stable net magnetic moment because there is no spare electron occupying the high-energy orbitals to excite high-spin state [79].

4.2.4 *The Effect of Size and Shape on Magnetism of MnO_2*

Shape anisotropy can increase the coercivity [80], where the magnetic spins are preferentially aligned along the long axis and their reversal to the opposite direction requires higher energies than that for spheres [81]. The shape of rods is advantageous because the nanorod will exhibit a greater dipole moment under the application of the same magnetic field [82] using the same volume of material as a corresponding nanosphere. When the size [81] of the magnetic particles decreases, they change from multidomain to single domain [83]. If the single domain particles become small enough, the magnetic moment in the domain fluctuates in direction due to thermal agitation, which leads to superparamagnetism.

Subhra et al. put forward that the shape of β-MnO_2 makes difference in magnetism [55]. They reported that one-dimensional nanostructures have increased anisotropies in both the shape anisotropy [42] and the magnetocrystalline anisotropy, compared to spherical nanoparticles, which exert influence on their magnetic properties [80, 84, 85]. β-MnO_2 nanorods exist as a superparamagnetic state at 77 K, as the coercive force and remnant magnetization are zero, shown in Fig. 4.11a. At low temperatures, such as 2 K, the *M-H* curve (Fig. 4.11b) displays a small hysteresis with coercive force of 435 Oe and remnant magnetization of 0.84 emu/g, indicating its ferromagnetic property below the blocking temperature (TB). The TB for nanorods has been determined to be 4 K obtained from temperature dependence magnetization studies between 0 K and 60 K with ZFC and FC in the applied field of 100 Oe,shown in Fig. 4.11c, whereas for nanospheres, the TB is calculated to be 40 K as shown in Fig. 4.11d. The nanorods remain superparamagnetic until 4 K, while for nanospheres it is only 40 K. The field dependence magnetization curves for the spherical nanoparticles at 77 and 5 K are given in Fig. 4.11e, f. The ZFC magnetization curve shows a maximum at low temperature that is characteristic of the superparamagnetism, and the temperature at which it occurs may be taken as blocking temperature. The FC and ZFC data show divergence at low temperatures as indeed was found in the case of other oxide nanoparticles.

4.3 The Applications of MnO_2 Magnetism

MnO_2 has been applied in various fields due to their excellent magnetic performance in magnetic resonance imaging [86, 87] enhancement, magnetic drug delivery, and magnetic memory storage, which is mainly applied in the medical diagnosis. MRI is one of the important techniques of medical diagnosis, shown in Fig. 4.12. Compared with tomography, MRI can achieve better imaging results in soft tissue when choosing the right contrast agent. MRI technology depends to a large extent on the different r1 (longitudinal relaxation rate) or r2 (transverse relaxation rate) [88] relaxation rates of heavy water in different tissues (such as normal tissues and tumors). There are many factors that usually affect the relaxation rate of contrast media. MRI is three dimensional on tumors. Imaging or the application of advanced or early diagnosis of tumors is limited by the low signal-to-noise ratio. Paramagnetic material has better nuclear magnetic resonance signal enhancement. Paramagnetic materials can be potential contrast agents for MRI. Paramagnetic ions can effectively enhance proton relaxation due to their unpaired electron spins [89]. After cancerous tissue selectively absorbs such contrast agents, the relaxation rate of heavy water is increased, resulting in nuclear magnetic resonance images of tumors and surrounding normal tissues. The extranuclear electron arrangement of manganese ion Mn(II) is special. The 3d-orbital of Mn^{2+} has five unpaired electrons [90]. This distribution is a semi-full shell layer, which has strong paramagnetism and can greatly improve heavy water. The r1 relaxation time depends on the paramagnetic response.

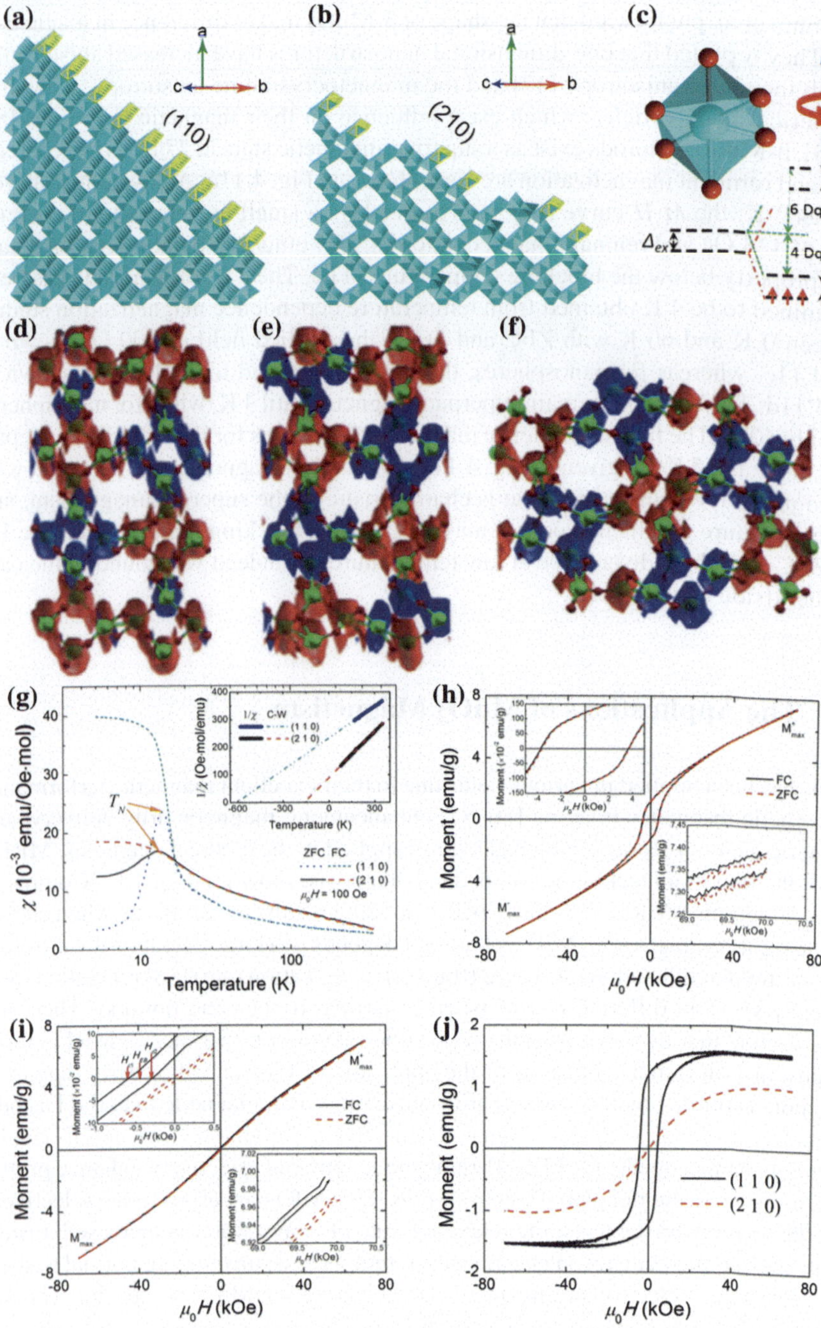
(a)
(110)
(b)
(210)
(c)
e_g
6 Dq
4 Dq
Δ_ex
t_2g
(d)
(e)
(f)
(g)
T_N
1/χ C-W
(1 1 0)
(2 1 0)
1/χ (Oe·mol/emu)
Temperature (K)
ZFC FC
μ0H = 100 Oe
χ (10^-3 emu/Oe·mol)
Temperature (K)
(h)
FC
ZFC
M+max
M−max
Moment (emu/g)
μ0H (kOe)
(i)
FC
ZFC
Moment (emu/g)
μ0H (kOe)
(j)
(1 1 0)
(2 1 0)
Moment (emu/g)
μ0H (kOe)

◂**Fig. 4.10** Origination of magentic moment of α-MnO_2 nanowires **a** (2 1 0) planes composed and **b** (1 1 0) planes composed of MnO_6 octahedra. The highlighted parts indicate the weak linked MnO_6 octahedral chains on the (2 1 0) and (1 1 0) plane when it is at the surface. **c** The thermoremanent magnetization (TRM) and isothermoremanent magnetization (IRM) dependences on applied magnetic fields at 5 K were measured in the field range of 0.4 to 50 kOe. The isosurface plots of the spin density of **d** O-terminated (1 1 0) surface, **e** Mn-terminated (1 1 0) surface and **f** (2 1 0) surface with the isosurface value being 0.02 electron/Å^3. Reproduced with permission from Li et al. [27] © Scientific Report, 2014

In practical applications, MnO_2 nanoparticles, in the ischemic hypoxic region of tumor tissue and endogenous H_2O_2 response, Mn^{2+}, specific part of the reaction to magnetic resonance-enhanced Mn^{2+}, significantly improve the tumor longitudinal relaxation rate rl and transverse relaxation rate r2, enhancing T_1 and T_2-weighted magnetic resonance imaging of tumor sites. Liu et al. [91] developed an acidic H_2O_2-responsive and O_2-evolving SiO_2-MB@MnO_2 core-shell photodynamic therapy (PDT) nanoplatform by using MnO_2 shell as a switchable shield, which prevents the premature release of loaded photosensitizer in core and elevate the O_2 concentration within tumor tissue in PDT. An intelligent and biocompatible bovine serum albumin-Ce_6-Si QDs-MnO_2 nanocomplex as a pH/H_2O_2-responsive photosensitizer nanocarrier constructed by Wang et al. [92] to modulate tumor hypoxia for fluorescence imaging-guided efficient PDT and a self-reinforcing chemodynamic therapy nanoagent based on MnO_2 with both Fenton-like Mn^{2+} delivery and glutathione (GSH) depletion properties reported by Lin et al. [93] to achieve MRI-monitored chemochemodynamic combination therapy, in which the MnO_2 shell undergoes a redox reaction with GSH to form glutathione disulfide and Mn^{2+}, resulting in GSH depletion-enhanced PDT. PDT is the most widely used and least side effect treatment in MRI. However, two problems still need to be solved during the PDT process. One is the photo-bleaching, and premature leakage and poor selectivity of photosensitizers severely affect the therapeutic efficacy. The other is altered vasculature and consequent chaotic blood flow often cause gradients of O_2 tension and acute hypoxia in fast-growing tumors during tumor progression, which poses a survival pressure to favor tumor initiation, metastasis and resistance to chemotherapy, and deteriorate the PDT efficacy. To overcome those problems, He et al. [94] developed a pH-/H_2O_2-activatable nanosystem with self-powered O_2 for O_2-elevated PDT and potential MRI, shown in Fig. 4.13. Manganese dioxide nanoparticles (MnO_2 NPs), capable of generating O_2 by reacting with H^+/H_2O_2, were grown in situ within the pores of large pore silica nanoparticles (LPMSNs). Chlorin e6 (Ce6) with high photosensitizing capability and low dark cytotoxicity was covalently linked to LPMSNs. The AS1411 aptamer, which has high affinity to the overexpressed nucleolin on the plasma membrane of most cancer cells, was anchored onto the surface of LPMSNs to endow the NPs with cancer selectivity. In this system, the MnO_2 NPs will serve as the reactants. Upon selective internalization into cancer cells, they will react rapidly with lysosomal H^+/H_2O_2 to generate massive O_2 in situ to enhance PDT. Moreover, in vitro experiments revealed that under stimulation, they can dissolve to release

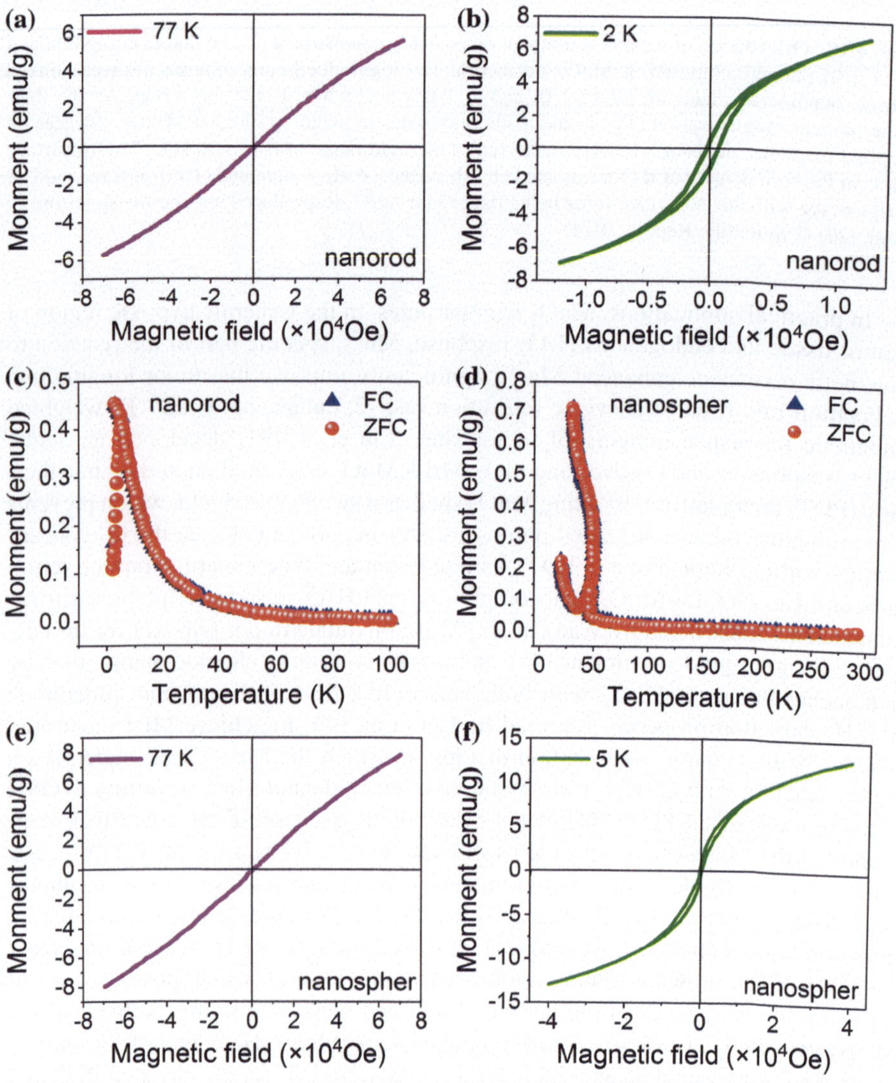

Fig. 4.11 Field dependence magnetization curves of β-MnO_2 nanorods at **a** 77 K and **b** 2 K. The M-T curve at an applied field of 100 Oe of **c** spherical and **d** rod-shaped β-MnO_2 nanoparticles. Field dependence magnetization curve of α-MnO_2 nanospheres at **e** 77 K and **f** 5 K. Reproduced with permission from Subhra et al. [55] © American Chemical Society, 2007

Mn^{2+} ions that enable the access of Mn^{2+} to the surrounding water molecules, acting as a "turn on" T_1 probe to localize the tumor sites in real time during therapy.

Recently, several new types of manganese dioxide contrast agents have drawn attention. Cao et al. [95] synthesized $MnO_2/Cu_{2-x}S$-siRNA nanoagent used in photothermal therapy (PTT) and PDT dual-modal therapy, in which reduced Mn^{2+} ions

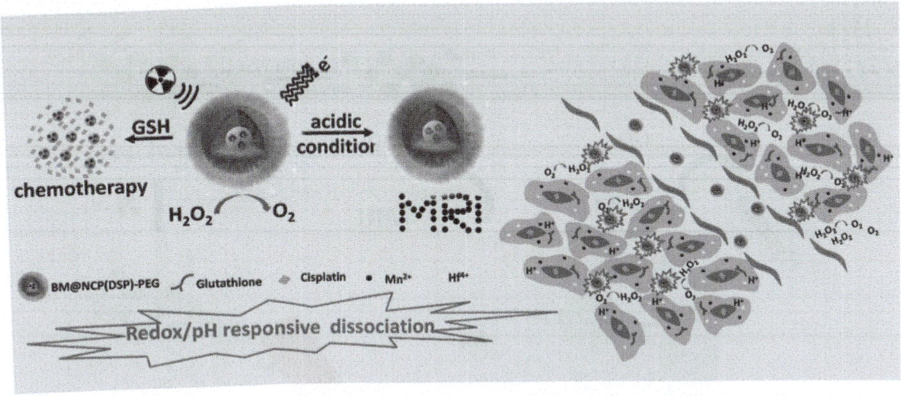

Fig. 4.12 Scheme illustrating the redox/pH-responsive behaviors of BM@NCP(DSP)-PEG composite nanoparticles in the tumor microenvironment. Such nanoparticles are responsive to reduced pH, reductive environment, and existence of tumor endogenous H_2O_2, enabling tumor-specific imaging, drug release, hypoxia relief, and enhanced cancer chemoradiotherapy. Reprinted with permission from Liu et al. [4] © WILEY-VCH Verlag GmbH & Co. KGaA, Weinhei, 2017

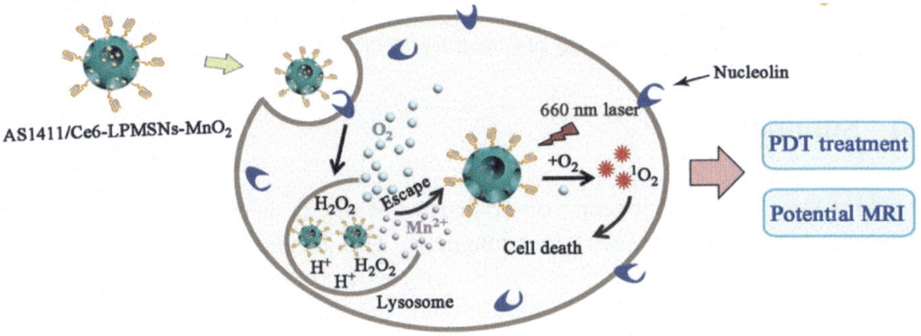

Fig. 4.13 Schematic illustration of H^+/H_2O_2-activated nanomaterials for potential MRI and O_2-elevated PDT. Reprinted with permission from He et al. [4] © The Royal Society of Chemistry, 2018

significantly enhance MRI contrast and the $Cu_{2-x}S$ acts a powerful photoacoustic (PA) and photothermal (PT) imaging agent, leading to tri-modal accurate tumor-specific imaging and detection. Tao et al. [96] designed polyethylene glycol (PEG) modified reduced nano-graphene oxide-manganese dioxide (rGO-MnO_2-PEG) nanocomposites to trigger oxygen generation from H_2O_2 to reduce the tumor hypoxic microenvironments, which used radioisotope labeled rGO-MnO_2-PEG as therapeutic agents for in vivo tumor radioisotope therapy (RIT), achieving excellent tumor killing and further enhancing the therapeutic efficiency of RIT, shown in Fig. 4.14.

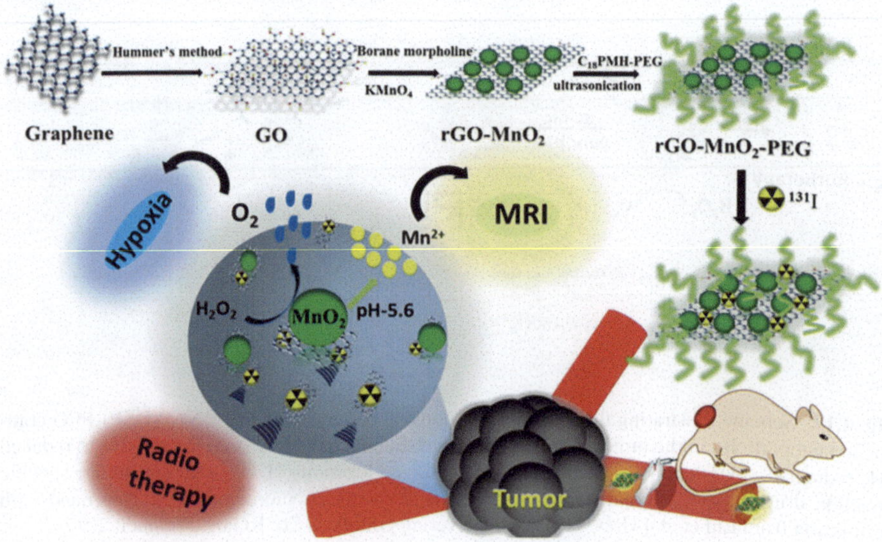

Fig. 4.14 Schematic illustration of the formation of 131I-rGO-MnO_2-PEG nanocomposites for reducing tumor hypoxia and enhancing radioisotope therapy efficiency. Reprinted with permission from Tao et al. [4] © The Royal Society of Chemistry, 2018

4.4 Summary

The Mn–Mn coupling between corner-sharing MnO_6 is antiferromagnetic, while between edge-sharing MnO_6 is ferromagnetic. The ground state of α-MnO_2 and β-MnO_2 is generally considered as antiferromagnetic properties, while δ-MnO_2 possesses the FM ground state. Electronic structure, morphologies, and the introduction of ions have great impacts on the magnetic properties of the MnO_2. Introducing ions into MnO_2 will change the interatomic distances and coordination number of Mn and O, leading to the various magnetism. Doping of K will induce the spin glass due to the triangular lattice arrangement of the magnetic Mn ions and the mixed valence state of the Mn^{3+} and Mn^{4+}. Fe doping reinforces the spin polarization of MnO_2 and increases the total magnetic moment, leading to the magnetic enhancement. The magnetism of MnO_2 also changes with exposed planes, size, and shape because of the strong surface magnetic effects on the particles.

Acknowledgements This work is financially supported by National Natural Science Foundation of China (Grant No. 51572166) and the Shanghai Key Laboratory of High Temperature Superconductors (No. 14DZ2260700). The authors thank the Analysis and Research Center of Shanghai University for their technical support. Wenxian Li acknowledges research support from the Program for Professors with Special Appointments (Eastern Scholar: TP2014041) at Shanghai Institutions of Higher Learning.

References

1. B.A. Pinaud, Z. Chen, D.N. Abram, T.F. Jaramillo, Thin films of sodium birnessite-type MnO_2: optical properties, electronic band structure, and solar photoelectrochemistry. J. Phys. Chem. C **115**, 11830–11838 (2011)
2. Y.L. Li, J.J. Wang, Y. Zhang, M.N. Banis, J. Liu, D.S. Geng, R.Y. Li, X.L. Sun, Facile controlled synthesis and growth mechanisms of flower-like and tubular MnO_2 nanostructures by microwave-assisted hydrothermal method. J. Colloid Interface Sci. **369**, 123–130 (2012)
3. A.M. Toufiq, F. Wang, Q. Li, Y. Li, Hydrothermal synthesis of MnO_2 nanowires: structural characterizations, optical and magnetic properties. Appl. Phys. A **116**, 1127–1132 (2013)
4. J.J. Liu, Q. Chen, W.W. Zhu, X. Yi, Y. Yang, Z.L. Dong, Z. Liu, Nanoscale-coordination-polymer-shelled manganese dioxide composite nanoparticles: a multistage redox/pH/H_2O_2-responsive cancer theranostic nanoplatform. Adv. Func. Mater. **27**, 1605926–1605936 (2017)
5. M.L. Song, T. Liu, C.R. Shi, X. Zhang, X. Chen, Bioconjugated manganese dioxide nanoparticles enhance chemotherapy response by priming tumor-associated macrophages toward M1-like phenotype and attenuating tumor hypoxia. ACS Nano **10**, 633–647 (2016)
6. C.S. Zhou, J.F. Wang, X. Liu, F.Y. Chen, Y.Y. Di, S.L. Gao, Q. Shi. Magnetic and thermodynamic properties of α, β, γ and δ-MnO_2. New J. Chem. (2018). http://doi.org/10.1039.c1038nj00896e
7. X. Zhang, X. Chang, N. Chen, K. Wang, L. Kang, Z.-H. Liu, Synthesis and capacitive property of δ-MnO_2 with large surface area. J. Mater. Sci. **47**, 999–1003 (2011)
8. M.M. Thackeray, Manganese oxides for lithium batteries. Prog. Solid State Chem. **25**, 1–71 (1997)
9. S. Ching, S.L. Suib, Synthetic routes to microporous manganese oxides. Comments Inorg. Chem. **19**, 263–282 (1997)
10. Y.F. Shen, R.P. Zerger, S.L. Suib, Manganese oxide octahedral molecular sieves: preparation, characterization, and applications. Cheminform **23**, 511–515 (1992)
11. S. Sithambaram, L.P. Xu, C.H. Chen, Y.S. Ding, R. Kumar, C. Calvert, S.L. Suib, Manganese octahedral molecular sieve catalysts for selective styrene oxide ring opening. Catal. Today **140**, 162–168 (2009)
12. M. Musil, B. Choi, A. Tsutsumi, Morphology and electrochemical properties of α-, β-, γ-, and δ-MnO_2 synthesized by redox method. J. Electrochem. Soc. **162**, A2058–A2065 (2015)
13. D. Tompsett, S. Parker, M.S. Islam, Surface properties of α-MnO_2: Relevance to catalytic and supercapacitor behaviour. J. Mater. Chem. A **2**, 15509–15518 (2014)
14. H. Sato, K. Wakiya, T. Enoki, Magnetic structure of β-MnO_2: X-ray magnetic scattering study. J. Phys. Soc. Jpn **70**, 37–41 (2001)
15. H. Sato, J. Yamaura, T. Enoki, Magnetism and electron transport phenomena of manganese oxide ion exchanger with tunnel structure. J. Alloy. Compd. **262**, 443–449 (1997)
16. G.C. Lv, X.B. Xing, L.B. Liao, P.F. An, H. Yin, L.F. Mei, Z.H. Li, Synthesis of birnessite with adjustable electron spin magnetic moments for the degradation of tetracycline under microwave induction. Chem. Eng. J. **326**, 329–338 (2017)
17. J.E. Greedan, N.P. Raju, A.S. Wills, Structure and magnetism in λ-MnO_2. Geometric frustration in a defect spinel. Chem. Mater. **10**, 3058–3067 (1998)
18. B. Yin, S.W. Zhang, H. Jiang, Phase-controlled synthesis of polymorphic MnO_2 structures for electrochemical energy storage. J. Mater. Chem. A **3**, 5722–5729 (2015)
19. Y.C. Zhang, T. Qiao, X.Y. Hu, W.D. Zhou, Simple hydrothermal preparation of γ-MnOOH nanowires and their low-temperature thermal conversion to β-MnO_2 nanowires. J. Cryst. Growth **280**, 652–657 (2005)
20. D.W. Su, H.J. Ahn, G.X. Wang, β-MnO_2 nanorods with exposed tunnel structures as high-performance cathode materials for sodium-ion batteries. NPG Asia Mater. **5**, e70–e76 (2013)
21. G. Zhu, H. Li, L. Deng, Z.-H. Liu, Low-temperature synthesis of δ-MnO_2 with large surface area and its capacitance. Mater. Lett. **64**, 1763–1765 (2010)

22. V. Ilyasov, B. Meshi, A. Ryzhkin, I. Ershov, I. Nikiforov, A. Ilyasov, Materials for spintronics: magnetic and transport properties of ultrathin (monolayer graphene)/MnO(001) and MnO(001) films. J. Mod. Phys. **2**, 1120–1135 (2011)
23. C. Franchini, R. Podloucky, J. Paier, M. Marsman, G. Kresse, Ground-state properties of multivalent manganese oxides: density functional and hybrid density functional calculations. Phys. Rev. B **75**, 195128–195139 (2007)
24. Q. Wang, Q. Sun, P. Jena, Y. Kawazoe, Antiferromagnetic coupling driven by bond length contraction near the $Ga_{1-x}Mn_xN$ film surface. Phys. Rev. Lett. **93**, 155501–155504 (2004)
25. J. Hafner, D. Hobbs, Understanding the complex metallic element Mn. II. Geometric frustration in β-Mn, phase stability, and phase transitions. Phys. Rev. B **68**, 014408–014422 (2003)
26. W.X. Li, X.Y. Cui, R.K. Zeng, G.D. Du, Performance modulation of α-MnO_2 nanowires by crystal facet engineering. Sci. Rep. **5**, 8987–8995 (2015)
27. W.X. Li, R.K. Zeng, Z.Q. Sun, D.L. Tian, S.X. Dou, Uncoupled surface spin induced exchange bias in alpha-MnO_2 nanowires. Sci. Rep. **4**, 6641–6647 (2014)
28. Q. Feng, H. Kanoh, K. Ooi, Manganese oxide porous crystals. J. Mater. Chem. **9**, 319–333 (1999)
29. S. Ching, J.L. Roark, Sol–gel route to the tunneled manganese oxide cryptomelane. Chem. Mater. **9**, 750–754 (1997)
30. M. Oku, K. Hirokawa, S. Ikeda, X-ray photoelectron spectroscopy of manganese-oxygen systems. J. Electron Spectrosc. Relat. Phenom. **7**, 465–473 (1975)
31. D.M. Sherman, The electronic structures of manganese oxide minerals. Am. Minera. **69**, 788–799 (1984)
32. Y.F. Li, S.C. Zhu, Z.P. Liu, Reaction network of layer-to-tunnel transition of MnO_2. J. Am. Chem. Soc. **138**, 5371–5380 (2016)
33. J.B. Goodenough, Theory of the role of covalence in the perovskite-type manganites [La, M(II)]MnO_3. Phys. Rev. **100**, 564–573 (1955)
34. Y.P. Duan, Z. Liu, H. Jing, Y.H. Zhang, S.Q. Li, Novel microwave dielectric response of Ni/Co-doped manganese dioxides and their microwave absorbing properties. J. Mater. Chem. **22**, 18291–18299 (2012)
35. J.Q. Tian, R. Ning, Q. Liu, A.M. Asiri, A.O. Al-Youbi, X.P. Sun, Three-dimensional porous supramolecular architecture from ultrathin g-C_3N_4 nanosheets and reduced graphene oxide: Solution self-assembly construction and application as a highly efficient metal-free electrocatalyst for oxygen reduction reaction. ACS Appl. Mater. Inter. **6**, 1011–1017 (2014)
36. N. Yamamoto, T. Endo, M. Shimada, Single crystal growth of α-MnO_2. Jpn. J. Appl. Phys. **13**, 723–724 (1974)
37. P.Y. Kuang, M.H. Liang, W.Y. Kong, Anion-assisted one-pot synthesis of 1D magnetic α- and β-MnO_2 nanostructures for recyclable water treatment application. New J. Chem. **39**, 2497–2505 (2015)
38. J.G. Zhao, J.Z. Yin, S.G. Yang, Hydrothermal synthesis and magnetic properties of α-MnO_2 nanowires. Mater. Res. Bull. **47**, 896–900 (2012)
39. J.B. Yang, X.D. Zhou, W.J. James, S.K. Malik, C.S. Wang, Growth and magnetic properties of $MnO_{2-\delta}$ nanowire microspheres. Appl. Phys. Lett. **85**, 3160–3162 (2004)
40. K.Y. Mulyukov, I.Z. Sharipov, G.F. Korznikova, The influence of crystalline grain size on the thermal stability of the $Er_{0.45}Ho_{0.55}Fe_2$, compound. Phys. Solid State. **44**, 1106–1110 (2002)
41. B.L. Zhu, L. Qiherima, X.J. Lv, Wang, Preparation and magnetic properties of α-MnO_2 nanoparticles. Key Eng. Mater. **492**, 264–267 (2011)
42. J.B. Yi, J. Ding, Y.P. Feng, G.W. Peng, G.M. Chow, Y. Kawazoe, B.H. Liu, J.H. Yin, S. Thongmee, Size-dependent magnetism and spin-glass behavior of amorphous NiO bulk, clusters, and nanocrystals: Experiments and first-principles calculations. Phys. Rev. B **76**, 224402–224405 (2007)
43. P. Umek, C. Bittencourt, A. Gloter, R. Dominko, Z. Jagličić, P. Cevc, D. Arčon, Local coordination and valence states of cobalt in sodium titanate nanoribbons. J. Phys. Chem. C **116**, 11357–11363 (2012)

44. J. Vicat, E. Fanchon, P. Strobel, The structure of $K_{1.33}Mn_8O_{16}$, and cation ordering in hollandite-type structures. Acta Crystallogr. **42**, 162–167 (1986)
45. H. Sato, T. Enoki, M. Isobe, Transport properties and magnetism of a helically hund-coupled conductor: β-MnO_2. Phys. Rev. B **61**, 3563–3570 (2000)
46. J. Luo, H.T. Zhu, J.K. Liang, Tuning magnetic properties of α-MnO_2 nanotubes by k^+ doping. J. Phys. Chem. C **114**, 8782–8786 (2010)
47. B. Tang, G. Wang, L. Zhuo, J. Ge, Novel dandelion-like beta-manganese dioxide microstructures and their magnetic properties. Nanotechnology **17**, 947–951 (2006)
48. H. Sato, Y. Kawamura, T. Ogawa, Critical phenomena in helical magnet beta-MnO_2: X-ray magnetic scattering study. Physica B **329**, 757–758 (2003)
49. K. Woo, H.J. Lee, Synthesis and magnetism of hematite and maghemite nanoparticles. J. Magn. Magn. Mater. **272–276**, E1155–E1156 (2004)
50. S.Y. Bae, H.W. Seo, H.C. Choi, Heterostructures of ZnO nanorods with various one-dimensional nanostructures. J. Phys. Chem. B **108**, 12318–12326 (2004)
51. G.H. Lee, S.H. Huh, J.W. Jeong, Anomalous magnetic properties of MnO nanoclusters. J. Am. Chem. Soc. **124**, 12094–12095 (2002)
52. A. Yoshimori, A new type of antiferromagnetic structure in the rutile type crystal. J. Phys. Soc. Jpn. **14**, 807–821 (1959)
53. M. Blume, Magnetic scattering of x rays (invited). J. Appl. Phys. **57**, 3615–3618 (1985)
54. D. Gibbs, D.E. Moncton, K.L. D'Amico, Magnetic x-ray scattering studies of holmium using synchro- tron radiation. Phys. Rev. Lett. **55**, 234–237 (1985)
55. S. Jana, S. Basu, S. Pande, Shape-selective synthesis, magnetic properties, and catalytic activity of single crystalline. J. Phys. Chem. C **111**, 16272–16277 (2007)
56. J.C. Ge, L.H. Zhuo, F. Yang, One-dimensional hierarchical layered K_xMnO_2 ($x < 0.3$) nanoarchitectures: Synthesis, characterization, and their magnetic properties. J. Phys. Chem. B **110**, 17854–17859 (2006)
57. H.T. Zhang, X.H. Chen, J.H. Zhang, Synthesis and characterization of one-dimensional $K_{0.27}MnO_2 \cdot 0.5H_2O$. J. Cryst. Growth. **280**, 292–299 (2005)
58. Z.P. Liu, R.Z. Ma, Y. Ebina, Synthesis and delamination of layered manganese oxide nanobelts. Chem. Mater. **19**, 6504–6512 (2007)
59. K. Dwight, N. Menyuk, Magnetic properties of and the canted spin problem. Phys Rev. **119**, 1470–1479 (1960)
60. H.T. Zhu, J. Luo, H.X. Yang, Birnessite-type MnO_2 nanowalls and their magnetic properties. J. Phys. Chem. C **112**, 17089–17094 (2017)
61. D.S. Zheng, The research on nano-magnetic properties of hexagonal nanosheets γ-MnO_2. J. Xi' An Univ Technol. **32**, 465–469 (2012)
62. G.C. Li, L. Jiang, H.T. Pang, H.R. Peng, Synthesis of γ-MnO_2 single-crystalline nanobelts. Mater. Lett. **61**, 3319–3322 (2007)
63. J. Park, E. Kang, C.J. Bae, Synthesis, characterization, and magnetic properties of uniform-sized MnO nanospheres and nanorods. J Phys. Chem. B **108**, 13594–13598 (2004)
64. G.H. Lee, S.H. Huh, J.W. Jeong, Anomalous magnetic properties of MnO nanoclusters. J. Am. Chem. Soc. **124**, 12094–12098 (2002)
65. G.L. Wang, B. Tang, L.H. Zhuo, J.C. Ge, M. Xue, Facile and selected-control synthesis of β-MnO_2 nanorods and their magnetic properties. Eur. J. Inorg. Chem. **2006**, 2313–2317 (2006)
66. M. Giot, L.C. Chapon, J. Androulakis, M.A. Green, P.G. Radaelli, A. Lappas, Magnetoelastic coupling and symmetry breaking in the frustrated antiferromagnet α-$NaMnO_2$. Phys. Rev. Lett. **99**, 247211–247214 (2007)
67. A. Zorko, S. El Shawish, D. Arčon, Z. Jagličić, A. Lappas, H. van Tol, L.C. Brunel, Magnetic interactions in α–$NaMnO_2$: Quantum spin-2 system on a spatially anisotropic two-dimensional triangular lattice. Phys. Rev. B **77**, 24412–24418 (2008)
68. C. Stock, L.C. Chapon, O. Adamopoulos, A. Lappas, M. Giot, J.W. Taylor, M.A. Green, C.M. Brown, P.G. Radaelli, One-dimensional magnetic fluctuations in the spin-2 triangular lattice α-$NaMnO_2$. Phys. Rev. Lett. **103**, 77202–77205 (2009)

69. Q. Feng, H. Kanoh, Y. Miyai, Hydrothermal synthesis of lithium and sodium manganese oxides and their metal ion extraction/insertion reactions. Cheminform **26**, 1226–1232 (1995)
70. X.F. Shen, Y.S. Ding, J. Liu, A magnetic route to measure the average oxidation state of mixed-valent manganese in manganese oxide octahedral molecular sieves (OMS). J. Am. Chem. Soc. **127**, 6166–6167 (2005)
71. S.L. Suib, L.E. Iton, Magnetic studies of manganese oxide octahedral molecular sieves: A new class of spin glasses. Chem. Mater. **6**, 429–433 (1994)
72. J. Luo, H.T. Zhu, F. Zhang, J.K. Liang, G.H. Rao, J.B. Li, Z.M. Du, Spin-glasslike behavior of k^+-containing α-MnO_2 nanotubes. J. Appl. Phys. **105**, 93925–93930 (2009)
73. J. Hui, D. Yu Ping, L. Zhuo, Z. Jia, L. Shun Hua, Influence of Fe-doping on the microstructure and electromagnetic performance of manganese oxides. Phys. B: Condens. Matter. **407**, 971–977 (2012)
74. J.B. Yi, J. Ding, Z.L. Zhao, B.H. Liu, High coercivity and exchange coupling of Ni/NiO nanocomposite film. J. Appl. Phys. **97**, K306–K309 (2005)
75. J. Luo, H.T. Zhu, J.K. Liang, Study on the magnetic properties of α-MnO_2 nanotubes by k ion doping. In: International symposium on phase diagram academic conference and phase diagram and material design
76. J. Lu, C.Y. Nan, L.H. Li, Q. Peng, Y.D. Li, Flexible SnS nanobelts: Facile synthesis, formation mechanism and application in Li-ion batteries. Nano Res. **6**, 55–64 (2012)
77. R.L. Stamps, Dynamic magnetic properties of ferroic films, multilayers, and patterned elements. Adv. Func. Mater. **20**, 2380–2394 (2010)
78. J.L. Tholence, R. Tournier, Susceptibility and remanent magnetization of a spin glass. Le J. de Phys. Colloques **35**, C4-229–C224-235 (1974)
79. P. Umek, A. Gloter, M. Pregelj, Synthesis of 3D hierarchical self-assembled microstructures formed from α-MnO_2 nanotubes and their conducting and magnetic properties. J. Phys. Chem. C **113**, 14798–14803 (2009)
80. N.M. Carneiro, W.C. Nunes, P.B. Rui, NiO nanoparticles dispersed in mesoporous silica glass. J. Phys. Chem. C **114**, 18773–18778 (2010)
81. W.S. Seo, H.H. Jo, K. Lee, B. Kim, S.J. Oh, J.T. Park, Size-dependent magnetic properties of colloidal Mn_3O_4 and MnO nanoparticles. Angew. Chem. Int. Ed. Engl. **43**, 1115–1121 (2004)
82. I. Khan, A. Khalil, F. Khanday, A.M. Shemsi, A. Qurashi, K.S. Siddiqui, Synthesis, characterization and applications of magnetic iron oxide nanostructures. Arabian J. Sci. Eng. **43**, 43–61 (2017)
83. N. Amin, S. Arajs, E. Matijevic, Magnetic properties of submicronic α-Fe_2O_3, particles of uniform size distribution at 300 k. Phys. Status Solidi C **104**, K65–K68 (1987)
84. F. Dumestre, C. Amiens, B. Chaudret, Synthesis of cobalt nanoparticles, nanorods and nanowires assisted by oleic acid and oleylamine based mixtures. MRS Proceedings. **735**, 165–170 (2002)
85. T. Hyeon, Y. Chung, J. Park, Synthesis of highly crystalline and monodisperse cobalt ferrite nanocrystals. J. Phys. Chem. B **106**, 6831–6833 (2002)
86. M. Zhang, L. Xing, H.T. Ke, Y.J. He, P.F. Cui, Y. Zhu, G. Jiang, J.B. Qiao, N. Lu, H. Chen, H.L. Jiang, MnO_2-based nanoplatform serves as drug vehicle and MRI contrast agent for cancer theranostics. ACS Appl. Mater. Interfaces **9**, 11337–11344 (2017)
87. Y.H. Zhang, W.X. Qiu, M. Zhang, L. Zhang, X.Z. Zhang, MnO_2 motor: A prospective cancer-starving therapy promoter. ACS Appl. Mater. Interfaces **10**, 15030–15039 (2018)
88. H.Q. Liu, X.L. Yu, B. Cai, S.J. You, Z.B. He, Q.Q. Huang, L. Rao, S.S. Li, C. Liu, W.W. Sun, W. Liu, S.S. Guo, X.Z. Zhao, Capture and release of cancer cells using electrospun etchable MnO_2 nanofibers integrated in microchannels. Appl. Phys. Lett. **106**, 93703–93708 (2015)
89. F. Xiao, Y.Q. Li, H.C. Gao, S.B. Ge, H.W. Duan, Growth of coral-like PtAu-MnO_2 binary nanocomposites on free-standing graphene paper for flexible nonenzymatic glucose sensors. Biosens. Bioelectron. **41**, 417–423 (2013)
90. X. Yang, D.G. He, X.X. He, K.M. Wang, Z. Zou, X.C. Li, H. Shi, J.R. Luo, X.X. Yang, Glutathione-mediated degradation of surface-capped MnO_2 for drug release from mesoporous silica nanoparticles to cancer cells. Part. Part. Syst. Charact. **32**, 205–212 (2015)

91. Z.F. Ma, X.D. Jia, J. Bai, MnO_2 gatekeeper: An intelligent and O_2-evolving shell for preventing premature release of high cargo payload core, overcoming tumor hypoxia, and acidic H_2O_2-sensitive MRI. Adv. Func. Mater. **27**, 1604258–1604269 (2017)
92. R.G. Wang, M.Y. Zhao, D. Deng, X. Ye, Intelligent and biocompatible photosensitizer conjugated silicon quantum dots-MnO_2 nanosystem for fluorescence imaging-guided efficient photodynamic therapy. Mater. Chem. B **6**, 4592–4601 (2018)
93. L.S. Lin, J.B. Song, L. Song, Simultaneous fenton-like ion delivery and glutathione depletion by MnO_2-based nanoagent enhances chemodynamic therapy. Angew. Chem. **57**, 4902–4906 (2018)
94. Z.M. He, Y. Xiao, J.R. Zhang, In situ formation of large pore silica-MnO_2 nanocomposites with H^+/H_2O_2 sensitivity for O_2-elevated photodynamic therapy and potential MR imaging. Chem. commun. **54**, 2962–2965 (2018)
95. Y. Cao, X.D. Meng, D.D. Wang, K. Zhang, W.H. Dai, H.F. Dong, X.J. Zhang, Intelligent $MnO_2/Cu_{2-x}S$ for multimode imaging diagnostic and advanced single-laser irradiated photothermal/photodynamic therapy. ACS Appl. Mater. Interfaces **10**, 17732–17741 (2018)
96. Y. Tao, L. Zhu, Y. Zhao, X. Yi, L. Zhu, F. Ge, X. Mou, L. Chen, L. Sun, K. Yang, Nano-graphene oxide-manganese dioxide nanocomposites for overcoming tumor hypoxia and enhancing cancer radioisotope therapy. Nanoscale **10**, 5114–5123 (2018)

Chapter 5
Hydrogel Responsive Nanomaterials for Colorimetric Chemical Sensors

Dandan Men, Honghua Zhang and Yue Li

Abstract The stimuli-responsive hydrogels are three-dimensional hydrophilic polymeric networks with a fascinating property that they will undergo an obvious and reversible volumetric variation in response to a small variation of external environmental stimuli. In particular, combining of the stimuli-responsive hydrogels with photonic crystals (PCs) or Au nanoparticles (NPs), the volumetric variation responded to external stimuli could be converted into a color change, thus creating a kind of colorimetric sensors. These colorimetric sensors attract more and more interest of researchers in different fields due to their simple operation and visualized read-out. Herein, after presenting a brief review on the basis concept, synthesis methods and sensitive mechanisms of the stimuli-responsive hydrogels, this chapter mainly focuses on their applications as colorimetric chemical sensors by combining with PCs. And some typical applications are proposed in detail, such as detecting pH value, ionic species, solvents, humidity, and biomolecules. In order to meet the increasing requirements of practical applications, the selectivity, response rate, and resolution ratio of these colorimetric sensors need to be improved in the near further.

5.1 Introduction

Hydrogels are three-dimensional (3D) hydrophilic polymeric networks with an ability to retain lots of water [1–9]. Hydrogels are constructed by natural and synthetic polymers [1]. Due to their good hydrophilicity, biocompatibility, and stability, hydrogels have been widely applied in many fields, such as biomedical fields [10–14],

---

D. Men · H. Zhang
Jiangxi Key Laboratory of Surface Engineering, Jiangxi Science and Technology Normal University, Nanchang 330013, People's Republic of China

Y. Li (✉)
Key Laboratory of Materials Physics and Anhui Key Laboratory of Nanomaterials and Nanotechnology, Institute of Solid State Physics, Hefei Institutes of Physical Science, Chinese Academy of Sciences (CAS), Hefei 230031, Anhui, People's Republic of China
e-mail: yueli@issp.ac.cn

Z. Sun and T. Liao (eds.), *Responsive Nanomaterials for Sustainable Applications*, Springer Series in Materials Science 297,
https://doi.org/10.1007/978-3-030-39994-8_5

electronic devices [15], and sensors [16–27]. Importantly, if they are synthesized from stimuli-responsive polymers, hydrogels will possess an amazing property that their volume can undergo a significant and reversible change in response to small alterations of specific external environmental stimuli without structural destruction [6, 7, 27–33]. Such property causes the hydrogels to be good candidates for sensors.

The stimuli-responsive hydrogels are known as "smart" or "intelligent" materials [34–36] and their volume changes are commonly referred as swelling–shrinking transitions in literatures [37–40]. However, most hydrogels are colorless and transparent [41, 42]. In this case, as a sensor, it is very difficult to detect the external stimuli only by means of measuring volume change [41]. Therefore, remarkable efforts have been made to immobilize or produce regular pattern on surfaces or insides of the hydrogels. Specifically, combining the stimulus-responsive hydrogels with photonic crystals (PCs), has created exciting interest of researchers because they can realize colorimetric detection [43–55]. The PCs are spatially periodic arrangements of materials with different dielectric constants. Because of the periodic arrangement, the PCs materials own an unique photonic band gap (PBG). This PBG can prohibit a specific wavelength of light located in the PBG from transmitting through the PCs, so generates a structural color [56–58]. When the PCs are integrated with a stimuli-responsive hydrogel, the volume change responding to external stimuli will alter the periodicities of the PCs, resulting in a variation of PBG, accompanied by a change of structural color [57–63]. If the forbidden wavelength lies in the region of visible light, the structural color could be recognized by naked eye, thus creating a colorimetric sensor. According to this mechanism, over the last decades, various colorimetric sensors based on RPCs have been prepared, such as temperature [64–69], light [70], magnetic field [71], pH [72–76], solvent [77, 78], and ions [79–83]. By conjugating appropriate recognition agents to the hydrogels matrix, these RPCs can respond to specific molecules [84–88]. Besides PCs, Au nanoparticles (NPs) also are combined with the stimuli-responsive hydrogels to construct the colorimetric sensors [89–93]. The performance of these colorimetric sensors can be improved by changing crosslinking density of hydrogel or increasing the concentration of functional groups.

Although most sensors based on stimulus-responsive hydrogels still require further development before practical application, they will become very promising materials in the near future. The next hydrogel-based chemical sensor demands that the hydrogel matrix can identify the specific analysts and their concentrations, respond to them with excellent selectivity as well as stability, and monitor them easily. This chapter provides an overview of recent researches in the fields of synthesis and application of the stimuli-responsive hydrogels for chemical sensors whose photonic properties are in response to external stimuli. After introducing the general development strategies, this chapter focuses on (1) briefly presenting the fabrication methods of hydrogels; (2) summarizing the sensitive mechanism of the stimuli-responsive hydrogels; and (3) giving an overview of the applications of the stimulus-responsive hydrogels as sensors. Herein, this chapter mainly focuses on the colorimetric chemical sensors based on the composites of stimuli-responsive hydrogels and PCs. Additionally, the composites of Au NPs and stimuli-responsive hydrogels as colorimetric chemical sensors also are slightly introduced.

5.2 Synthesis of Hydrogel

Hydrogels are hydrophilic polymers with a 3D network structures crosslinked mainly through chemical crosslinking and physical crosslinking [1, 7]. A crosslinking step is required to overcome the solubility and improve mechanical strength. Generally, for the chemical crosslinking, there are two methods [7]. The first one is adding a crosslinker in a reaction solution to simultaneously link polymer chains during their synthesizing process. For example, polyacrylic acid (PAAc) as a typical pH-sensitive hydrogel was synthesized via a free radical polymerization of a reaction solution containing monomer (acrylic acid, AAc) and crosslinker (*N*,*N*-methylene bisacrylamide) [72]. The second one is applying crosslinker to interconnect two polymer chains. For instance, 3D chitosan (CS) networks as basic hydrogels were synthesized through interconnecting the CS chains by glutaraldehyde [43, 94]. Recently, a kind of transparent and multistimuli-responsive hydrogels were fabricated within a few seconds by crosslinking the CS chains with a variety of transition metal ions at appropriate pH value [95]. Upon mixing, these metal ions underwent ultrafast complexation with the CS chains (within 2 s), resulting in forming interwoven network structure, as shown in Fig. 5.1. The facile preparation process and short gelation time lead it to be a promising approach for designing advanced hydrogel materials.

Compared with the chemical crosslinking, physically crosslinked hydrogels have attracted increasing attentions owing to their relatively facile production process. However, in comparison with their counterparts prepared via the chemical crosslinking, the physically crosslinked hydrogels tend to be frailer because relatively weak interactions [1].

5.3 Sensitive Mechanism of Hydrogel

For the chemical sensor based on hydrogel, three primary sensing mechanisms were demonstrated [96, 97]: (1) immobilizing ions on the hydrogel [98–101]; (2) changing the crosslinking density [102–104]; (3) variation the solubility of the hydrogel polymer [88, 105, 106].

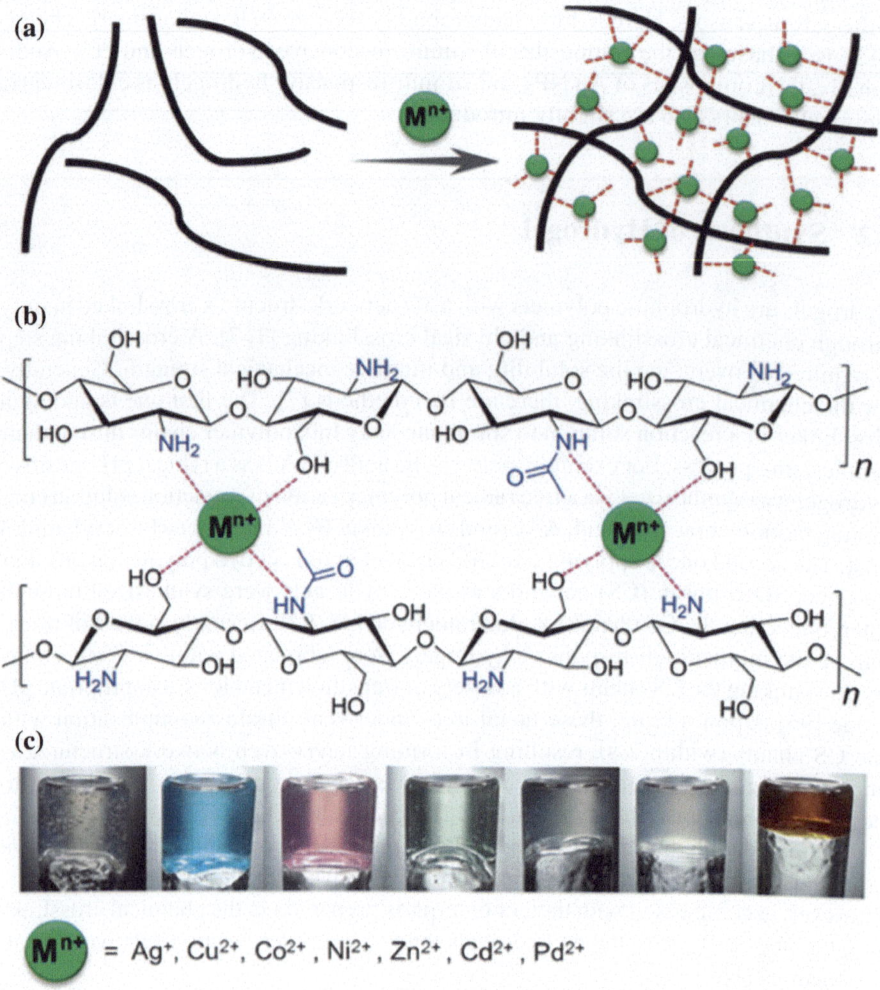

Fig. 5.1 **a** Schematic illustration of crosslinking the CS chains with metal ions in water via fast complexation. **b** Chemical structures of the CS interwoven networks crosslinked by metal ions; **c** Photographs of hydrogels formed by crosslinking the CS chains with different metal ions (from left to right: Ag^{+}, Cu^{2+}, Co^{2+}, Ni^{2+}, Zn^{2+}, Cd^{2+}, and Pd^{2+}, respectively) [95]

5.3.1 *Immobilizing Ions on the Hydrogel*

Immobilizing ions on the hydrogel network will result in an increase of osmotic pressure within the hydrogel networks owing to increased Donnan potential arising from introducing mobile counterions, leading to a swell [98–101]. This sensing motif has been demonstrated for monitoring pH [43, 44, 72], Pb^{2+} [79], K^{+} [101] and glucose [98, 107]. Taking a pH-sensitive hydrogel as an example, this sensitive motif is explained in detail as follows. The well-known property of hydrogels with weak acidic or weak basic pendant groups in their polymeric structure has been utilized as a pH sensor with good reversibility [7]. When the pH value of surrounding environment is above the pKa of the ionizable groups, the anionic hydrogels, such as with carboxylic or sulfonic pendant groups, are ionized that forces the hydrogel to swell. On the other hand, the cationic hydrogels, such as with amine pendant groups, are protonated at the pH value below the pKb of the ionizable groups, thus swelling occurs at lower pH. PAAc as a typical pH-responsive hydrogel has been widely applied in the area of drug delivery and pH sensor. It deprotonates in a basic surrounding condition as following:

$$[\mathrm{RCOOH}]_{\mathrm{hydrogel}} + [\mathrm{OH}^-]_{\mathrm{aq}} \rightarrow [\mathrm{RCOO}^-]_{\mathrm{hydrogel}} + \mathrm{H_2O};$$

In an acidic ambient as following:

$$[\mathrm{RCOO}^-]_{\mathrm{hydrogel}} + [\mathrm{H}^+]_{\mathrm{aq}} \rightarrow [\mathrm{RCOOH}]_{\mathrm{hydrogel}}$$

In a basic surrounding condition, as the pH value increases, the carboxy pendant groups on the PAAc backbone are ionized to carboxylate anions, accompanied by an adequate generation of mobile counterions inside the hydrogel, which results in an increasing Donnan osmotic pressure, thus, causing a corresponding swell of the hydrogel. However, in an acidic solution, the charge density and the mobile counterions within the PAAc hydrogel strongly decrease, resulting in a reduction in Donnan osmotic pressure and hence a shrink of hydrogel. Additionally, approximately at pKa of the ionizable groups, the ionization begins and the volume change appears obviously. However, when the ionization of the ionizable component is completed, further increasing pH only increases the ionic strength, which causes the hydrogel to shrink. The general swelling behavior of the acidic, basic, and amphiphilic hydrogels at different pH are shown in Fig. 5.2. From it, it can be seen that the swelling of the

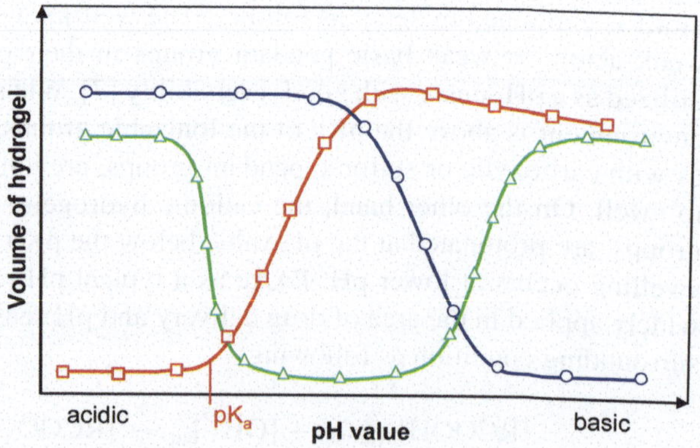

Fig. 5.2 Phase transition behavior of acidic (□), basic (○), and amphiphilic (Δ) hydrogels [7]

acidic hydrogels appears in basic solutions because of the deprotonation. By contrast, the basic hydrogels swell in acidic solutions because the protonation occurs in acidic solution. Unlike them, the amphiphilic hydrogels show two phases transitions both in basic and acidic condition.

5.3.2 Changing the Crosslinking Density

If an analyte molecule has the ability to bind with more than one recognition group, it can form additional crosslinks within the hydrogel, thus causing hydrogel to shrink. Based on this general principle, various chemical sensors related to hydrogel were fabricated. For example, an ammonia-sensitive hydrogel was developed by covalently coupling 3-aminophenol as a recognition agent to poly (2-hydroxyethyl acrylate) (PHEA) hydrogel backbone. When it is exposed to ammonia, an ammonia molecule can react with two recognition groups to generate additional crosslinking. The creation of new crosslinking leads the hydrogel to shrink [108]. In addition to ammonia sensor, other sensors for detecting including glucose [104], avidin and lectin protein concanavalin. A [109] have also been prepared by changing the crosslink density of the hydrogel.

5.3.3 Variation in the solubility of the Hydrogel Polymer

In this motif, altering the hydrogel polymer to increase its solubility can result in the free energy of mixing of the hydrogel polymer with the surrounding medium becoming more favorable, thereby swells the hydrogel [96]. For instance, at high ionic strength condition, as pH increases, the swelling behavior of poly(2-hydroxyethyl acrylate-co-acrylic acid) (Poly(HEA-co-AA)) was attributed to increasing solubility of carboxylates compared to carboxyl acid groups, rather than increasing Donnan potential [75]. Additionally, since the solubility of phenolates in water exceeds that of phenol, another pH sensor that works in high ionic strength medium was fabricated by attaching nitrophenol groups onto the polyacrylamide (PAAm) backbone. At higher pH, the phenol groups dissociate to phenolates, leading to increase solubility and resulting in swell of hydrogel [105].

5.4 Chemical Sensors Based on Stimuli-Responsive Hydrogel

There are many ways for the chemical sensors responding to specific analytes. One of the most convenient and obvious readout methods is changing their optical properties, especially a visually perceptible color variation. Relevant types of colorimetric sensors based on the composites of stimuli-responsive hydrogels and PCs or Au NPs are discussed below.

5.4.1 pH Sensor

The pH-sensitive hydrogel constructed by a 3D crosslinked polymer backbone carrying with weak acidic or basic groups [7, 110, 111]. The weak acidic or basic groups contribute to the pH sensitivity. PAAc is a typical pH-sensitive hydrogel, but high water solubility limited its further application. In order to overcome this shortcoming, AAc is usually copolymerized with other monomers. For example, poly(acrylamide-co-acrylic acid) (P(AAm-co-AAc)) was synthesized by using hydrophilic AAm monomer as a basic monomer, AAc as a functional monomer, and *N*,*N*-methylene bisacrylamide as a crosslinker [111]. What is more, both mechanical property and sensitivity could be adjusted by changing the ratio between the PAAm and PAAc or crosslinking density.

5.4.1.1 pH Sensor Based on Hydrogel/PCs Composites

Recently, various colorimetric pH sensors were developed by embedding the PCs into a pH-responsive hydrogel [57, 74]. Generally, 3D and 2D PCs have been employed to construct hydrogel/PCs composites [57]. 3D hydrogel/PCs composites are mainly formed by two methods, as shown in Fig. 5.3a, b. In the first one, the 3D PCs is prepared primarily, and then the responsive materials are infiltrated into the interstices of PCs prepared in advance to lock the PCs (Fig. 5.3a). By this method, Song et al. fabricated a pH sensor [74]. When the pH increases from 3.3 to 6.7, the diffraction peaks shift from 510 to 590 nm, meanwhile, the color changes from blue to orange, as shown in Fig. 5.3c, d. In the second approach, the 3D RPCs are prepared by

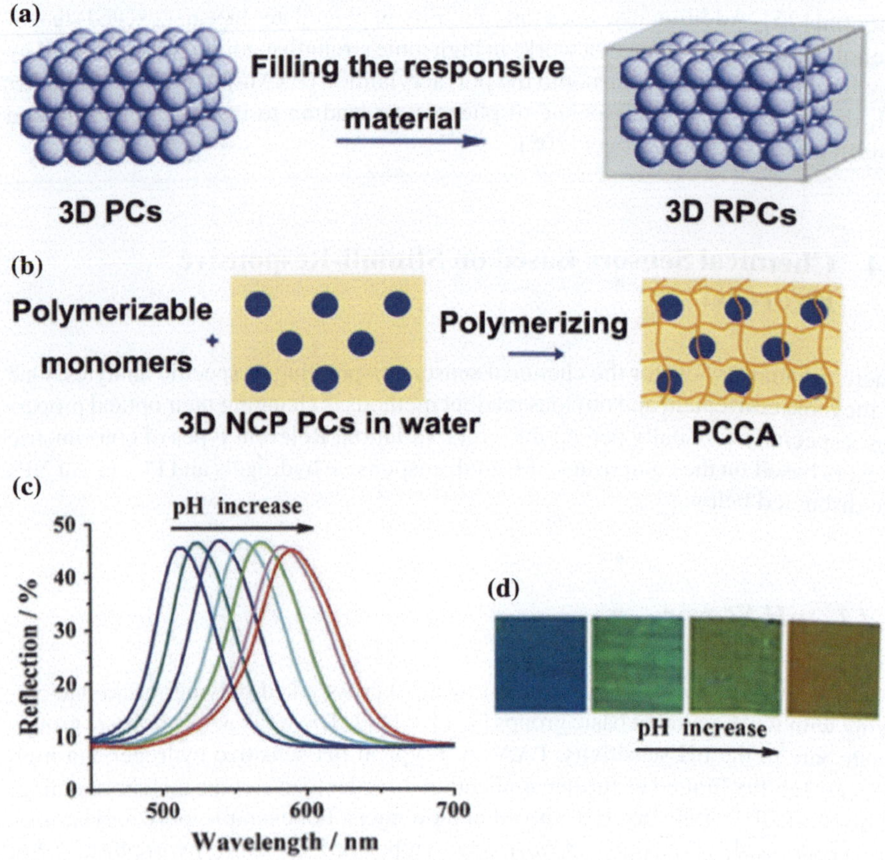

Fig. 5.3 **a**, **b** Two preparation route of 3D RPCs [57]. Diffraction spectra (**c**) and color variation (**d**) of the pH sensor based on 3D hydrogel/PCs composite [74]

electrostatic self-assembly of microspheres in a solution containing monodispersed microparticles and polymerizable monomers, followed by polymerization, as shown in Fig. 5.3b. This method was proposed by Asher et al. and the prepared RPCs were called polymerized colloidal crystal array (PCCA) [79, 112]. In this method, the formed 3D PCs are non-close-packed (NCP) structures, and the polymerizable monomers should be nonionic to avoid disrupting charge stabilized PCs. Due to this, compared to the first method, its functionalized process is relatively complex. For example, a hydrolysis step is required to produce ionizable groups (–COOH) for the pH-sensitive PCCA [112]. Moreover, the microspheres in the 3D RPCs can be removed to form inverse-opal structure.

Due to fast, simple, and efficient self-assembly process of high-quality 2D PCs on a large area, a lot of research has been devoted to fabricate 2D hydrogel/PCs composites to construct visual sensors [109]. 2D hydrogel/PCs composite film was fabricated by directly in situ filled responsive materials into the dried 2D PCs. By this method, Asher et al. successfully prepared CS/2D PCs composite for detecting pH value [94]. However, current reported 2D RPCs composite sensors are mainly limited to 2D PS microspheres arrays, which have relatively small scattering cross section, causing the diffraction signal too weak to be detected by instrument. Therefore, high reflective mirror, such as Al, Au or Ag mirror [57, 94, 104], was applied to increase their diffraction signal during the processes of diffraction measurement. However, as a visual sensor, high diffraction intensity is an important factor. For improving the diffraction intensity, Li et al. attached the 2D PS microsphere arrays onto double surfaces of the CS film to detect pH (Fig. 5.4) [43]. It exhibits stronger diffraction intensity than that attached 2D PS microspheres arrays only on one surface due to multi-diffraction effect. Furthermore, with increase of pH value, the diffraction peaks gradually red shift (Fig. 5.4c), accompanied with a corresponding color change (Fig. 5.4d). However, although the diffraction intensity is improved, it is not high enough and the high reflective mirror (Al mirror) is still needed. To further increase the diffraction intensity, 2D Au nanosphere arrays were chosen as the PCs to replace 2D PS microspheres arrays to fabricate pH sensor (Fig. 5.5a) [72]. Compared with the sensor based on 2D PS microspheres arrays, it displays obvious iridescent color (Fig. 5.5b) and intense diffraction signals on quartz slide (Fig. 5.5c), which is helpful

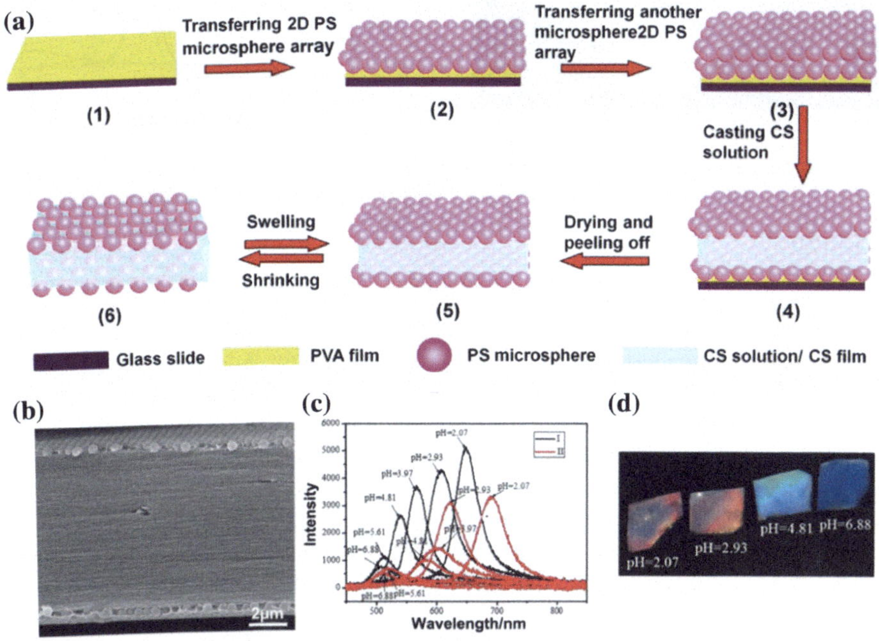

Fig. 5.4 Preparation route (**a**) and SEM image (**b**) of the CS film-attached 2D PS microsphere arrays on both surfaces. **c** Comparison of diffraction spectra of the CS film attached 2D PS microsphere arrays on both surfaces (I) and only one surface (II) in different pH values. **d** Photographs of the CS film attached 2D PS microsphere arrays on both surfaces at different pH values [43]

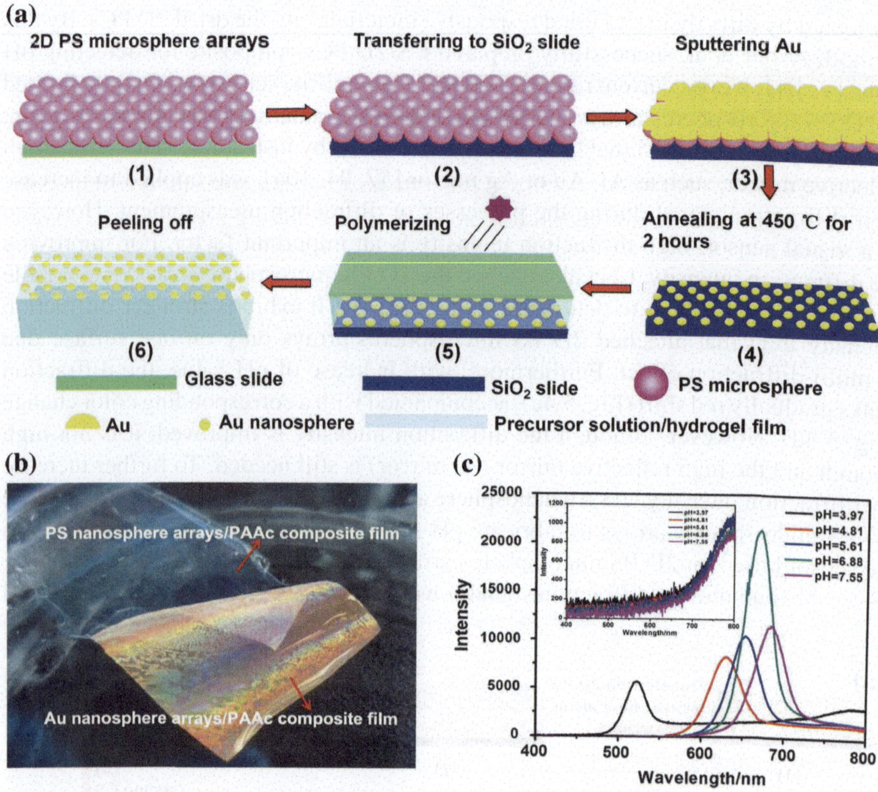

Fig. 5.5 **a** Preparation route of the 2D Au nanosphere arrays/PAAc composite film. **b** Photograph of the 2D Au and PS nanosphere arrays/PAAc composite film in water. **c** Diffraction spectra of the 2D Au and PS nanosphere arrays (inset)/PAAc composite films on quartz slide in different pH values [72]

for their practical applications in visual detection. The high diffraction intensity was attributed to large scattering cross section of Au nanosphere. Additionally, this sensor with intense diffraction signal can be expanded to monitor other analysts by attaching appropriate identification agent to the hydrogel matrix.

The above sensors based on PCs/hydrogel composites may be unsuitable for microscale detection owing to their bulk sizes. Additionally, as colorimetric sensors, they may suffer from color uneveness. Guan et al. had fabricated a new class of RPCs for microenvironment sensing with high resolution by a hydrogen bond-guided template polymerization method [75]. The new RPCs are composed of individual one-dimensional (1D) periodic structure of Fe_3O_4 NPs and responsive hydrogel shells with a thickness of tens-of-nanometer. It is worth noting that since their submicrometer size, both lateral resolution (around 2 μm) and response rate (less than 40 ms), are 2–3 orders of magnitude smaller than that of the previous RPCs films. Figure 5.6a depicts the preparation process of the

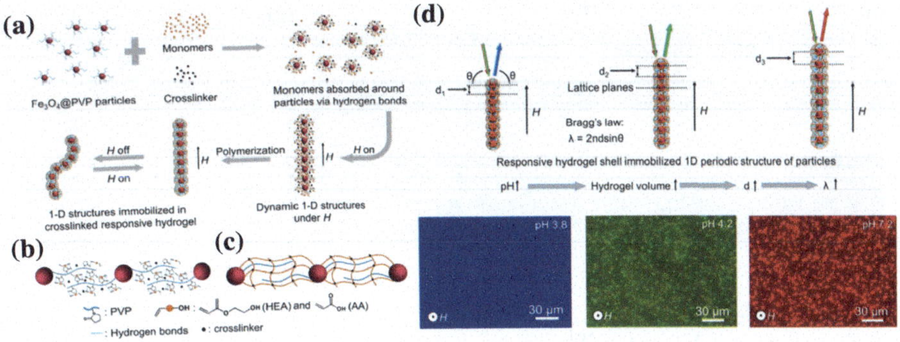

Fig. 5.6 **a** Preparation route of the Fe_3O_4 NPs@PVP@Poly(HEA-co-AA) photonic nanochains. Schematic illustration of hydrogen bonds between monomers and PVP chains (**b**) and immobilizating the nanochains in the crosslinked Poly(HEA-co-AA) hydrogel shell (**c**). **d** Schematic illustration of structural color changing process triggered by pH and dark-field optical microscopy images of the Fe_3O_4 NPs@PVP@Poly(HEA-co-AA) photonic nanochains in different pH values [75]

Fe_3O_4@polyvinylpyrrolidone(PVP)@Poly(HEA-co-AA) photonic nanochains. In the process, monomers were concentrated in the vicinity of magnetic Fe_3O_4 NPs via hydrogen bonds between monomers in the precursor solution and PVP fixed on the surfaces of the Fe_3O_4 NPs (Fig. 5.6b), thus insuring the polymerization reaction appears mainly close the NPs. Then, these Fe_3O_4 NPs were assembled linearly into dynamic nanochains with periodic structure by magnetic field. Afterward, these dynamic nanochains were in situ locked in the crosslinked Poly(HEA-co-AA) hydrogel (Fig. 5.6c). Figure 5.6d describes the color-changing process and dark-field optical microscopy images of this Fe_3O_4@PVP@Poly(HEA-co-AA) photonic nanochains as a pH sensor in different pH values. Furthermore, this method can be extended to fabricate other sensors, such as temperature and solvent, by coupling different functional groups to the hydrogel shell.

5.4.1.2 pH Sensor Based on Hydrogel/Au Dimer Composites

As aggregation changes, the color of Au NPs will change from red to purple, blue, yellow, and even black because of interparticle plasmonic coupling effects [113–117]. This color change provides an excellent platform for constructing colorimetric sensors to detect target analytes. Au dimer, consisted of two Au NPs, is a simplest plasmonic coupling structure. A noteworthy feature of such Au dimer is that it has two distinguishing extinction peaks which named local surface plasmon resonance (LSPR) peak and plasmonic coupling peak, respectively. Interestingly, the plasmonic coupling peak position could be tuned by changing the interparticle gap between the two Au NPs. According to this concept, Li et al. developed a novel colorimetric pH sensor by embedding Au dimers in CS hydrogel film, as shown in Fig. 5.7a, b [90]. With decreasing pH value, the plasmonic coupling peak positions blue shift gradually

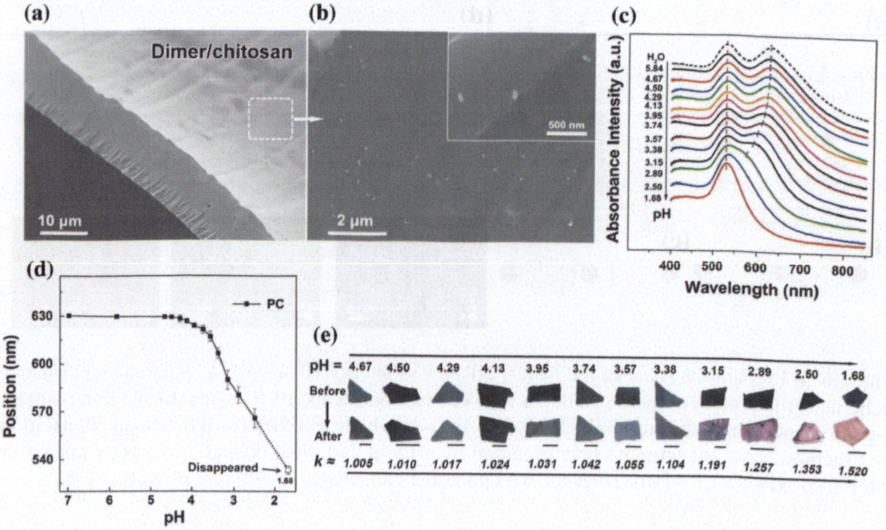

Fig. 5.7 **a**, **b** SEM images of the CS composite film embedded with Au dimers with different magnifications. **c** The extinction spectra of the CS films embedded with Au dimers at different pH values. **d** The positions of the plasmonic coupling peaks of the spectra in (**c**). **d** The photographs of the CS composite film embedded with Au dimers before and after soaking in buffer solutions with different pH values [90]

and the color changes from dark blue to bright red due to increasing interparticle gap, caused by the swell of the CS hydrogel matrix (Fig. 5.7c–e). Furthermore, the sensitivity of the composite film could be adjusted by crosslinking time, and thus will lead to a large shift range of plasmonic coupling peak for different applications.

5.4.1.3 pH Sensor Based on Au NPs@Hydrogel or Hydrogel@Au NPs

Compared with hydrogel film, the hydrogel microbeads can realize small-amount detection with a relatively fast response rate. A pH sensor constructed by Au NPs@P(AAm-co-AA) hydrogel microbeads were successfully created via encapsulating homogeneous Au NPs in the P(AAm-co-AA) hydrogel microbeads by droplet microfluidics technology [41]. These composite microbeads exhibit a good pH-responsive behavior, as shown in Fig. 5.8. When the pH increases from 2 to 9, the dimater of the Au NPs@P(AAm-co-AA) composite microbeads gradually increases. As the pH is above 9, the hydrogel composite microbeads shrink again owing to increase ionic strength (Fig. 5.8a–n). Thus, the pH can be detected via measuring the change in size of the microbead by the microspectrometer system, which leads this microbead to be a visual microsensor. Opposite to the changing trend of the size, their absorbance intensity firstly declines and then increases with increasing

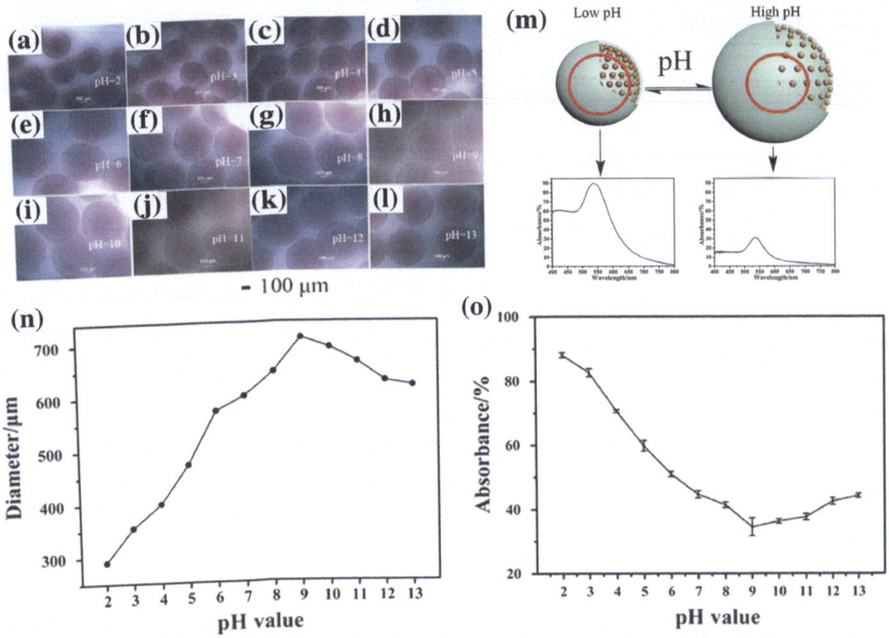

Fig. 5.8 **a–l** Microscopy images, **m** sensitive mechanism of absorbance intensity and **n** fluctuation curve of the diameters of the Au NPs@hydrogel microbeads in different pH values. The red circles in **m** represent the measurement area of absorption spectra. **o** Fluctuation curve of the absorbance intensity at ca. 540 nm [41]

pH value, as shown in Fig. 5.8o. This is because the number of the Au NPs per unit volume firstly declines and then increases due to hydrogel swelling and shrinking.

In addition, these P(AAm-co-AA) hydrogel microbeads embedded with Au NPs are also employed as a sensor for glucose by adding glucose oxidase in the hydrogel matrix. When they are exposed to glucose, glucose will be converted to gluconic acid. The generated gluconic acid can lower the pH value of solution, thus causing hydrogel to shrink, accompanied with gradually increase in optical absorbance. In addition to encapsulating the Au NPs in the inside of the hydrogel microbeads, Li et al. also attached the Au nanospheres on their surface through electrostatic self-assembly method [99]. In these hydrogel microbead@Au nanospheres, the nanogaps between adjacent Au NPs can be dynamically tuned by volume change derived from the change of pH value, leading them to be highly sensitive dynamic surface-enhanced Raman scattering (SERS) substrates.

5.4.2 Ion Sensor

Ion-sensitive hydrogel is constructed by introducing certain functional groups along the polymer chains. For example, by modifying hydrogel with crown ether, an Pb^{2+} [118], K^+ [101], and Be^{2+} [24] sensors were obtained because the crown ether groups can bind with certain metal cations to form crown ether-cation complexes. These complexes behave as ionized side chains to result in an ionic hydrogel, thus swelling the hydrogel, red shifting the diffraction, and finally causing a color change. The swell of ionic hydrogel is mainly due to increasing osmotic pressure. Additionally, repulsion between ionized side chains and increased solubility of the crown ether-cation side groups compared with the neutral uncomplexed side groups also cause the gel to swell [119]. Asher et al. prepared a Pb^{2+} sensor by embedding 3D PCs into a Pb^{2+}-sensitive hydrogel [79]. Figure 5.9a, b depicts the sensitive mechanism and optical properties of this sensor in different Pb^{2+} concentrations. From it, it can be seen with increasing Pb^{2+} concentration, the diffraction wavelengths red shift. However, in higher Pb^{2+} concentration, the diffraction wavelength blue shifts again because the Donnan potential is attenuated in high ionic strength solution. As a result, the hydrogel swelling is decrease and the diffraction peak blue shifts. This phenomenon indicates the response of this sensor is significantly influenced by ionic strength. Moreover, by covalently immobilizing benzo-9-crown-3 (B9C3) groups to

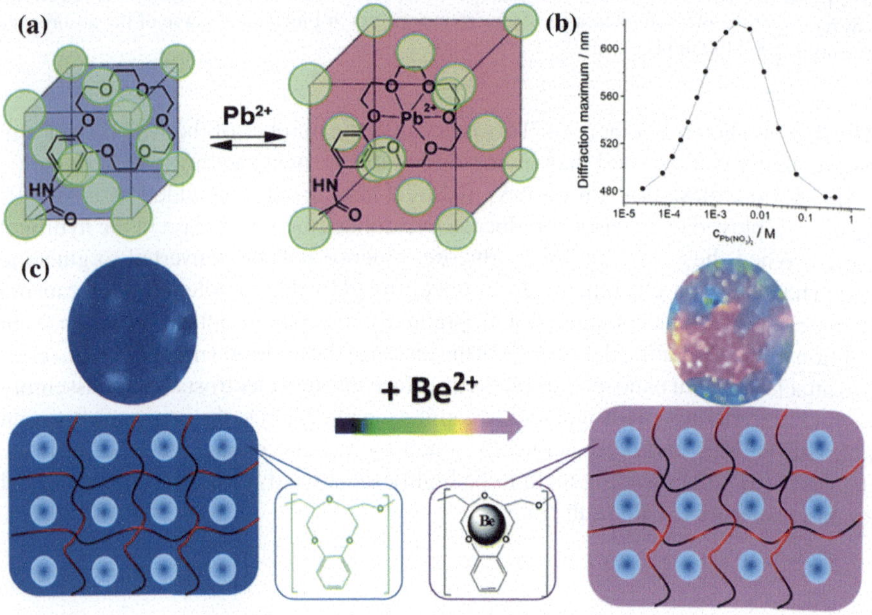

Fig. 5.9 Sensitive mechanism (**a**) and optical properties (**b**) of the Pb^{2+} responsive photonic hydrogels [79]. **c** Sensitive mechanism and optical photographs of the Be^{2+} responsive photonic hydrogels [24]

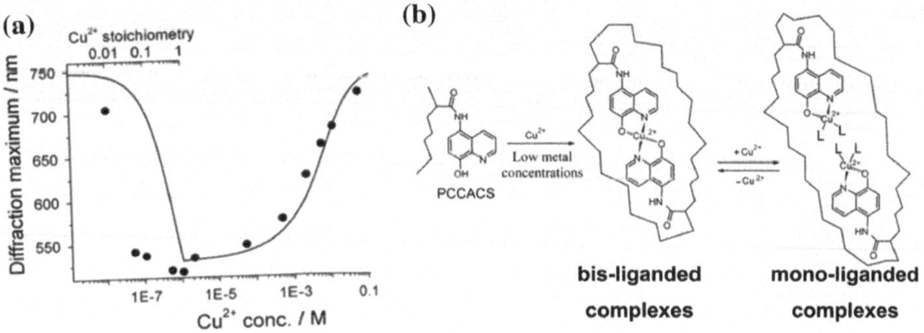

Fig. 5.10 a Effect of the Cu^{2+} concentrations on the diffraction wavelength of the 8-hydroxyquinoline-functionalized PCCA. **b** Sensitive mechanism of the 8-hydroxyquinoline-functionalized PCCA to Cu^{2+} [103]

the hydrogel, a Be^{2+} responsive photonic hydrogel is prepared, as shown in Fig. 5.9c [24]. The grafted B9C3 can strongly and selectively chelate with Be^{2+} ions in solution, thus causing the hydrogel to swell and the diffraction color to change, just like the above-mentioned Pb^{2+} sensor.

Besides immobilizing ions on the hydrogel, changing crosslinking density of hydrogel is also an effective method for synthesizing ion-responsive hydrogel. For example, a PCCA sensor for Cu^{2+} was prepared by grafting 8-hydroxyquinoline as recognition agents into hydrogel matrix [103]. Interestingly, at low concentrations of metal ion (<μM), this sensor responds to increasing ion concentration by blue shifting the diffraction peaks, as shown in Fig. 5.10a. This is because the metal cations formed bis-liganded complexes with two 8-hydroxyquinolines that crosslink and shrink the hydrogel matrix. However, at higher cation concentration, these bis-liganded complexes will be broken into mono-liganded complexes, which lead the hydrogel to swell and the diffraction peak to red shift (Fig. 5.10b). However, the 8-hydroxyquinoline-modified PCCA sensor is not a nonspecific metal cation sensor because it displayed high association constants to a wide range of metal ions in aqueous solution, such as Ni^{2+}, Co^{2+}, and Zn^{2+}. Therefore, it is desirable to develop a sensor for heavy metal ion with high selectivity.

Single-stranded DNA (or RNA) molecules with specific sequences as aptamers can bind to a variety of molecules or ions with good sensitivity and selectivity. For instance, Hg^{2+} has high affinity for thymine–thymine (T-T) base pairs in DNA and it can form stable T-Hg^{2+}-T structures, resulting in change of the aptamer's conformation [82, 120–122]. By functionalizing the hydrogel matrix of the hydrogel/PCs composite by these aptamers, in the appearance of target analyst, the specific binding of heavy metal ions and these aptamers can crosslink single-stranded aptamers, which triggers hydrogel to shrink to alter the periodicity of the PCs, as shown in Fig. 5.11a. Based on this unique feature, Gu et al. fabricated a Hg^{2+} sensor-based hydrogel/PCs composite by modifying the hydrogel matrix with 5′-NH_2-$(CH_2)_6$-TTCTTTCTTCCCCTTGTTTGTT-$(CH_2)_6$-NH_2-3′ (Fig. 5.11b) [82]. In the presence

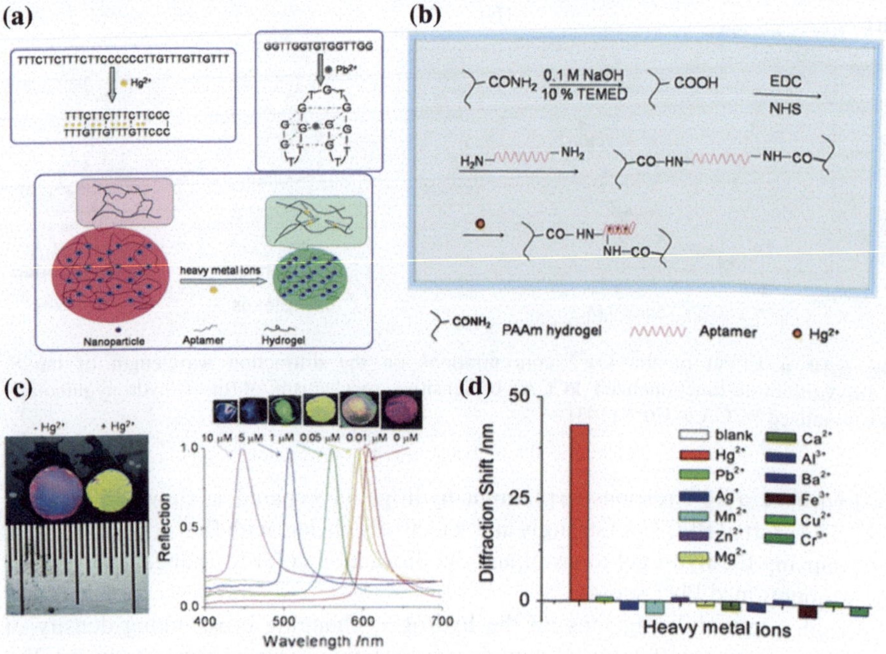

Fig. 5.11 **a** Corresponding aptamer sequences for Hg^{2+} and Pb^{2+}, and sensitive mechanism of the aptamer to heavy metal ions. **b** Synthetic route of aptamer crosslinked hydrogel. **c** Volume change, color change, and optical property of the aptamer-functionalized hydrogel/composites in different Hg^{2+} concentrations. **d** The effect of various metal ions on the diffraction shift (1 μM for Hg^{2+} and 100 μM for other metal ions) [82]

of Hg^{2+}, the sensor reveals a prominent volume change, color variation, and diffraction peak shift (Fig. 5.11c). Importantly, this senor displays high selectivity to its target analyte, as shown in Fig. 5.11d. This good selectivity can improve the possibility of practical application for this sensor because the Hg^{2+} generally coexist with other metals ions. Later, they further developed PCs microcapsules for detecting heavy metal ions [123]. Compared to the former, the PCs microcapsules can realize label-free multiplex detecting. These PCs microcapsules are composed of close-packed opal PCs cores and stimulus-responsive inverse-opal PCs shells functionalized with specific sequences. When used in targets analysis, the PCs cores of the microcapsules provide stable diffraction peaks for encoding, while the hydrogel shells specifically recognize and react with the targets, causing the shells to shrink, as shown in Fig. 5.12a. With the increasing concentration of Hg^{2+}, the diffraction peak gradually blue shifts due to the formed T-Hg^{2+}-T structures, but the encoding peaks are stable without obvious shift (Fig. 5.12b). Moreover, both above methods could be extended for detecting other heavy metal ions with good selectivity, such as Pb^{2+} and Ag^{+} by selecting specific aptamers.

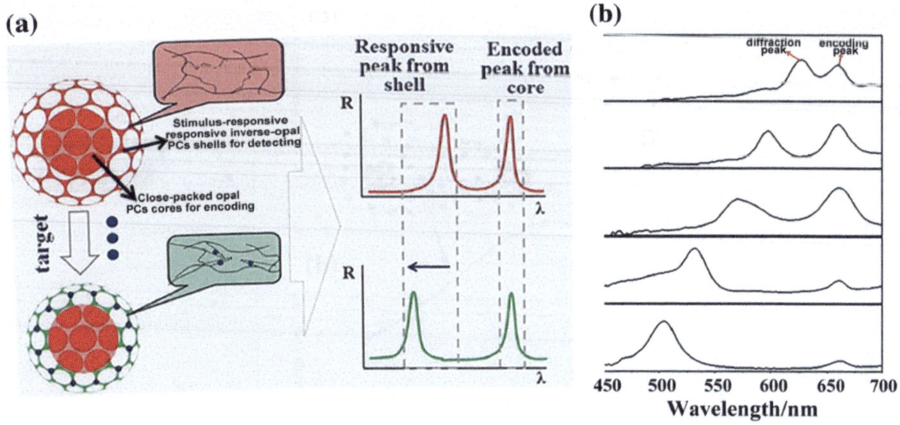

Fig. 5.12 **a** Encoding and sensitive mechanism of the microcapsules for heavy metal ions. **b** Optical properties of the microcapsules in different Hg^{2+} concentrations [123]

5.4.3 Surfactant Sensor

Amphiphilic surfactants are organic compounds that comprise of a hydrophilic, ionic or nonionic group bound to a hydrophobic group [124]. When a slightly hydrophobic nonionic water-soluble polymer is immersed in the solution containing ionic surfactants, strong cooperatively associating will take place owing to hydrophobic interactions between the polymers and the surfactant hydrophobic tails [124–127]. The association between the slightly hydrophobic nonionic polymer and ionic surfactant is regarded as micelles formation [125]. The concentration where the surfactant forms polymer-bound micelles is called critical aggregation concentration (CAC). The CAC is lower than the critical micelle concentration (CMC) of the pure surfactant. Figure 5.13a displays the swelling equilibrium isotherm of the slightly hydrophobic nonionic polymer soaked in a quasi-infinite solution of ionic surfactant. At low surfactant concentration, the surfactant distributes evenly between the gel and the external solution, and no micelles are formed. Thus, the volume does not change obviously. However, above the CAC, a significantly increase in volume is observed with increasing surfactant concentration. This phenomenon can be explained by the fact that the polymer-bound micelles are formed inside the polymer, thus binding charged surfactant to the polymer. The amount of bound surfactant increases as the surfactant concentration increases, which causes the hydrogel to gradually swell. After a swelling maximum, the swelling degree begins to decrease. The reason for the decrease is that above the CMC, the added surfactant molecules mainly form "free" ionic micelles. These "free" ionic micelles distributing between the gel network and the outside solution will increase the ionic strength of solution, thus reduces the swelling degree. A visible sensor for surfactants was prepared by attaching 2D PS NPs arrays onto a poly(N-isopropylacrylamide) (PNIPAAm) hydrogel, as shown in Fig. 5.13b [127]. PNIPAAm has side hydrophobic isopropyl groups attached to

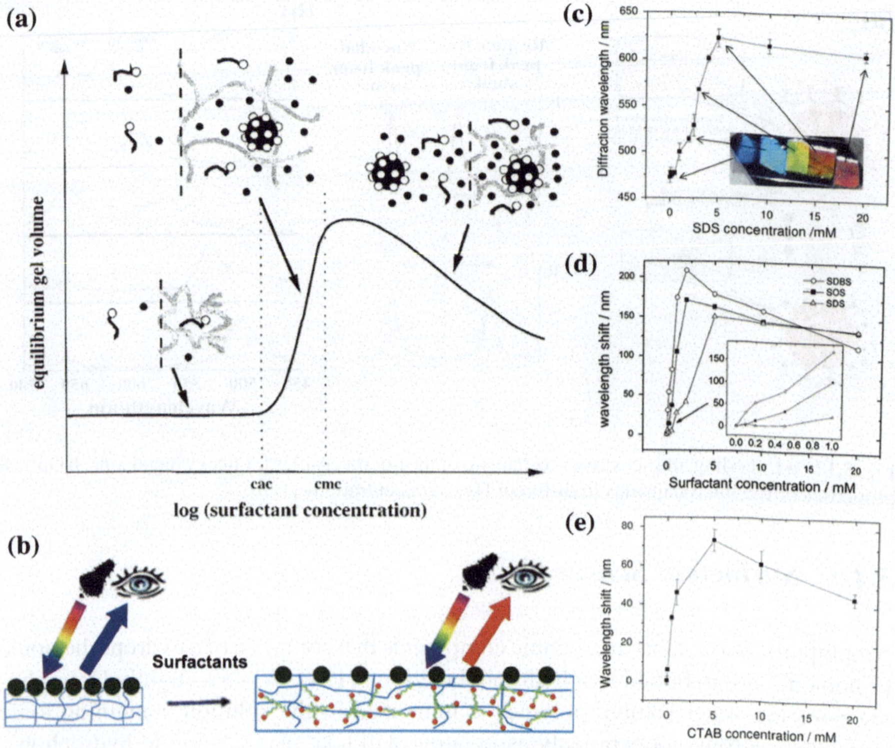

Fig. 5.13 **a** Equilibrium swelling isotherm of a slightly hydrophobic nonionic polymer soaked in a large volume of solution containing ionic surfactant that can bind to the polymer [125]. **b** Schematic illustration of the surfactant sensor based on 2D PCs and PNIPAAm. **c** The effect of SDS concentrations on diffraction wavelength. **d**, **e** The effect of SDS, SDBS, SOS, and CTAB concentrations on diffraction shift [127]

hydrophilic amide groups. Surfactants in aqueous solution can blind to it through hydrophobic interactions. Figure 5.13c shows the diffraction spectra of this sensor on the sodium dodecyl sulfate (SDS) concentration in aqueous solution. In pure water, it diffracts at 470 nm. Increasing the SDS concentration to 5 mM, the diffraction wavelength red shifts to 622 nm. However, at higher SDS concentrations (10 mM), the diffraction wavelength blue shifts due to increasing ionic strength originated from the formed "free" ionic micelles. Additionally, this sensor can be extended to detect other surfactants, such as sodium dodecylbenzene sulfonate (SDBS) and sodium C14-16 olefin sulfonate (SOS), as shown in Fig. 5.13d. Among these surfactants, this sensor is most sensitive to SDBS due to increasing hydrophobic binding affinity with increasing alkyl chain length. Besides anionicsurfactants, the sensor also responds to cationic surfactant, as shown in Fig. 5.13e. It is worth noting that the shift of diffraction peak of this sensor in the cationic cetyl trimethylammonium bromide (CTAB) solution is less than in anionic surfactants, possibly since smaller hydrophilicity of trimethylammonium ionic head groups of the CTAB. Moreover, by incorporating hydrophobic monomers into the hydrogel, the sensitivity can be increased through increasing hydrophobic binding affinity.

5.4.4 Solvent Sensor

Hydrogel is highly hydrophilic and tends to swell when it comes into contact with moisture. But this trend is restrained by adding ethanol in water. Based on this mechanism, Gao et al. prepared a solvent sensor based on PAAm hydrogel with an inverse-opal structure (IOPAAm) [77]. When it was immersed in mixed solution of ethanol and water, an obviously monochromatic color variation appears. The color changes from reddish orange, through green and finally to blue within a few seconds with increasing ethanol content (Fig. 5.14). Meanwhile, the diffraction peaks show a corresponding blue shift. The sensor has the advantages of rapid response in structural color and diffraction wavelength responding to environment variations due to highly porous structures. However, the procedure for removing the PC template may restrict the preparation of samples with a large area.

Besides PAAm hydrogel, PNIPAM also displays solvent sensibility [128, 129]. Below the volume phase transition temperature (LCST), the PNIPAM polymer chains exhibit cononsolvency behavior to collapse in the presence a second solvent in pure water such as alcohols or tetrahydrofuran. If the second solvent is a good solvent for PNIPAM, further adding its content will recover the PNIPAM to swollen state [128, 130]. Based on this, Guan et al. developed a solvent sensor by locking magnetic nanochains with 1D periodic structure in PNIPAM hydrogel balls [78]. The content of methanol or ethanol in the water can be monitored via a diffraction peak shift and color variation of the hydrogel/PCs composite balls, as shown in Fig. 5.15. As methanol or ethanol content increases, the diffraction peaks display abrupt blue shift and red shift at a critical methanol or ethanol content. This phenomenon is because the polymer chains undergo a reentrant coil–globule–coil transition with increasing methanol or ethanol content. Wu et al. attributed this reentrant transition to the formation of different water–methanol complexes [130]. Briefly, in a solution containing PNIPAM, water and methanol, there are at least three kinds of intermolecular interactions: water-PNIPAM, methanol-PNIPAM, and water-methanol. Among these three kinds of intermolecular interactions, the water–methanol interaction is the strongest. As a result of the strong interaction, water–methanol complexes are formed. They are poor solvents for PNIPAMm and can coexist with free water or methanol. In low methanol content, there is a lot of uncomplexed free water because methanol molecules are not enough to complex with all water molecules. In this case, these free water molecules associate on the polymer chains to cause the polymer chains in a coil state. With the increase of methanol content, more water–methanol complexes are formed, accompanied by a decrease of solvent quality, resulting in a contraction of the PNIPAM chains. When methanol content increases to a certain value, the water and methanol are nearly completely complexation, resulting in a sharp coil-to-globule transition. With the further increase of methanol content, the free

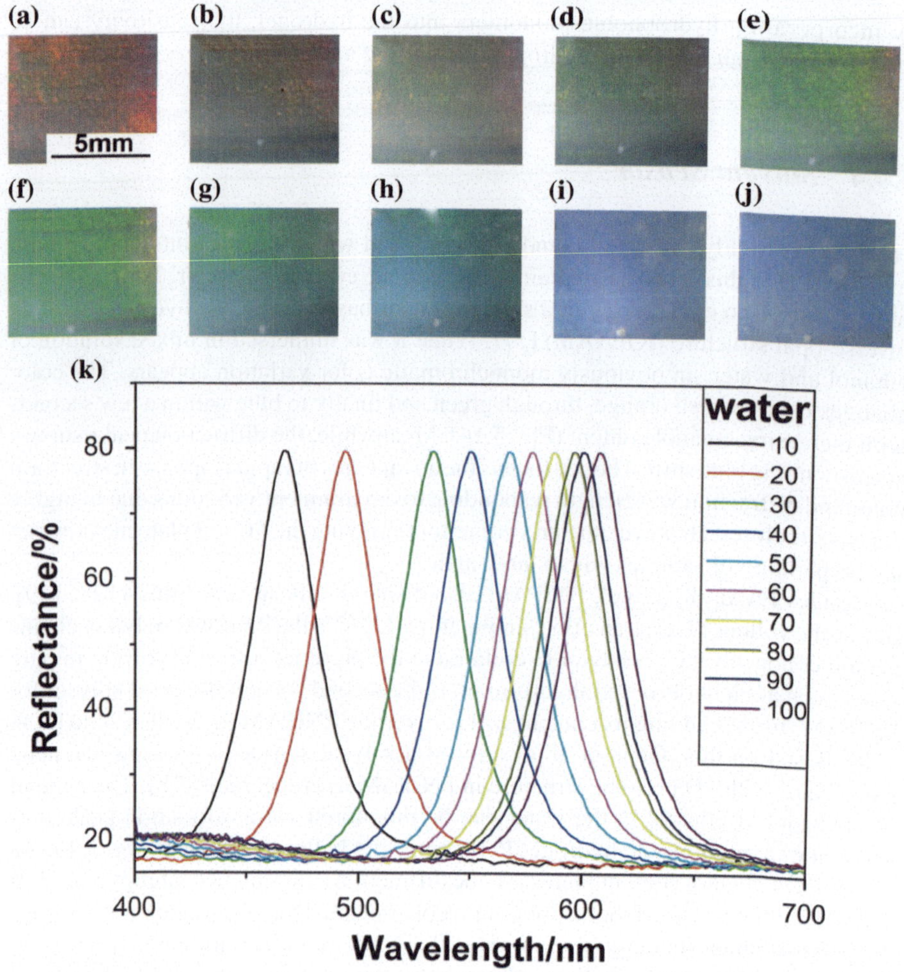

Fig. 5.14 Photographs of the IOPAM at different ethanol/water ratios: **a** 0/100, **b** 10/90, **c** 20/80, **d** 30/70, **e** 40/60, **f** 50/50, **g** 60/40, **h** 70/30, **i** 80/20, **j** 90/10. **k** The reflection spectra of the IOPAM at different ethanol/water ratios [77]

methanol molecules gradually increase, leading the PNIPAM to a coil state again by associating on the polymer chains. Later, Tanaka et al. ascribed the coil–globule–coil transition to a competition of the forming PNIPAM–water (p-w) hydrogen bonds and PNIPAM–methanol (p-m) or PNIPAM–alcohol (p-a) hydrogen bonds [128]. They thought that if both types of hydrogen bonds are cooperative, the total number of bound molecules shows a sharp minimum, thus the polymer chains are in a globule state. While there are competitive, the chains will be in a coil state. As a result, the polymer chain undergoes a very interesting reentrant coil–globule–coil transition.

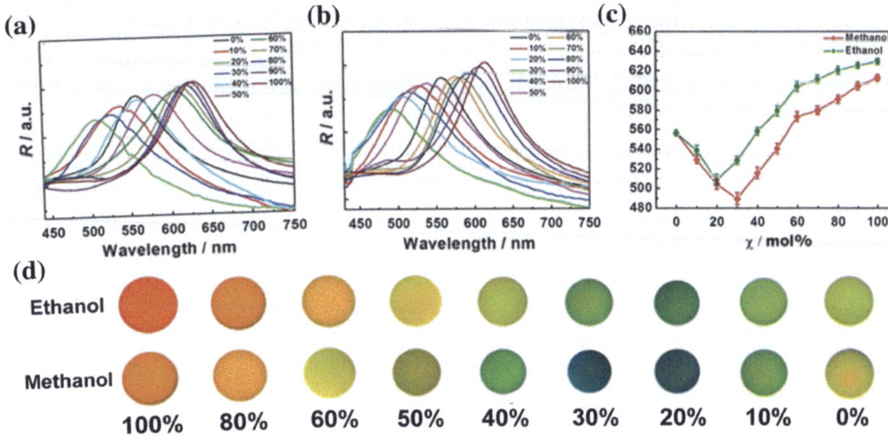

Fig. 5.15 Reflection spectra of the solvent sensor based on PNIPAM hydrogel balls embedded with 1D periodic structures in different ethanol (**a**) and methanol (**b**) contents in water. Effect of ethanol and methanol contents on the diffraction peaks (**c**) and color (**d**) [78]

5.4.5 *Humidity Sensor*

By means of hydrophilicity of the PAAm hydrogel, Song et al. prepared an humidity sensor by infiltrating AAm solution into voids of PCs template and subsequently photopolymerizing without a calcination or dissolution procedure to remove the PCs template [131]. For this sensor, the diffraction peak red shifted from 390 to 630 nm with increasing humidity from 20 to 100%, as shown in Fig. 5.16a. The shift range of diffraction peak achieves approached 240 nm, which spanned almost the whole visible range. In addition, the shift of spectral diffraction peak was accompanied with a series of color variations (Fig. 5.16b). The large range of change of diffraction

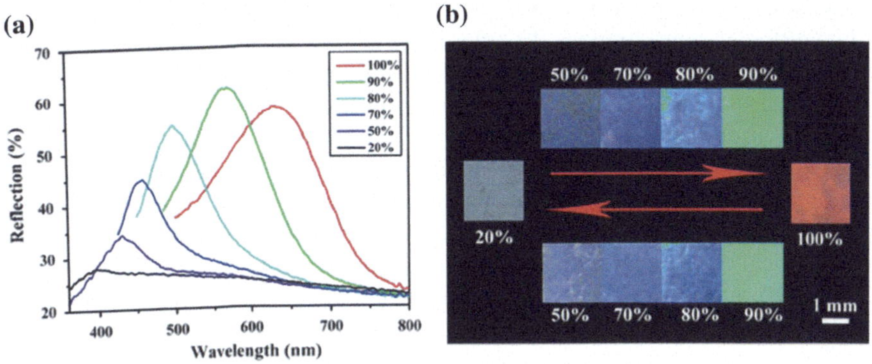

Fig. 5.16 **a** Reflection spectra and (**b**) colors of the humidity sensor in various humidity environments [131]

wavelength and color made this sensor could monitor a very broad humidity range. Later, Yin et al. developed a fast fabrication method for RPCs-based humidity sensor by producing PCs structures through instant magnetic assembly of $Fe_3O_4@SiO_2$ NPs in a in reaction solution of poly(ethylene glycol) acrylates, then immobilized the structures by photopolymerization [132]. Compared with the processes referred in above, which involves colloidal NPs assembly, responsive material infiltration and polymerization, this process is more simple and fast. These convenient and visual humidity sensors are highly desired because they need no specific apparatus and power for operating.

5.4.6 *Glucose Sensor*

It is well known that phenylboronic acids (PBAs) have substantial affinities to diol-containing molecules and can form a reversible covalent complex with them [85, 87, 104, 106, 107]. In particular, the PBAs exhibit high operational stability compared with glucose oxidase and lectin [85, 96]. Interestingly, glucose can bind with PBAs to generate 1:1 PBA–glucose complexes and 2:1 PBA–glucose complexes, as shown in Fig. 5.17a. In low ionic strength aqueous solutions, only 1:1 PBA–glucose complexes are formed and force hydrogel to swell [104]. The formation of 1:1 PBA–glucose complexes may be because the PBA groups distance is too far to form the 2:1 PBA–glucose complex in low ionic strength aqueous. Oppositely, in high ionic strength condition (ca.150 mM, as typical as physiological condition), the hydrogels are in the shrinking state, which makes the PBA groups distance be near enough to form 2:1 PBA–glucose complexes. The formed 2:1 PBA–glucose complexes will increase the crosslinking density of hydrogel, thus leads the hydrogel to shrink [104]. However, at higher glucose concentration, the 2:1 PBA–glucose complexes may be converted into 1:1 PBA–glucose complexes, leading hydrogel to swell again, as shown in Fig. 5.17b, c [85, 133]. This phenomenon will cause interfere for practical application because this sensor may display same volume and same diffraction peak at two different concentrations. In order to solve this problem, Braun et al. used a volume resetting agent to minimize the competing effects of these two complexes, yielding sensor materials that show linear responses under physiological condition [87]. In their design, polyvinyl alcohol (PVA) as a volume resetting agent is added into a preformed PBA-functionalized hydrogel to form PBA–PVA–PBA crosslinking, which shrinks the hydrogel. In contrast to the volume shrinkage caused by the forming 2:1 PBA–glucose complexes in conventional PBA-modified hydrogels at physiological condition, the formed PBA–PVA–PBA crosslinking will be replaced by 1:1 PBA–glucose complexes, causing the hydrogel only to swell with increasing glucose concentration (Fig. 5.17d, e). This because PBAs have lower affinities to 1,3-diols (e.g., PVA) than 1,2-diols (e.g., glucose), and, therefore, 1,2-diol target molecules can displace 1,3-diols to form a preformed 1,2-diol-PBA complex (1:1 PBA-glucose complex), causing the hydrogel to swell.

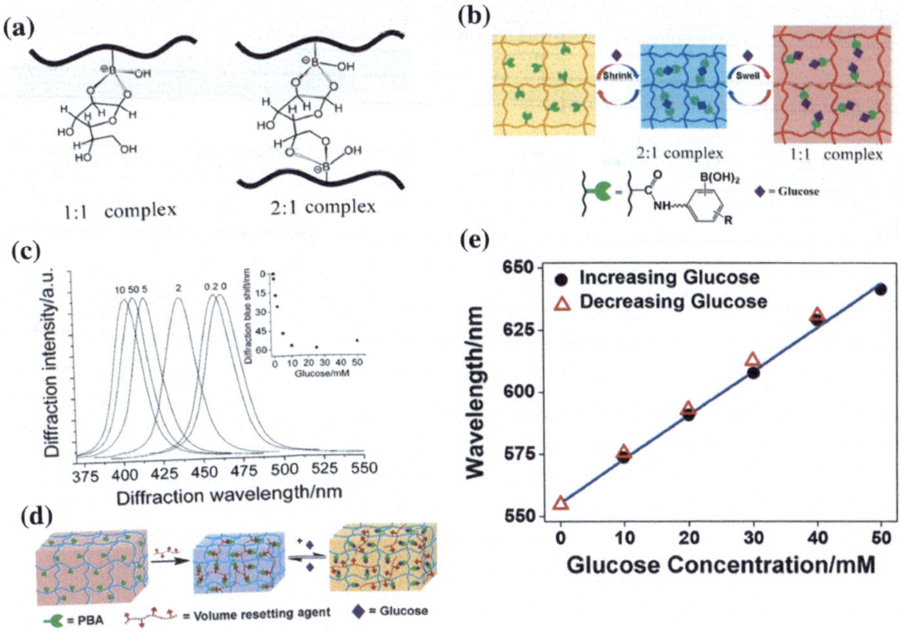

Fig. 5.17 **a** Schematic illustration of the 1:1 and 2:1 PBA–glucose complex. **b** Interactions of the PBA-modified hydrogel with glucose and their effect on the hydrogel volume with the increasing glucose concentration in high ionic strength condition [85]. **c** Diffraction spectra of sensor based on PBA-modified PCCA in different glucose concentrations in high ionic strength condition [133]. **d** Design route of the glucose-sensitive hydrogel with a linear response. **e** Diffraction wavelength versus the glucose concentrations [87]

The above sensors are constructed by responsive hydrogel matrix and insensitive PCs. Their sensitive mechanism is based on swell or shrink of hydrogel film induced by external stimulus, further resulting in a red or blue shift of diffraction wavelength and a corresponding color change of PCs. Contrarily, Zhang et al. constructed a new glucose sensors by embedding P(NIPAM–PBA) microgel sphere arrays in a PAAm hydrogel matrix [97]. In this system, the hydrogel matrix is nonresponsive, instead, the spheres constructed PCs are stimulus-responsive. In the presence of glucose, it does not show a significant wavelength change, instead have an intensity response to glucose concentration. This change of diffraction intensity is caused by decreasing refractive index of hydrogel microspheres with increasing degree of swelling. Additionally, an important feature of this microgels sensor is it had a much faster response rate than bulk hydrogel film.

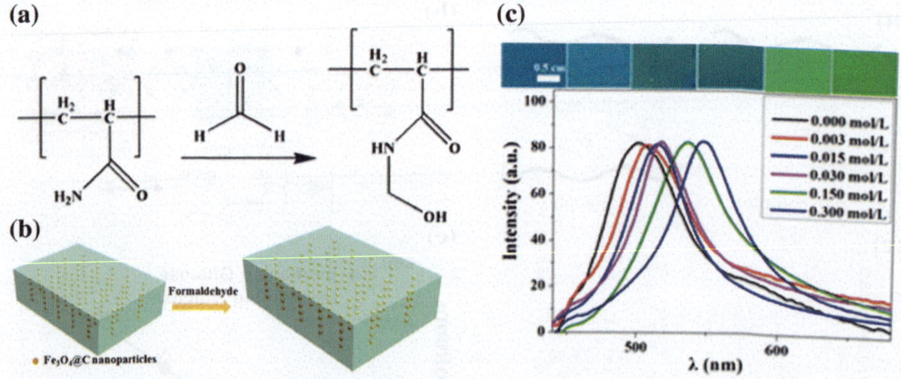

Fig. 5.18 **a** Reaction of PAAm and formaldehyde. **b** Schematic illustration of the variation of lattice spacing of PCs induced by formaldehyde. **c** Photographs and corresponding reflection spectra of the PAAm hydrogels/PCs composites in the aqueous solution with different formaldehyde concentrations [84]

5.4.7 Aldehydes Sensor

Variation of Flory–Huggins interaction parameter between the polymer and solvent would result in a change of hydrogel volume [105, 106]. For instance, amide groups of PAAm can effectively undergo a hydroxymethylation reaction with aldehyde in an aqueous solution of Na_2CO_3 [84]. Once the hydroxymethylation reaction takes place, the polar hydroxyl groups are produced in the hydrogels, as shown in Fig. 5.18a. The generated polar hydroxyl group will lead to a concomitant change of polymer–water interaction parameter and enhance interaction between the polymer and water, which expands the hydrogels. According to this mechanism, Zhu et al. prepared aldehyde sensor by combining carbon-encapsulated Fe_3O_4 nanoparticles PCs and PAAm hydrogels, as shown in Fig. 5.18b. This sensor exhibits obvious reflection spectra shift and structural color variation after treated with formaldehyde aqueous solution. It also can be used for distinguishing different aldehydes by different shift of reflection peaks. In addition, such RPCs sensors based on variation of Flory–Huggins interaction parameter-induced swell can be expanded to detect to other analyst, such as pH [88] and kinases [105].

5.4.8 Hydrogel Sensor Based on Strong Polyelectrolytes

So far, most responsive hydrogel materials for chemical sensors were constructed by weak polyelectrolytes. Their response to pH and metal ions mainly depend on changing charge quantity bounded in the hydrogel matrix. However, the response could be significantly reduced by ionic strength [41, 79, 99]. For strong polyelectrolytes hydrogels, the interference derived from the ionic strength can be shield or reduced due to high charge density in the hydrogels. Additionally, the strong polymer electrolytes are capable to rapidly exchange ions in solution and the bond energy of ion–ion interaction can compare with covalent. Importantly, the aggregation and conformation of the strong polyelectrolytes hydrogels could be changed by counterions with different degrees of solvation, charge densities, and chemical structures. The changed aggregation and conformation will affect volume changes of hydrogel [81]. Based on these mechanisms, Zhang et al. prepared anions, cations, and amino acid sensor by the strong polyelectrolyte hydrogels with an inverse-opal structure, as shown in Fig. 5.19a [81]. The volume of this sensor changes with the exchanging anions, thus causing a shift of diffraction signal. Taking the cation-sensitive sensor as an example, it exhibits different diffraction peaks (Fig. 5.19b) and an obvious color change (Fig. 5.19b) in different cations aqueous. The sensitive mechanism to cations of this sensor is not only due to the different salvation but also tightly coupled with the integration between cations and poly(ethylene glycol) which is a crosslinker. The ionic strength dependence of this sensor was characterized by varying the salt concentration (first by increasing and later by decreasing, Fig. 5.19e). It shows that the anions concentration below 2×10^{-4} M has little effect on the diffraction wavelength. However, this sensor does not exhibit good selectivity. As an extension of their work, they functionalized the photonic polyelectrolyte sensors with N-containing,

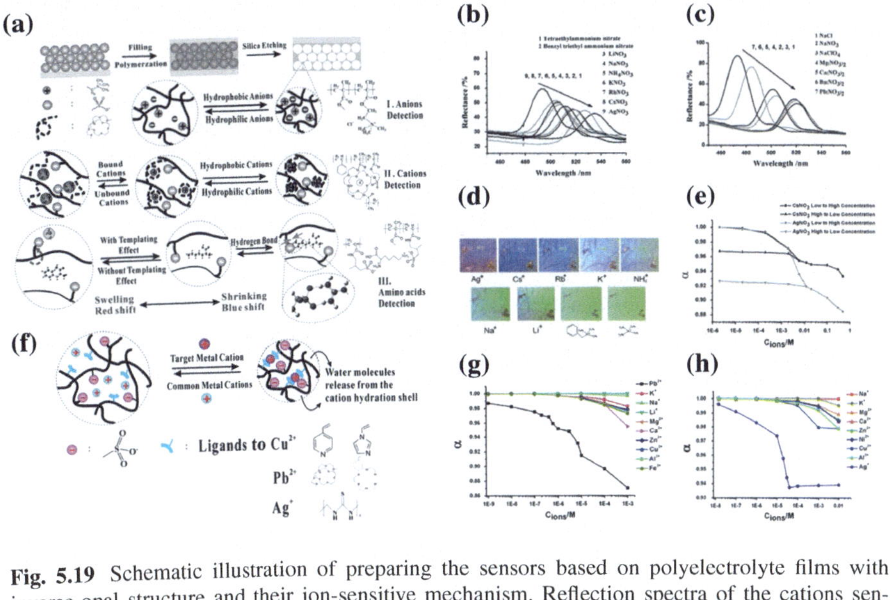

Fig. 5.19 Schematic illustration of preparing the sensors based on polyelectrolyte films with inverse-opal structure and their ion-sensitive mechanism. Reflection spectra of the cations sensor in 0.02 M diverse monovalent cations aqueous solutions (**b**) and their corresponding colors (**c**). **d** Reflection spectra of the cations sensor in 0.02 M diverse divalent cations aqueous solutions. **e** Effect of the anions concentration on the relative diffraction maximum ($\alpha = \lambda/\lambda_{max}$) of the cations sensor [81]. **f** Schematic illustration of sensitive mechanism of the cations sensor responded to specific ions. Effects of other anions on the monitoring Pb^{2+} (g) and Ag^{+} (**h**) [82]

O-containing, and S-containing ligands for detecting Cu^{2+}, Pb^{2+}, and Ag^{+} with good sensibility and selectivity (Fig. 5.19f) [80]. From Fig. 5.19g, h, it can be seen that these sensor responds to target metal ions by blue shifting their diffraction peaks. And other ions show little effect in the concentration below 10 μM for Pb^{2+} and 100 μM for Ag^{+}, respectively. Due to their high sensitivity and ionic strength-shielding, these photonic polyelectrolyte sensors may implement practical applications in the near future in ions coexist environment, such as drinking water and waste water.

5.5 Conclusion and Outlook

The recent progress of the stimuli-responsive hydrogels has been demonstrated in this chapter. The stimuli-responsive hydrogels are promising materials for chemical sensor because of their exciting property that can undergo an obvious and reversible volumetric variation in response to a small change of external environmental stimuli. In particular, combination of the stimuli-responsive hydrogels and PCs or Au NPs, the response to external stimuli can be converted into color change perceived by naked eye without any necessary treatments or complex instruments. Many examples of such sensors are discussed in this chapter. Additionally, the potential of the stimuli-responsive hydrogels is not fully exploited, which offers opportunities for the development of new and functional materials to construct new chemical sensor for detecting more target analyte to satisfy more demands in practical applications. Furthermore, the future research directions in the field of the chemical sensor based on stimuli-responsive hydrogels may include further developing microsensors for micro detection; improving selectivity, sensitivity, and resolution of the sensor. Moreover, in addition to chemical sensors, the stimuli-responsive hydrogels can be used in other fields such as selective sorbents, drug delivery systems and memory devices.

Acknowledgements The authors acknowledge the financial support from the National Science Fund for Distinguished Young Scholars (Grant No. 51825103), the National Key Research and Development Program of China (Grant No. 2017YFA0207101), and the Natural Science Foundation of China (Grant Nos. 51771188, 51571189, and 51701054).

References

1. M. Mahinroosta, Z.J. Farsangi, A. Allahverdi et al., Hydrogels as intelligent materials: a brief review of synthesis, properties and applications. Mater. Today Chem. **8**, 42–55 (2018)
2. T.E. Brown, B.J. Carberry, B.T. Worrell et al., Photopolymerized dynamic hydrogels with tunable viscoelastic properties through thioester exchange. Biomaterials **178**, 496–503 (2018)
3. B.H. Cipriano, S.J. Banik, R. Sharma et al., Superabsorbent hydrogels that are robust and highly stretchable. Macromolecules **47**(13), 4445–4452 (2014)
4. M.T.I. Mredha, Y.Z. Guo, T. Nonoyama et al., A facile method to fabricate anisotropic hydrogels with perfectly aligned hierarchical fibrous structures. Adv. Mater. **30**(9), 1704937 (2018)

5. Y. Zhou, M. Layani, S.C. Wang et al., Fully printed flexible smart hybrid hydrogels. Adv. Funct. Mater. **28**(9), 1705365 (2018)
6. I. Tokarev, S. Minko, Stimuli-responsive hydrogel thin films. Soft Matter **5**(3), 511–524 (2009)
7. A. Richter, G. Paschew, S. Klatt et al., Review on hydrogel-based pH sensors and microsensors. Sensors **8**(1), 561–581 (2008)
8. S. Basu, H.S. Samanta, J. Ganguly, Green synthesis and swelling behavior of Ag-nanocomposite semi-IPN hydrogels and their drug delivery using dolichos biflorus linn. Soft Mater **16**(1), 7–19 (2018)
9. T. Montheil, C. Echalier, J. Martinez et al., Inorganic polymerization: an attractive route to biocompatible hybrid hydrogels. J. Mater. Chem. B **6**(21), 3434–3448 (2018)
10. S.L. Vega, M.Y. Kwon, K.H. Song et al., Combinatorial hydrogels with biochemical gradients for screening 3D cellular microenvironments. Nat. Commun. **9**, 614 (2018)
11. S.J. Buwalda, K.W.M. Boere, P.J. Dijkstra et al., Hydrogels in a historical perspective: from simple networks to smart materials. J. Control. Rel. **190**, 254–273 (2014)
12. A.M. Rosales, K.S. Anseth, The design of reversible hydrogels to capture extracellular matrix dynamics. Nat. Rev. Mater. **1**(2), 15012 (2016)
13. B.H. Hu, C.O. Owh, P.L. Chee et al., Supramolecular hydrogels for antimicrobial therapy. Chem. Soc. Rev. **47**(18), 6917–6929 (2018)
14. S.Y. Chin, Y.C. Poh, A.C. Kohler et al., Additive manufacturing of hydrogel-based materials for next-generation implantable medical devices. Sci. Robot. **2**(2), 6451 (2017)
15. Q.F. Rong, W.W. Lei, M.J. Liu, Conductive hydrogels as smart materials for flexible electronic devices. Chemistry–A European Journal **24**(64), 16930–16943 (2018)
16. H.Y. Peng, W. Wang, F.H. Gao et al., Ultrasensitive diffraction gratings based on smart hydrogels for highly selective and rapid detection of trace heavy metal ions. J. Mater. Chem. C **6**(42), 11356–11367 (2018)
17. Q.S. Chen, W.H. Shi, M.F. Cheng et al., Molecularly imprinted photonic hydrogel sensor for optical detection of L-histidine. Microchim. Acta **185**(12), 557 (2018)
18. R. Wu, S.H. Zhang, Q. Zhang et al., Volumetric hydrogel sensor enables visual and quantitative detection of sulfion. Sensor. Actuat. B: Chem. **282**, 750–755 (2019)
19. Z.Y. Lei, Q.K. Wang, S.T. Sun et al., A bioinspired mineral hydrogel as a self-healable, mechanically adaptable ionic skin for highly sensitive pressure sensing. Adv. Mater. **29**(22), 1700321 (2017)
20. M.Q. Li, H.W. Liao, Q.L. Deng et al., Preparation of an intelligent hydrogel sensor based on g-C_3N_4 nanosheets for selective detection of Ag^+. J. Macromol. Sci. Part A **55**(5), 408–413 (2018)
21. C. Chen, Z.Q. Dong, J.H. Shen et al., 2D photonic crystal hydrogel sensor for tear glucose monitoring. ACS Omega **3**(3), 3211–3217 (2018)
22. J. Nam, I.B. Jung, B. Kim et al., A colorimetric hydrogel biosensor for rapid detection of nitrite ions. Sensor. Actuat. B: Chem. **270**, 112–118 (2018)
23. H. Liu, M.X. Li, C. Ouyang et al., Biofriendly, stretchable, and reusable hydrogel electronics as wearable force sensors. Small **14**(36), 1801711 (2018)
24. J.J. Qin, B.H. Dong, L.X. Cao et al., Photonic hydrogels for the ultratrace sensing of divalent beryllium in seawater. J. Mater. Chem. C **6**(15), 4234–4242 (2018)
25. R. Wu, S.H. Zhang, J. Lyu et al., A visual volumetric hydrogel sensor enables quantitative and sensitive detection of copper ions. Chem. Commun. **51**(38), 8078–8081 (2015)
26. I. Willner, Stimuli-controlled hydrogels and their applications. Acc. Chem. Res. **50**(4), 657–658 (2017)
27. M. Sun, R.B. Bai, X.Y. Yang et al., Hydrogels: hydrogel interferometry for ultrasensitive and highly selective chemical detection. Adv. Mater. **30**(46), 1870352 (2018)
28. H.Z. Kang, A.C. Trondoli, G.Z. Zhu et al., Near-infrared light-responsive core-shell nanogels for targeted drug delivery. ACS Nano **5**(6), 5094–5099 (2011)
29. T. Jing, H.R. Du, Q. Dai et al., Magnetic molecularly imprinted nanoparticles for recognition of lysozyme. Biosensor. Bioelectro **26**(2), 301–306 (2010)

30. B.G. Kabra, S.H. Gehrke, R.J. Spontak, Microporous, responsive hydroxypropyl cellulose gels. 1. Synthesis and microstructure. Macromolecules **31**(7), 2166–2173 (1998)
31. W.A. Laftah, S. Hashim, A.N. Ibrahim, Polymer hydrogels: a review. Polym. Plast. Technol. **50**(14), 1475–1486 (2011)
32. I. Tokarev, S. Minko, Stimuli-responsive porous hydrogels at interfaces for molecular filtration, separation, controlled release, and gating in capsules and membranes. Adv. Mater. **22**(31), 3446–3462 (2010)
33. H. Li, T.Y. Ng, Y.K. Yew et al., Modeling and simulation of the swelling behavior of pH-stimulus-responsive hydrogels. Biomacromol **6**(1), 109–120 (2005)
34. A. Döring, W. Birnbaum, D. Kuckling, Responsive hydrogels-structurally and dimensionally optimized smart frameworks for applications in catalysis, micro-system technology and material science. Chem. Soc. Rev. **42**(17), 7391–7420 (2013)
35. R.V. Kulkarni, S.A. Biswanath, Electrically responsive smart hydrogels in drug delivery: a review. J. Appl. Biomater. Funct. Mater. **5**(3), 125–139 (2007)
36. A.L. Navarro-Verdugo, F.M. Goycoolea, G. Romero-Meléndez et al., A modified Boltzmann sigmoidal model for the phase transition of smart gels. Soft Mater. **7**(12), 5847–5853 (2011)
37. Y. Qiu, K. Park, Environment-sensitive hydrogels for drug delivery. Adv. Drug Deliver. Rev **53**(3), 321–339 (2001)
38. Y. Ogawa, K. Ogawa, E. Kokufuta, Swelling-shrinking behavior of a polyampholyte gel composed of positively charged networks with immobilized polyanions. Langmuir **20**(7), 2546–2552 (2004)
39. F. Ganji, F.S. Vasheghani, F.E. Vasheghani, Theoretical description of hydrogel swelling: a review. Iran. Polym. J. **19**(5), 375–398 (2010)
40. J. Kim, M.J. Serpe, L.A. Lyon, Hydrogel microparticles as dynamically tunable microlenses. J. Am. Chem. Soc. **126**(31), 9512–9513 (2004)
41. H.L. Li, D.D. Men, Y.Q. Sun et al., Optical sensing properties of Au nanoparticle/hydrogel composite microbeads using droplet microfluidics. Nanotechnology **28**, 405502 (2017)
42. X.L. Xiong, C.C. Wu, C.S. Zhou et al., Responsive DNA-based hydrogels and their applications. Macromol. Rapid Commun. **34**(16), 1271–1283 (2013)
43. D.D. Men, H.H. Zhang, Y. Li et al., Optical sensor based on hydrogel films with 2D colloidal arrays attached on both the surfaces: anti-curling performance and enhanced optical diffraction intensity. J. Mater. Chem. C **3**, 3659–3665 (2015)
44. D.D. Men, F. Zhou, Y. Li et al., Gold nanoshell arrays-based visualized sensors of pH: Facile fabrication and high diffraction intensity. J. Mater. Res. **32**(4), 717–725 (2017)
45. D.D. Men, L.F. Hang, Y. Li et al., 3-Acrylamidophenylboronic acid-modified hydrogel film attached to a gold nanosphere array to detect hydrofluoric acid with good selectivity and recyclability. Chem. Nano Mat. **4**(2), 165–169 (2018)
46. X.G. Han, Y.D. Liu, Y.D. Yin, Colorimetric stress memory sensor based on disassembly of gold nanoparticle chains. Nano Lett. **14**(5), 2466–2470 (2014)
47. C. Fenzl, S. Wilhelm, T. Hirsch et al., Optical sensing of the ionic strength using photonic crystals in a hydrogel matrix. ACS Appl. Mater. Interfaces **5**(1), 173–178 (2013)
48. L. Nucara, V. Piazza, F. Greco et al., Ionic strength responsive sulfonated polystyrene opals. ACS Appl. Mater. Interfaces **9**(5), 4818–4827 (2017)
49. J.P. Couturier, M. Sütterlin, A. Laschewsky et al., Responsive inverse opal hydrogels for the sensing of macromolecules. Angew. Chem. Int. Ed. **54**(22), 6641–6644 (2015)
50. G.B. Huang, Y.B. Yin, Z. Pan et al., Fabrication of 3D photonic crystals from chitosan that are responsive to organic solvents. Biomacromol **15**(12), 4396–4402 (2014)
51. C. Fenzl, T. Hirsch, O.S. Wolfbeis, Photonic crystals for chemical sensing and biosensing. Angew. Chem. Int. Ed. **53**(13), 3318–3335 (2014)
52. J.Y. Xu, C.X. Yan, C. Liu et al., Photonic crystal hydrogel sensor for detection of nerve agent. IOP Conf. Ser.: Mater. Sci. Eng. **167**(1), 012024 (2017)
53. H. Xu, J.Y. Zhang, Y.S. Xu et al., Down's syndrome screening with hydrogel photonic barcodes. Sensor. Actuat. B: Chem. **255**, 2690–2696 (2018)

54. Y.S. Xu, H. Wang, C.X. Luan et al., Porous hydrogel encapsulated photonic barcodes for multiplex microRNA quantification. Adv. Funct. Mater. **28**(1), 1704458 (2018)
55. Y.J. Zhao, X.W. Zhao, Z.Z. Gu, Photonic crystals in bioassays. Adv. Funct. Mater. **20**(18), 2970–2988 (2010)
56. K.I. MacConaghy, C.I. Geary, J.L. Kaar et al., Photonic crystal kinase biosensor. J. Am. Chem. Soc. **136**(19), 6896–6899 (2014)
57. D.D. Men, D.L. Liu, Y. Li, Visualized optical sensors based on two/three-dimensional photonic crystals for biochemicals. Sci. Bul. **61**(17), 1358–1371 (2016)
58. J.P. Ge, Y.D. Yin, Responsive photonic crystals. Angew. Chem. Int. Ed. **50**(7), 1492–1522 (2011)
59. P. Lova, G. Manfredi, D. Comoretto, Advances in functional solution processed planar 1D photonic crystals. Adv. Opt. Mater. 1800730 (2018)
60. J. Sevilla, A. Andueza, Optical sensing based on photonic crystal structures. Fiber Opt. Sens. **21**, 223–240 (2017)
61. C.I. Aguirre, E. Reguera, A. Stein, Tunable colors in opals and inverse opal photonic crystals. Adv. Funct. Mater. **20**(16), 2565–2578 (2010)
62. H. Wang, K.Q. Zhang, Photonic crystal structures with tunable structure color as colorimetric sensors. Sensors **13**(4), 4192–4213 (2013)
63. Y.J. Zhao, Z.Y. Xie, H.C. Gu et al., Bio-inspired variable structural color materials. Chem. Soc. Rev. **41**(8), 3297–3317 (2012)
64. S.J. Jeon, M.C. Chiappelli, R.C. Hayward, Photocrosslinkable nanocomposite multilayers for responsive 1D photonic crystals. Adv. Funct. Mater. **26**(5), 722–728 (2016)
65. W.D. Zhao, M.H. Quan, Z.Q. Cao et al., Visual multi-triggered sensor based on inverse opal hydrogel. Colloid. Surf. A: Physicochem. Eng. **554**, 93–99 (2018)
66. Z. Hu, X. Lu, J. Gao, Hydrogel opals. Adv. Mater. **13**(22), 1708–1712 (2001)
67. Y. Takeoka, M. Watanabe, Tuning structural color changes of porous thermosensitive gels through quantitative adjustment of the cross-linker in pre-gel solutions. Langmuir **19**(22), 9104–9106 (2003)
68. K. Ueno, K. Matsubara, M. Watanabe et al., An electro-thermochromic hydrogel as a full-color indicator. Adv. Mater. **19**(19), 2807–2812 (2007)
69. M.C. Chiappelli, R.C. Hayward, Photonic multilayer sensors from photo-crosslinkable polymer films. Adv. Mater. **24**(45), 6100–6104 (2012)
70. J. Wang, Y. Hu, R. Deng et al., Multiresponsive hydrogel photonic crystal microparticles with inverse-opal structure. Langmuir **29**(28), 8825–8834 (2013)
71. W.T. Wang, X.Q. Fan, F.H. Li et al., Magnetochromic photonic hydrogel for an alternating magnetic field-responsive color display. Adv. Opt. Mater. **6**(4), 1701093 (2018)
72. D.D. Men, F. Zhou, Y. Li et al., A functional hydrogel film attached with a 2D Au nanosphere array and its ultrahigh optical diffraction intensity as a visualized sensor. J. Mater. Chem. C **4**, 2117–2122 (2016)
73. J. Shin, P.V. Braun, W. Lee, Fast response photonic crystal pH sensor based on templated photo-polymerized hydrogel inverse opal. Sensor. Actuat. B: Chem. **150**(1), 183–190 (2010)
74. E.T. Tian, Y. Ma, L.Y. Cui et al., Color-oscillating photonic crystal hydrogel. Macromol. Rapid Commun. **30**(20), 1719–1724 (2009)
75. W. Luo, Q. Cui, K. Fang et al., Responsive hydrogel-based photonic nanochains for microenvironment sensing and imaging in real time and high resolution. Nano Lett. (2018). https://doi.org/10.1021/acs.nanolett.7b04218
76. Y.J. Lee, P.V. Braun, Tunable inverse opal hydrogel pH sensors. Adv. Mater. **15**(7–8), 563–566 (2003)
77. J.Y. Wang, Y. Cao, Y. Feng et al., Multiresponsive inverse-opal hydrogels. Adv. Mater. **19**(22), 3865–3871 (2007)
78. W. Luo, J.D. Yan, Y.L. Tan et al., Rotating 1-D magnetic photonic crystal balls with a tunable lattice constant. Nanoscale **9**(27), 9548–9555 (2017)
79. A.V. Goponenko, S.A. Asher, Modeling of stimulated hydrogel volume changes in photonic crystal Pb^{2+} Sensing materials. J. Am. Chem. Soc. **127**, 10753–10759 (2005)

80. W. Hong, W.H. Li, X.B. Hu et al., Highly sensitive colorimetric sensing for heavy metal ions by strong polyelectrolyte photonic hydrogels. J. Mater. Chem. **21**(43), 17193–17201 (2011)
81. W. Hong, X.B. Hu, B.Y. Zhao et al., Tunable photonic polyelectrolyte colorimetric sensing for anions, cations and zwitterions. Adv. Mater. **22**(44), 5043–5047 (2010)
82. B.F. Ye, Y.J. Zhao, Y. Cheng et al., Colorimetric photonic hydrogel aptasensor for the screening of heavy metal ions. Nanoscale **4**(19), 5998–6003 (2012)
83. C. Price, J. Carroll, T.L. Clare, Chemoresistive and photonic hydrogel sensors of transition metal ions via Hofmeister series principles. Sensor. Actuat. B: Chem. **256**, 870–877 (2018)
84. X. Jia, T. Zhang, J. Wang et al., Responsive photonic hydrogel-based colorimetric sensors for detection of aldehydes in aqueous solution. Langmuir **34**(13), 3987–3992 (2018)
85. C.J. Zhang, M.D. Losego, P.V. Braun, Hydrogel-based glucose sensors: effects of phenylboronic acid chemical structure on response. Chem. Mater. **25**, 3239–3250 (2013)
86. Y.X. Yuan, Z.L. Li, Y. Liu et al., Hydrogel photonic sensor for the detection of 3-Pyridinecarboxamide. Chem. Eur. J. **18**, 303–309 (2012)
87. C.J. Zhang, G.G. Cano, P.V. Braun, Linear and fast hydrogel glucose sensor materials enabled by volume resetting agents. Adv. Mater. **26**, 5678–5683 (2014)
88. K.I. MacConaghy, D.M. Chadly, M.P. Stoykovich et al., Optically diffracting hydrogels for screening kinase activity in vitro and in cell lysate: impact of material and solution properties. Anal. Chem. **87**(6), 3467–3475 (2015)
89. Y.S. Huang, Y.L. Ma, Y.H. Chen et al., Target-responsive DNAzyme cross-linked hydrogel for visual quantitative detection of lead. Anal. Chem. **86**(22), 11434–11439 (2014)
90. D.L. Liu, L.L. Fang, Y. Li et al., Ultrasensitive and stable Au dimer-based colorimetric sensors using the dynamically tunable gap-dependent plasmonic coupling optical properties. Adv. Funct. Mater. **28**(18), 1707392 (2018)
91. J.H. Kim, B.W. Boote, J.A. Pham et al., Thermally tunable catalytic and optical properties of gold-hydrogel nanocomposites. Nanotechnology **23**(27), 275606 (2012)
92. J.J. Zhang, L. Mou, X.Y. Jiang, Hydrogels Incorporating Au@ polydopamine nanoparticles: robust performance for optical sensing. Anal. Chem. **90**(19), 11423–11430 (2018)
93. S. Lim, J.E. Song, J.A. La et al., Gold nanospheres assembled on hydrogel colloids display a wide range of thermoreversible changes in optical bandwidth for various plasmonic-based color switches. Chem. Mater. **26**(10), 3272–3279 (2014)
94. J.T. Zhang, L.L. Wang, D.N. Lamont et al., Fabrication of large-area two-dimensional colloidal crystals. Angew. Chem. Int. Ed. **51**(25), 6117–6220 (2012)
95. Z.F. Sun, F.C. Lv, L.J. Cao et al., Multistimuli-responsive, moldable supramolecular hydrogels cross-linked by ultrafast complexation of metal ions and biopolymers. Angew. Chem. Int. Ed. **54**(27), 7944–7948 (2015)
96. M.M.W. Muscatello, S.A. Asher, Poly (vinyl alcohol) rehydratable photonic crystal sensor materials. Adv. Funct. Mater. **18**(8), 1186–1193 (2008)
97. Y. Liu, Y.J. Zhang, Y. Guan, New polymerized crystalline colloidal array for glucose sensing. Chem. Commun. (14), 1867–1869 (2009)
98. A.K. Yetisen, N. Jiang, A. Fallahi et al., Glucose-sensitive hydrogel optical fibers functionalized with phenylboronic acid. Adv. Mater. **29**(15), 1606380 (2017)
99. H.L. Li, D.D. Men, Y. Li et al., Surface enhanced Raman scattering properties of dynamically tunable nanogaps between Au nanoparticles self-assembled on hydrogel microspheres controlled by pH. J. Colloid Interface Sci. **505**, 467–475 (2017)
100. X. Fei, T. Lu, J. Ma et al., Bioinspired polymeric photonic crystals for high cycling pH-sensing performance. ACS Appl. Mater. Interfaces **8**(40), 27091–27098 (2016)
101. H. Saito, Y. Takeoka, M. Watanabe Simple and precision design of porous gel as a visible indicator for ionic species and concentration. Chem. Commun. 0, 2126–2127 (2003)
102. Z.Y. Cai, A. Sasmal, X.Y. Liu et al., Responsive photonic crystal carbohydrate hydrogel sensor materials for selective and sensitive lectin protein detection. ACS Sens. **2**(10), 1474–1481 (2017)
103. S.A. Asher, A.C. Sharma, A.V. Goponenko et al., Photonic crystal aqueous metal cation sensing materials. Anal. Chem. **75**(7), 1676–1683 (2003)

104. F. Xue, Z.H. Meng, F.Y. Wang et al., A 2-D photonic crystal hydrogel for selective sensing of glucose. J. Mater. Chem. A **2**(25), 9559–9565 (2014)
105. A.C. Sharma, T. Jana, R. Kesavamoorthy et al., A general photonic crystal sensing motif: creatinine in bodily fluids. J. Am. Chem. Soc. **126**(9), 2971–2977 (2004)
106. F. Horkay, I. Tasaki, P.J. Basser, Osmotic swelling of polyacrylate hydrogels in physiological salt solutions. Biomacromol **1**(1), 84–90 (2000)
107. D. Nakayama, Y. Takeoka, M. Watanabe et al., Simple and precise preparation of a porous gel for a colorimetric glucose sensor by a templating technique. Angew. Chem. Int. Ed. **115**(35), 4329–4332 (2003)
108. K.W. Kimble, J.P. Walker, D.N. Finegold et al., Progress toward the development of a point-of-care photonic crystal ammonia sensor. Anal. Bioanal. Chem. **385**(4), 678–685 (2006)
109. Z.Y. Cai, N.L. Smith, J.T. Zhang et al., Two-dimensional photonic crystal chemical and biomolecular sensors. Anal. Chem. **87**(10), 5013–5025 (2015)
110. A. Bal, B. Özkahraman, Z. Özbaş, Preparation and characterization of pH responsive poly (methacrylic acid-acrylamide-N-hydroxyethyl acrylamide) hydrogels for drug delivery systems. J. Appl. Polym. Sci. **133**(13), 43226 (2016)
111. S. Nesrinne, A. Djamel, Synthesis, characterization and rheological behavior of pH sensitive poly (acrylamide-co-acrylic acid) hydrogels. Arab. J. Chem. **10**(4), 539–547 (2017)
112. K. Lee, S.A. Asher, Photonic crystal chemical sensors: pH and ionic strength. J. Am. Chem. Soc. **122**(39), 9534–9537 (2000)
113. Y. Zhang, Y.M. Guo, X.Y. Jiang et al., Nanomaterials for ultrasensitive protein detection. Adv. Mater. **25**(28), 3802–3819 (2013)
114. J.S. Sun, Y.L. Xianyu, X.Y. Jiang, Point-of-care biochemical assays using gold nanoparticle-implemented microfluidics. Chem. Soc. Rev. **43**(17), 6239–6253 (2014)
115. Y.P. Chen, Y.L. Xianyu, X.Y. Jiang, Surface modification of gold nanoparticles with small molecules for biochemical analysis. Acc. Chem. Res. **50**(2), 310–319 (2017)
116. D.L. Liu, F. Zhou, Y. Li et al., Black gold: plasmonic colloidosomes with broadband absorption self-assembled from monodispersed Au nanospheres by using a reverse emulsion system. Angew. Chem. Int. Ed. **54**(33), 9596–9600 (2015)
117. D.L. Liu, C.C. Li, Y. Li et al., Capillary gradient-induced self-assembly of periodic Au spherical nanoparticle arrays on an ultralarge scale via a bisolvent system at air/water interface. Adv. Mater. Interfaces **4**(10), 1600976 (2017)
118. P.J. Yan, F. He, W. Wang et al., Novel membrane detector based on smart nanogels for ultrasensitive detection of trace threat substances. ACS Appl. Mater. Interfaces **10**(42), 36425–36434 (2018)
119. J.H. Holtz, J.S.W. Holtz, C.H. Munro et al., Intelligent polymerized crystalline colloidal arrays: novel chemical sensor materials. Anal. Chem. **70**(4), 780–791 (1998)
120. Y. Wang, F. Yang, X.R. Yang, Colorimetric detection of mercury(II) Ion using unmodified silver nanoparticles and mercury-specific oligonucleotides. ACS Appl. Mater. Interfaces **2**(2), 339–342 (2010)
121. X.F. Ding, L.T. Kong, J. Wang et al., Highly sensitive SERS detection of Hg^{2+} ions in aqueous media using gold nanoparticles/graphene heterojunctions. ACS Appl. Mater. Interfaces **5**, 7072–7078 (2013)
122. J.H. Huang, X. Gao, Z.G. Li et al., Graphene oxide-based amplified fluorescent biosensor for Hg^{2+} detection through hybridization chain reactions. Anal. Chem. **86**, 3209–3215 (2014)
123. B.F. Ye, H.B. Ding, Y. Cheng et al., Photonic crystal microcapsules for label-free multiplex detection. Adv. Mater. **26**(20), 3270–3274 (2014)
124. E. Kokufuta, Y.Q. Zhang, T. Tanaka et al., Effects of surfactants on the phase transition of poly (N-isopropylacrylamide) gel. Macromolecules **26**(5), 1053–1059 (1993)
125. J. Sjöström, L. Piculell, Simple gel swelling experiments distinguish between associating and nonassociating polymer-surfactant pairs. Langmuir **17**(13), 3836–3843 (2001)
126. W. Xue, I.W. Hamley, Thermoreversible swelling behaviour of hydrogels based on N-isopropylacrylamide with a hydrophobic comonomer. Polymer **43**(10), 3069–3077 (2002)

127. J.T. Zhang, N. Smith, S.A. Asher, Two-dimensional photonic crystal surfactant detection. Anal. Chem. **84**(15), 6416–6420 (2012)
128. F. Tanaka, T. Koga, H. Kojima et al., Temperature and tension-induced coil-globule transition of poly (N-isopropylacrylamide) chains in water and mixed solvent of water/methanol. Macromolecules **42**(4), 1321–1330 (2009)
129. F.M. Winnik, H. Ringsdorf, J. Venzmer, Methanol-water as a co-nonsolvent system for poly (N-isopropylacrylamide). Macromolecules **23**(8), 2415–2416 (1990)
130. G. Zhang, C. Wu, The water/methanol complexation induced reentrant coil-to-globule-to-coil transition of individual homopolymer chains in extremely dilute solution. J. Am. Chem. Soc. **123**(7), 1376–1380 (2001)
131. E.T. Tian, J.X. Wang, Y.M. Zheng et al., Colorful humidity sensitive photonic crystal hydrogel. J. Mater. Chem. **18**(10), 1116–1122 (2008)
132. R.Y. Xuan, Q.S. Wu, Y.D. Yin et al., Magnetically assembled photonic crystal film for humidity sensing. J. Mater. Chem. **21**(10), 3672–3676 (2011)
133. V.L. Alexeev, A.C. Sharma, A.V. Goponenko et al., High ionic strength glucose-sensing photonic crystal. Anal. Chem. **75**(10), 2316–2323 (2003)

Chapter 6
Interfacial Responsive Functional Oxides for Nanoelectronics

Pankaj Sharma and Jan Seidel

Abstract Functional interfaces can be considered 'materials within materials' that are active at the nanoscale and are responsive to an applied external stimulus. At these interfaces, the material properties are not only distinct from the bulk, but new functionalities or phases can emerge. Notable examples of such functional interfaces include structural phase boundaries and domain walls in ferroelectric oxides. In this chapter, nano-mechanical–electromechanical properties of the morphotropic phase boundaries in bismuth ferrite thin films are reviewed. Specifically, using scanning probe microscopy techniques, factors governing nano-mechanical and electromechanical properties of morphotropic bismuth ferrite are identified and potential applications in oxide nanoelectronics are discussed.

6.1 Introduction

For long-time interfaces and boundaries in materials were regarded as defects that are detrimental to a (bulk) material's functionalities and potential applications. However, this viewpoint has changed over time, from being '*detrimental*' to offering new aspects and being '*responsive functional nanomaterials*' in their own right. The most prominent example, while restricting ourselves to the field of complex oxides, is the exciting discovery of a high-mobility electron liquid confined at the interface of two band insulators [1]. This functional interface is responsive to various external stimuli and its electronic density can be tuned via a multitude of means [2] enabling all oxide nanoelectronics concepts [3]. Another example of a responsive functional interface in

P. Sharma · J. Seidel (✉)
School of Materials Science and Engineering, The University of New South Wales, Sydney, NSW 2052, Australia
e-mail: jan.seidel@unsw.edu.au

P. Sharma
e-mail: pankaj.sharma@unsw.edu.au

ARC Centre of Excellence in Future Low-Energy Electronics Technologies, The University of New South Wales, Sydney, NSW 2052, Australia

Z. Sun and T. Liao (eds.), *Responsive Nanomaterials for Sustainable Applications*, Springer Series in Materials Science 297,
https://doi.org/10.1007/978-3-030-39994-8_6

ferroelectric oxides with great potential for sustainable nanoelectronics applications is the recently reported morphotropic phase boundaries in $BiFeO_3$. These nanoscale interfaces and their properties have been widely characterized using scanning probe microscopy (SPM) techniques. SPM techniques involve a sharp probe (radius few tens of nanometre) that interact through interatomic forces with the sample and are used to gain a nanoscopic insight into materials properties. Besides imaging the surface and spatially resolved properties, the tip of an SPM can be used to tailor and control materials functionalities at the same confined length scales. Furthermore, the sharp tip acts as a nano-indenter that can be used to deliver a mechanical impulse to trigger mechanical-activated phenomena such as phase transitions, polarization reversal [4], ion migration [5, 6] and metal–insulator transitions [7]. This chapter reviews recent pivotal developments in this research field and gives an outlook on future prospects.

6.2 Morphotropic Phase Boundaries

Ferroelectric oxides are important technological materials that exhibit a *spontaneous* polarization (electrical dipole moment), with many of them possessing a perovskite structure (general formula, ABO_3, where A and B are metal cations) [8]. The spontaneous polarization can be reversed or transitioned between equivalent states (two or more) by a suitably oriented electric field. Ferroelectrics have a non-centrosymmetric crystal structure (*piezoelectricity*) and therefore display a coupling between electrical and mechanical properties. When such materials are subjected to an external stress, electrical charges are generated, or conversely, they undergo a shape change (strain) on the application of an external electric field. This interlinking of electromechanical properties makes ferroelectrics highly suitable for '*sensing* and *actuation*' applications. An especially large coupling is achieved in materials that possess *morphotropic phase boundaries (MPBs)*. MPBs are actually transition regions in the phase diagram of ferroelectric oxides, where the crystal structure becomes very susceptible to external stimuli. This is due to a flattened energy landscape between competing for structural polymorphs at the MPB. The term *morphotropic* was originally coined to refer to phase transitions due to changes in the composition, but of late has acquired a more general meaning. Under this approach, the crystal structure changes at the MPB, and the electromechanical coupling attains a high value due to the associated polarization rotation. Initially, these MPBs were realized in composition-driven intricate solid solutions [9–11] that contain lead (Pb). It is only recently that these MPBs were realized in chemically simple oxide compounds, e.g. in $PbTiO_3$ [12, 13] and $BiFeO_3$ [14]. Here, we will consider the MPBs in the $BiFeO_3$ (BFO) system. The discovery of MPBs in BFO is quite special on two counts: (i) it provides an opportunity to unravel microscopic physical origins of the enhanced electromechanical couplings in chemically simple compounds and (ii) presents a lead-free recyclable alternative for ferroelectric devices.

The MPB in BFO is achieved using thin-film strain engineering. In epitaxially grown thin films, by imposing a compressive strain of about –4.3%, the ground state of the BFO is tuned to a novel tetragonal-like phase (*T-like*) with a large *c*/*a* ratio of ~1.25 [14]. With increasing film thickness, strain relaxation takes place and intermediate strain values drive BFO to the brink of an MPB, where a rhombohedral-like phase (*R-like*) starts to emerge to relieve elastic strain. This *R-like* phase coexists with tetragonal-like phase forming mixed-phase regions. These mixed-phase regions appear as corrugated nanoscale stripes that are embedded in the parent *T-like* phase matrix. Detailed structural studies reveal that the crystal structure in the corrugated nanoscale regions is highly distorted and differs greatly from their parent counterparts (i.e. the novel *T-like* phase and bulk-like relaxed rhombohedral phase of BFO) [15]. Therefore, from now on the tetragonal- and rhombohedral-like phases in the mixed-phase regions will be referred to as the T'- and R'-*like* phases, respectively. The T'- and R'-*like* phases form due to strain relaxation and are low symmetry intermediary phases that connect the novel *T*-like phase and relaxed *R*-phase (bulk-like) of the BFO. At this stage, it is important to note that in the mixed-phase regions, MPBs are real physical boundaries [16], in contrast to a *transition region* in the phase diagram for the traditional lead-based complex solid solutions. These physical MPBs are typically a few unit cells wide and spatially segregate T'- and R'-*like* phases. The crystal structure changes smoothly across these boundaries without the formation of any misfit dislocations and/or defects. At these MPBs, enhanced electromechanical responses have been observed [17] prompting investigation into the underlying microscopic origins [18–21] associated with structural phase transformations, motion of phase boundaries and nano-mechanical properties. In the following section, we will explore these aspects at the nanoscale in detail.

6.2.1 Nanoscale Structural Transformations

The competing structural polymorphs in morphotropic BFO are responsive to an applied external stimulus (i.e. electric field, strain/thickness, stress and chemical alloying) [17, 22–25] and can be transitioned into one another. The structural and polarization transformations can be inferred by X-ray diffraction (XRD) [15], scanning probe (SPM) [14, 15, 17] and transmission electron microscopy [17] techniques. However, precise control and visualization of the structural transformations at the nanoscale have been lacking. This has been addressed recently using a scanning probe technique called *static piezoresponse* [26]. In this technique, a nanoscale tip monitors the external field-induced static surface displacement of the electro-active ferroelectric film. Figure 6.1a displays a typical nanoscale surface displacement (δz) versus DC bias hysteretic curve for the morphotropic BFO film. The response is a distinct *butterfly-shaped* nonlinear hysteretic curve revealing a large electric field-induced reversible strain of about 6% strain $\varepsilon = \delta Z_{\max}/L$, $\delta Z_{\max}$ is the maximum surface deformation; L is film thickness, which is 60 nm. A closer look at the obtained spectroscopic curve elucidates concomitant phase transitions, i.e. both structural and

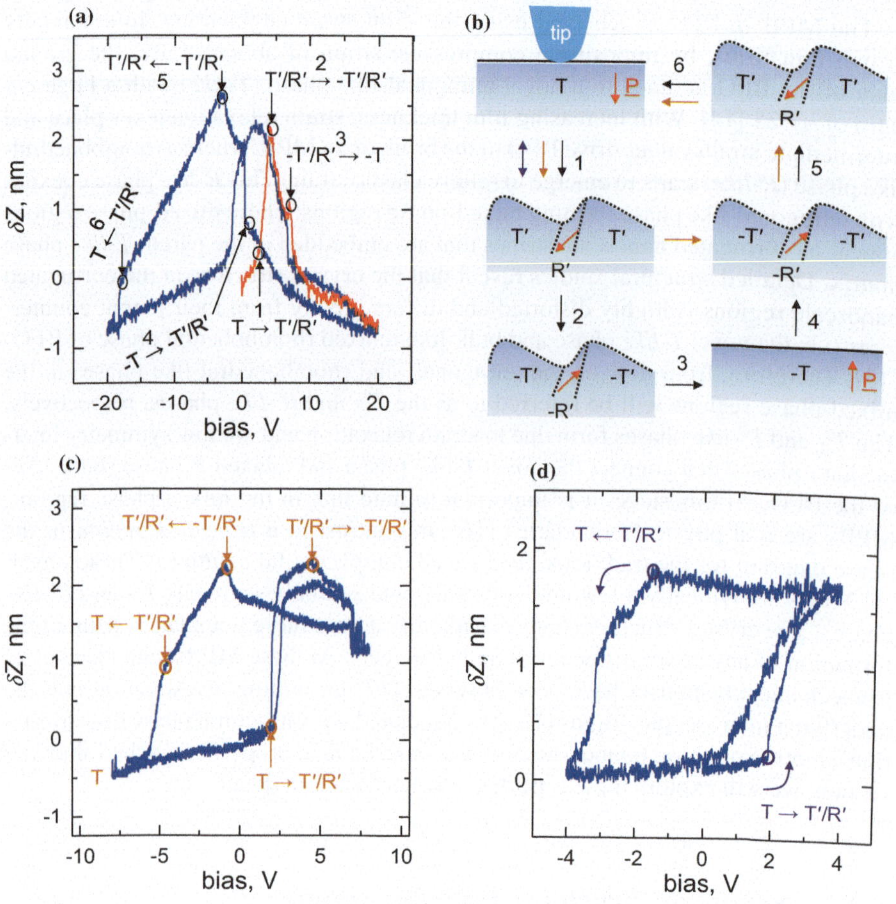

Fig. 6.1 Nanoscale phase transformations and pathways in morphotropic BFO. **a** Local surface displacement as a function of applied external bias (±20 V). **b** Schematic illustration of nanoscale phase transformations and bias-dependent pathways. **c**, **d** Surface displacement versus electrical bias in a range of ±8 V (**c**) and ±4 V (**d**). In **b**, black/orange/purple arrows show sequence of transitions for data shown in **a**/**c**/**d**, respectively. The initial quarter cycle in **a** is shown in red. Data in **a**, **c** and **d** was acquired at a film's location with an initial *T-like* phase [26]

that of polarization. The structural transformations between well-established competing polymorphs (i.e. *T-* and *T′/R′-mixed phase*) are decoded because of significant change in the *out-of-plane* lattice constant and consequently the surface morphology. Therefore, in the acquired response the structural transitions manifest as an abrupt jump from ~0.5 to 1 nm (event # 1, 3, 4 and 6, Fig. 6.1a). The electronic transformations associated with a flip in the direction of the ferroelectric polarization on the other hand are marked by reversal in the sign of the strain (i.e. where δz reverses direction, event # 2 and 5, Fig. 6.1a). Also, the polarization of the *T′/R′-mixed phase* is tilted away from the surface normal compared to the nearly vertical orientation in the *T-like* phase [15, 20, 27]. Therefore, the structural transitions are accompanied by polarization rotation away from the surface normal. The sequence of all transitions (1–6) as inferred from Fig. 6.1a is schematically sketched in Fig. 6.1b.

A few interesting features of observed nanoscale phase transformations are as follows: (i) *T*- to *T′/R′-mixed-phase* structural transition takes place at low bias compared to the inverse transformation, (ii) *T*- to *T′/R′-mixed-phase* structural transition precedes polarization reversal, (iii) Both *T*- to *T′/R′-mixed-phase* and *T′/R′-mixed-phase* to *T-phase* structural phase transformations can be realized by applying bias of the same polarity depending upon its history and magnitude, (iv) morphotropic phase boundary motion takes place due to reversible structural transitions between the completing polymorphs, and (v) the initial *T-like* phase after being subjected to a relatively high bias is no longer stable in the absence of bias and indicates a strong tendency towards the formation of nanoscale mixed-phase region with phase boundaries (after the initial quarter cycle, the subsequent *T*- to *T′/R′-mixed-phase* structural transition occurs at ~0 V). This is possibly due to charge injection from the tip. Their observations underpin the origins of enhanced piezoelectric coupling in morphotropic BFO thin films.

In addition, the number and sequence of these nanoscale phase transformations mediating electromechanical response can be controlled in a ‘dial-in’ manner by tuning the applied external bias. Consequently, the electrically recoverable strain can be regulated. For instance, in the bias range of ±20 V, all possible transitions #1–6 (Fig. 6.1a, b) were realized in a sequential fashion. However, by reducing the applied to ±10 V the pathway of phase transitions can be altered (Fig. 6.1c) to achieve a field-induced stain of ~4%. The new sequence of transitions is highlighted by orange colour arrows in Fig. 6.1b. In this case, four (i.e. two structural transitions and two polarization reversal events) transitions take place in a successive manner. On restricting the applied bias even further (Fig. 6.1d), only two structural transitions take place without any polarization reversal resulting in electrically recoverable strain of about ~3%. These quantitative strain measurements identify structural transitions, MPB kinetics and polarization rotation as the key factors behind the associated observation of large electromechanical responses.

6.2.2 *Elasticity Mapping Across MPBs*

Across physical MPBs, the crystal structure changes, and therefore, it is natural to expect a change in the associated nano-mechanical properties, e.g. elastic modulus. In the mixed-phase regions, the competing crystal phases (i.e. *T′-, and R′-like phase*) appear with a periodicity of few tens of nm. Therefore, unambiguous high-resolution elucidation of nano-mechanical properties across these boundaries has been a challenge. Lately, significant progress in this direction has been made with the use of SPM techniques, e.g. force–volume mapping [28] and ultrasonic force microscopy [29]. Figure 6.2 shows a spatially resolved map of elastic modulus across phase boundaries acquired using force–volume imaging. The force–volume mapping involves the

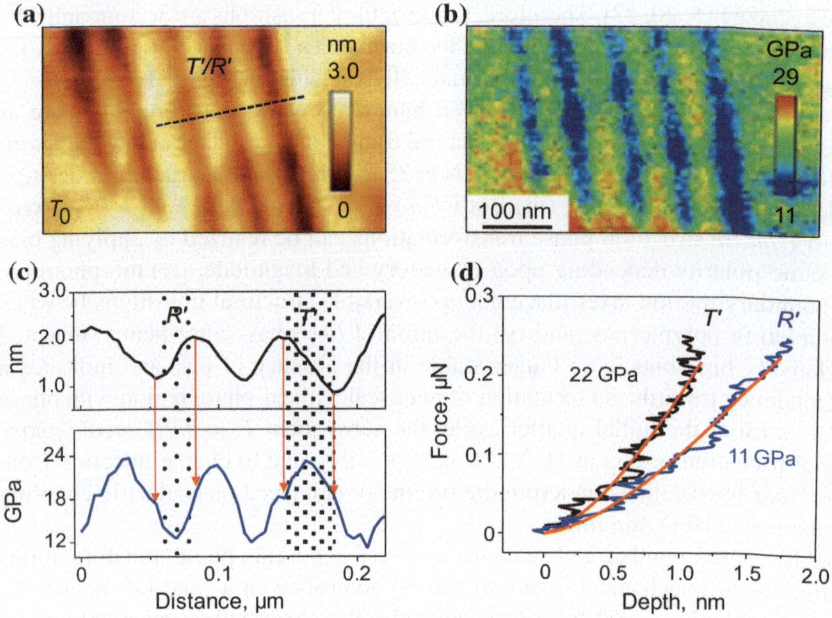

Fig. 6.2 Mapping of the elastic modulus across MPBs. **a**, **b** Topography of La-doped BFO (**a**), and the corresponding 2D map of the elastic modulus (**b**). **c** Cross-sectional profiles of images in **a**, **b** along the dashed-line in **a**. **d** Spectroscopic force versus depth curves on T'- and R'-like phases in the mixed-phase region. Red line in **d** shows fit using Hertzian model. The red arrows in **c** denote structural phase boundaries [28]

acquisition of spectroscopic force–distance curves over a grid of *M X N* points and their subsequent analysis in the elastic regime to quantify elastic modulus. The map shows a distinguishable modulation of elasticity across the physical MPBs (Fig. 6.2a, b). A one-to-one correlation between elasticity and crystalline phases can be established directly by performing a cross-sectional analysis. The asymmetric sawtooth profile of the morphology enables structural phase identification [15] (Fig. 6.2c, top plot). Namely, the R'*-like phase* shows a higher slope compared to the T'*-like phase*. The phase boundaries (i.e. R'/T' *and* T'/R') on the other hand manifest as extrema in this cross-sectional profile. In comparison with the corresponding elasticity profile, the high and low modulus values coincide with T'*- and* R'*-like* phases, respectively. This distinct elastic behaviour of structural phases is also illustrated by force–deformation spectroscopic curves (Fig. 6.2d). A comparatively larger deformation is seen for the R'*-like* phase, and the elastic modulus varies by a factor of ~2 across MPBs. Therefore, the R'*-like* phase is mechanically more compliant than the T'*-like* phase.

The observed trend is consistent with results from ultrasonic force microscopy and modelling results from thermodynamic analysis. A detailed X-ray diffraction strain analysis [29] suggests relatively large misorientation and spacing distortion of the *c*-axis for R'*-like* phase. This implies that the crystal structure of the R'*-like* phase is more ambiguous and thus malleable. Another argument that buttresses this position is based on a structural aspect. As noted before, the T'- and R'-*like* phases are intermediary phases between the T-like and the relaxed R phase of the BFO. Prior high-resolution XRD studies [15] have shown that the crystal structure of the T'-*like phase* (in the mixed-phase region) is the same as the strained *T-phase* but is slightly *tilted* with respect to the surface normal. Therefore, to traverse the large structural disparity between *T-like* and bulk-like R-phase of the BFO, the R'-*like* phase must vary more dramatically [29]. This together with the smaller spatial extent of the R'-*like phase* compared to the T'-*like phase* (Fig. 6.2c, top plot) suggests that the former must be more pliant elastically. However, the opposite behaviour has also been reported for thick (>100 nm) mixed-phase BFO films [30] with modulus values much higher than seen in Fig. 6.2.

Besides, the elastic modulation across phase boundaries correlates with the observation of large piezoelectric responses across such structures [28, 29]. A high piezoelectric response has been observed at phase boundaries compared to either of the crystalline phases one each side of the boundary. This enhancement of the electromechanical response has been attributed to flexoelectricity [29], i.e. a coupling between polarization and strain gradients. The flexoelectric effect scales inversely with length and therefore contributes substantially at the nanoscale. This is precisely the case at phase boundaries, where crystal lattice varies significantly over a length scale of a few nm to few tens of nm. Consequently, at phase boundaries distinct '*kinks*' are seen in the modulus profile (shown by red arrows in Fig. 6.2c). These '*kinks*' point to the presence of strain gradients at phase boundaries and implies flexoelectricity-induced high piezoelectricity at these structures. Further, the slope of the modulus profile for the R'-*like phase* is higher than the one for the T'-*like phase*—suggesting the presence of a larger strain gradient in the R'-*like phase*. This agrees with detected higher piezoresponse for the mechanically compliant R'-*like phase* relative to the neighbouring stiff phase.

6.2.3 *Mechanical Injection of MPBs and Phase Transition Yield Strength*

The manipulation and rewritable dynamic control of MPBs are critical for their potential applications in nanoelectronics. This can be achieved by the application of external stimuli [19, 23, 31, 32]. Here, we will explore the application of nanomechanical force for the injection of MPBs in BFO. Such an approach is supported theoretically by a thermodynamic analysis [28]. This analysis predicts that the structural transition in morphotropic BFO can be accomplished solely with the use of mechanical force, but thus far the magnitudes of applied external mechanical stimuli were below the threshold needed to stimulate phase transition.

To characterize mechanical force-induced phase transition and yield strength, the force–distance curves are acquired over an extended force regime (Fig. 6.3a). These

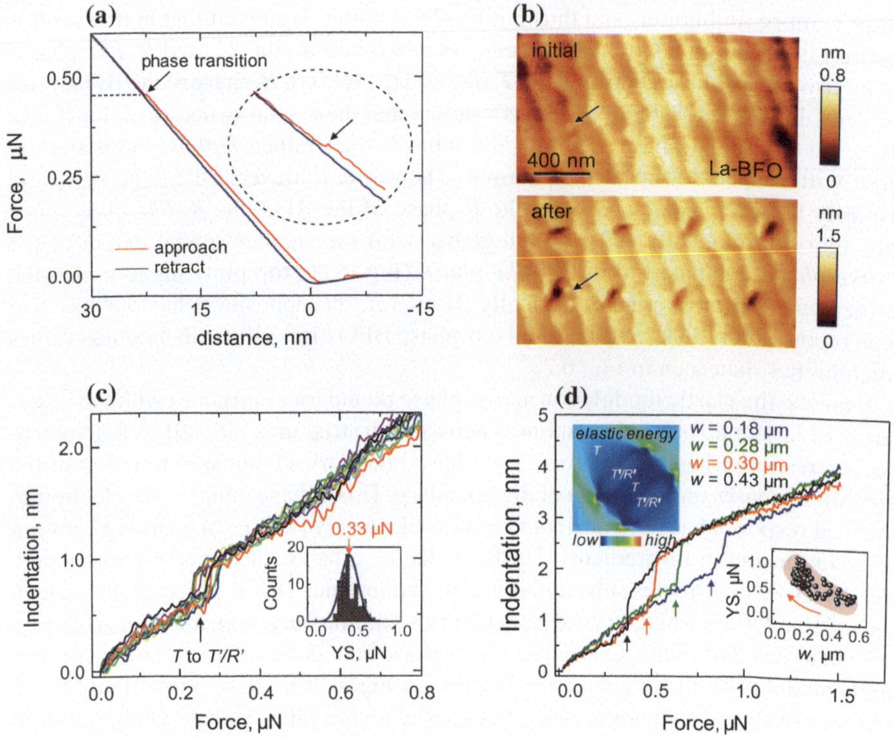

Fig. 6.3 Mechanical force-induced phase transition and transition yield strength. **a** A force–distance spectroscopic curve acquired on *T-like phase*. Inset in **a** shows zoom-in on the phase-transition region. **b** Topography images of the La-BFO before (top image) and after (bottom image) force-induced phase transitions at several spatial locations. **c** Local indentation versus force curves acquired at several distinct locations on *T-like phase*. Inset in **c** shows the histogram of phase transition yield strength (YS). The blue line in inset **c** represents a Gaussian fit. **d** Indentation versus force curves on *T-like phase* of different widths (w) enclosed within mixed-phase striped regions. Insets in **d** bottom right show scaling of phase transition yield strength (YS) with respect to width of enclosed *T-like phase*. Inset in **d** top left shows phase-field modelled map of elastic energy for a configuration in which the *T*-like phase is enclosed within two mixed-phase stripes. The results presented in this figure are acquired on a La-doped BFO film [28]

spectroscopic curves are acquired on the initial *T-like phase* of BFO (Fig. 6.3b, top image). The nano-mechanical force-induced phase transformation from *T*- to *T′/R′-like* phase appears as tip '*sink-in*' (i.e. discontinuity at ~0.44 µN, Fig. 6.3a) during force ramp-up [28, 33]. As a result, a gap arises between approaches and retracts traces for force magnitudes (in contact) below the phase transition. The tip '*sink-in*' represents the onset of the phase transition and the mechanical force needed to achieve it is the phase transition yield strength. These structural transitions can be explicitly visualized in the morphology (Fig. 6.3b). For instance, in Fig. 6.3b (bottom image) mechanical force caused local structural transitions are visible at eight (i.e. local force–distance curves are acquired at these locations) distinct spatial locations.

In the morphology, these transitions appear as a well-defined isolated dark stripe (i.e. *T′/R′-like* phase). Therefore, using mechanical stress, it is feasible to induce the phase transformation and insert localized isolated MPBs on demand. This is in contrast to external electric fields, where it is challenging to create isolated MPBs. This is due to field lines typically extending up to several hundred nm's away from the tip–sample contact. Here, the nanoscale SPM tip exerts highly localized mechanical pressures (tip radius of curvature ~30 nm) and results in structural transitions directly underneath the tip. These mechanical written MPBs are still responsive to electric or thermal stimulus and can be restored back to the initial state.

Next, the force–distance curves are analysed as indentation (depth) versus force to simplify the identification of transition yield strength (Fig. 6.3c). In these curves, at transition yield strength, there arises a distinct '*step-like feature*' in the indentation. Clearly, there is a slight spread in value of transition yield strength. To gain statistical information and get the average transition yield strength, these measurements are repeated. The observed distribution of yield strength is shown as inset in Fig. 6.3c. The distribution peaks at 0.330 ± 0.020 μN, representing the average phase transition yield strength of the *T-like phase* (60 nm La-doped BFO film). The phase transition yield strength is commonly given in units of pressure and can be calculated as: $F_{\text{th}}/\pi r^2$, where F_{th} is threshold mechanical force for phase transition and r is the radius of curvature of the tip (~30 nm). Thus, the average transition yield strength of the *T-like phase* for a morphotropic La-doped BFO film is ~116 $\pm$ 7 MPa. Additionally, the effect of doping on transition yield strength has also been explored in detail [34]. Relative to the pure BFO, the transition yield strength decreases by over 50% with isovalent Lanthanum dopant, whereas it increases by ~15% with aliovalent calcium.

Further, the scaling behaviour of transition yield strength with respect to in-plane dimensions of *T-like phase* has been addressed. For this, spectroscopic force–distance measurements are performed on *T-like phase* stripes of different widths (w) that are embedded within the mixed-phase regions (see top left inset, Fig. 6.3d). The transition yield strength or the threshold force increases (from 0.3 to 0.8 μN, Fig. 6.3d, also see inset) with decreasing width of *T-like phase*. A stronger pinning of *T-like phase* with decreasing width by the surrounding mixed-phase stripes has been argued as a key factor governing this increase. This scaling behaviour can also be understood by considering the changes in the elastic energy. Phase-field simulations (inset top left, Fig. 6.3d) show that the elastic energy of the *T-like phase* decreases with width, i.e. the elastic energy of the *T-like phase* surrounded by *T′/R′-like phase* stripes is low compared to the enclosing *T-like phase*. Considering this, the stable phase regions, i.e. *T*- and *T′/R′*-mixed phase, can be represented by a double-well potential. In such a situation, the energy barrier is proportional to the transition yield strength. With decreasing width, the elastic energy of the *T-like phase* decreases, resulting in a deeper well for the phase. Consequently, the barrier height for phase transformation increases and a high mechanical force (transition yield strength) is required to trigger this transition.

Precise injection of individual phase boundaries using a force stimulus also opens a possibility to dynamically regulate electronic characteristics of morphotropic BFO

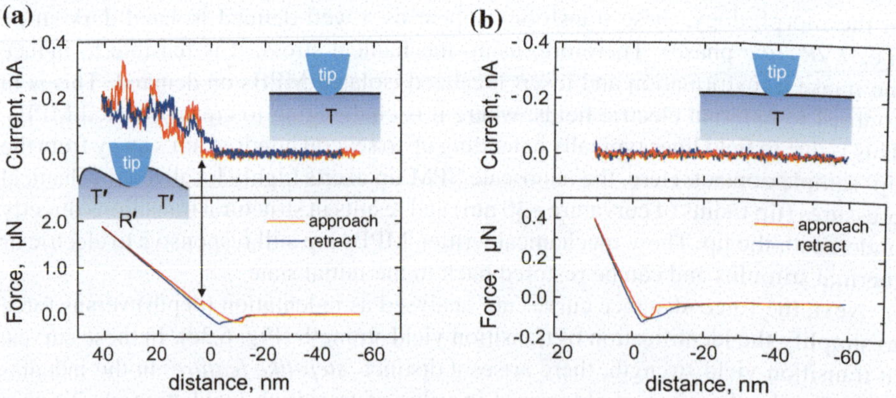

Fig. 6.4 Mechanical control of electrical properties of morphotropic La-doped BFO. **a, b** A simultaneous recorded electrical current (top plot) during acquisition of force–distance curve (bottom plot) with force values ramped up to above (**a**), and below (**b**), the phase transition yield strength of *T-like phase*. During these measurements, tip was maintained at a fixed DC bias of −3 V (below ferroelectric switching threshold), and electrical current is recorded across BFO between tip and bottom electrode [28]

at the nanoscale [28]. This concept is achieved by using a conductive SPM tip to both mechanically press and simultaneously monitor electrical current across BFO (Fig. 6.4). Thus, during acquisition of force–distance curves the tip is maintained at a fixed low DC bias, which is below the critical bias needed to switch polarization and inject MPBs. As before, a nano-mechanical force drives structural transition at a threshold value of ~0. 31 μN (Fig. 6.4a, top plot). Concurrently, at the structural transition the conductivity increases dramatically in a non-volatile fashion (Fig. 6.4a, bottom plot). This sharp increase in the conductivity arises from the creation of the stable phase boundaries which depending upon their type (i.e. *T′/R′* and *R′/T′*) have been shown to display distinct electrical properties [20, 21, 28]. A simultaneous fast '*step-like*' increase in electrical current implies the electronic nature of conduction at these isosymmetric phase boundaries. Also, to note, no such changes in the conductivity are seen if the applied mechanical force is below the local threshold needed to induce the phase transition (Fig. 6.4b).

6.2.4 Elastic Anomalies During Phase Transitions

Thus far, we reviewed elastic properties of the crystalline phases in morphotropic BFO. The elastic anomalies associated with structural phase transformations between thermodynamically stable competing crystalline phases of BFO remain unexplored. The phase transitions and the coupled elastic properties can be characterized using a contact resonance-based piezoresponse force microscopy (PFM) approach called as dual AC resonance tracking (DART) [35]. DART-PFM detects dynamic piezoresponse and involves excitations of the electromechanical active surface at frequencies close to the tip–sample contact resonance. SPM tip in contact with the surface is electrically driven with a linear combination of sinusoidal voltages, one below and one above the contact resonance frequency. The approach increases the sensitivity of the standard PFM by Q-factor of the resonance and allows the resonance frequency to be tracked. The contact resonance frequency is a function of local mechanical properties of the sample such as elastic modulus. One of the most prominent features of any structural transition is the accompanying mechanical softening. The elastic tunability during phase transition therefore can form the basis for the detection of such transitions using dynamic contact resonance-based atomic force microscopy techniques, e.g. DART-PFM [36]. The elastic softening during transition decreases AFM tip–sample contact stiffness, which, in turn, will shift the tip–sample contact resonance frequency dynamically to lower values. Therefore, a shift or decrease in the contact resonance frequency will signify the occurrence of structural phase transformation and the associated elastic softening of the underlying structural phase in the probed nanoscopic volume [36]. As a result, changes in tip–sample resonance frequency, provided the tip–sample contact area remains unchanged, will signify elastic variations in the underlying materials volume. Such changes in elastic properties manifested in the form of a resonance frequency shift can be quantified with the help of appropriate models for cantilever dynamics and tip–sample contact mechanics [36–38].

In addition, the spectroscopic mode of the DART-PFM, in which a low amplitude high-frequency (near resonance as described above) AC voltage is superimposed on a slowly varying DC waveform, allows simultaneous recording of resonance frequency, dynamic and static piezoresponse as a function of the applied DC bias. The simultaneous recorded spectroscopic signals for a morphotropic La-doped BFO thin film (thickness = 22 nm) are shown in Fig. 6.5. Figure 6.5a shows static piezoresponse and visualizes electrically induced reversible phase transformation between competing for stable crystalline phases (i.e. *T-like* and *T′/R′-mixed phase*) of the BFO. As before, for instance in Fig. 6.1d, at phase transitions the static piezoresponse shows a distinct jump. The hysteretic static piezoresponse in Fig. 6.1d is flipped compared to that shown in Fig. 6.5a because of the reversed polarity of the applied field (in Fig. 6.1, the bias is supplied to the bottom electrode of the sample and tip is grounded, whereas in Fig. 6.5, the tip is biased, and bottom electrode of the sample is grounded). Note that the applied bias range in Fig. 6.5 is chosen to be below the polarization switching threshold of the BFO film. From Fig. 6.5a, the electrically recoverable strain can be calculated as, $\varepsilon = 7.5\%$, using equation $\varepsilon = \delta Z_{max}/L$, where L is film thickness, which is 22 nm. This giant piezoelectricity is attributed to nanoscopic phase transitions and the accompanying slight polarization rotation between the competing crystalline phases. Further, the La-doped BFO film thickness is right at the brink where *T′/R′-mixed-phase* areas in the as-prepared state can spontaneously appear (we will explore this in more detail in the next section).

The simultaneously recorded bias-dependent evolution of contact resonance frequency and high-frequency PFM response is shown in Fig. 6.5b–d. The resonance

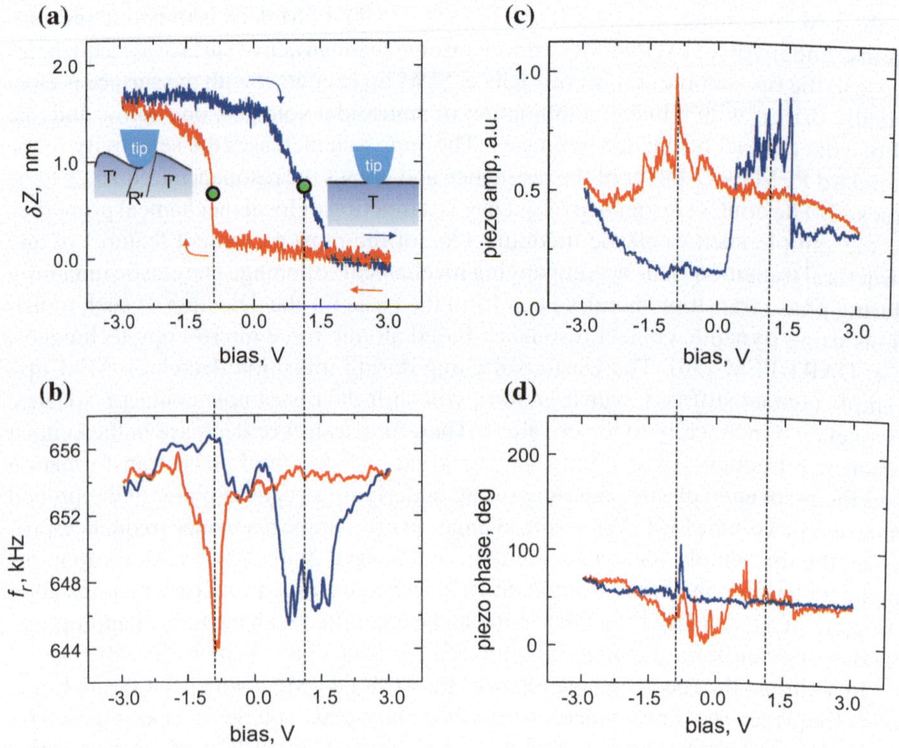

Fig. 6.5 Elastic anomalies during morphotropic structural phase transitions. **a** Spectroscopic local surface displacement as a function bias. **b–d** A simultaneously recorded tip–sample contact resonance frequency (**b**), dynamic piezoresponse amplitude (**c**) and phase response (**d**), versus applied bias. Inset in **a** shows a schematic illustration of nanoscale phase transformations [36]

frequency displays a hysteretic behaviour and shows noticeable sharp changes. Interestingly, these sharp changes emerge during crystalline phase transformations only. During phase transition, the resonance frequency decreases by ~10 kHz due to mechanical softening of the La-BFO. Once the transition is complete, the resonance frequency increases to nearly the same value as before. The increase in resonance frequency shows the hardening of the La-BFO film. The acquired high-frequency electromechanical response (Fig. 6.5c) shows enhanced piezoelectric coupling during nanoscale structural transitions. We reiterate that the polarization does not switch within the applied bias window of ±3 V. This conclusion is supported by observing no significant piezoresponse phase changes (Fig. 6.5d). Therefore, intrinsic rather than extrinsic lattice contributions such as softening of the crystal lattice, activated morphotropic phase boundary kinetics and pathways for the evolution of polarization among competing structural phases govern the enhanced piezoelectricity of the (22 nm thick) La-BFO during crystalline phase transitions. The extrinsic contributions include polarization switching and ferroelectric domain wall motion.

In order to explicitly visualize polarization switching and rule out extrinsic factors, similar spectroscopic curves can be acquired over an extended bias range (Fig. 6.6a–d). In these curves, the response is acquired for one half of a bias cycle (from −5.5 to

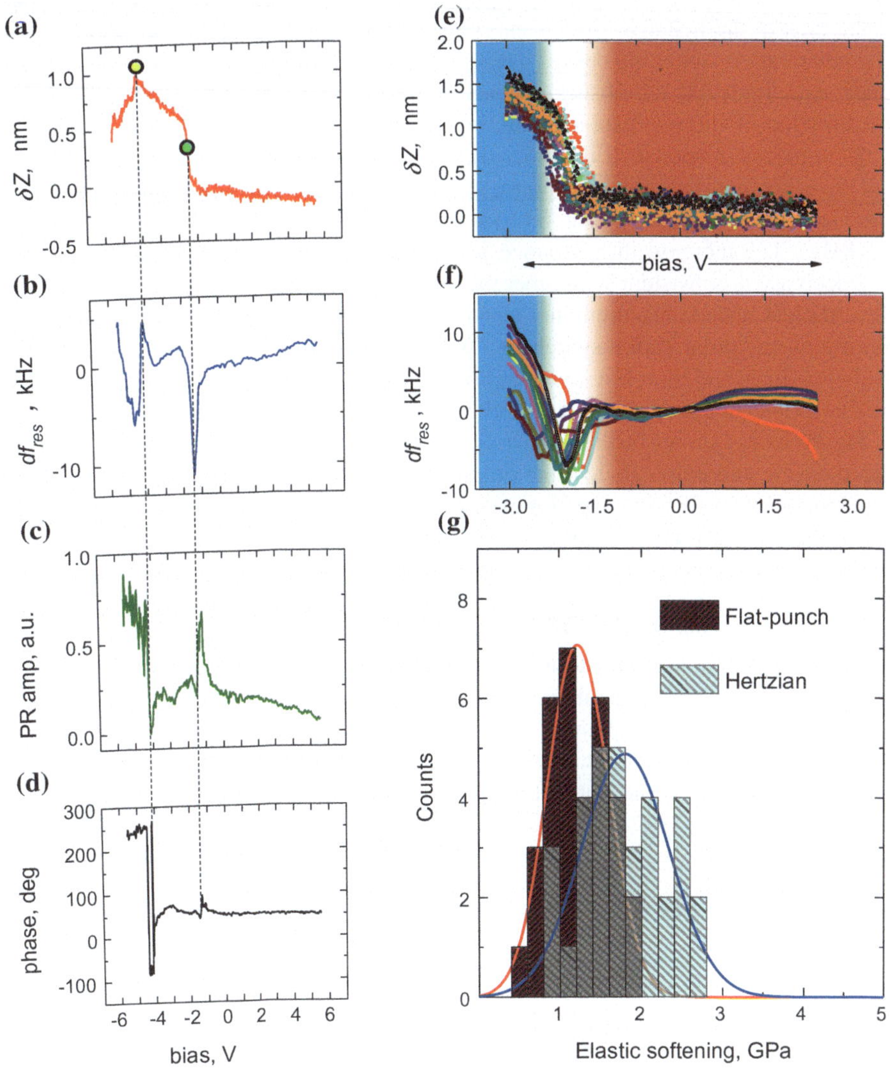

Fig. 6.6 Estimation of elastic softening during the structural transition. **a**–**d** A simultaneously acquired local surface displacement (**a**), contact resonance frequency shift (**b**), dynamic piezoresponse amplitude (**c**) and phase response (**d**), versus applied bias. **e**, **f** Colour-coded curves of local surface displacement (**e**) and contact resonance frequency shift (**f**), versus bias acquired at different spatial locations. **g** Histogram of the calculated elastic softening assuming Hertzian contact and flat-punch tip geometry. In **g**, solid red and blue lines indicate Gaussian fits. The shift in contact resonance frequency is defined as: $df_{res} = [f_{res}(V) - f_{res}(V = 0)]$ [36]

+5.5 V). Clearly, the phase transition from *T*-like phase to *T′/R′-mixed phase* takes place at ~−1.5 V. At the phase transition, as before, static piezoresponse shows a sharp jump and a concomitant dip in the resonance frequency occurs due to mechanical softening. At the same time, the high-frequency piezoresponse amplitude shows a peak and the corresponding phase signal remains unaffected. With further increase of the bias in the negative direction, polarization switching occurs at about −4 V. The polarization switching causes a reversal in sign of the acquired strain both in the quasi-static limit and at high-frequencies: (i) for the quasi-static response (Fig. 6.6a), the direction of surface displacement, i.e. δZ, reverses direction, (ii) whereas in the high-frequency response (Fig. 6.6c, d), the amplitude goes to a minimum value and the corresponding phase response shows a 180° flip. The simultaneously monitored resonance frequency (Fig. 6.6b) shows a small anomaly in the form of a peak (~3 kHz). The increase in resonance just before the switching threshold followed by a decrease is consistent with other previously reported experimental studies [39]. Nevertheless, these measurements reveal distinct onset biases for crystalline phase transitions and polarization switching. The phase transition takes place at comparatively low bias, and therefore contributions of extrinsic factors in the observed giant piezoresponse in the applied bias window of ±3 V can be ruled out (Fig. 6.5).

Figure 6.6e, f shows further statistics on bias-induced phase transitions and associated elastic anomalies (below the polarization switching threshold). Using appropriate models [36, 38], the actual mechanical softening of La-BFO *during* the phase transition can be estimated quantitatively (Fig. 6.6g). From Gaussian fits to calculated modulus change values, the elastic modulus softens by ≈1.93 ± 0.532 GPa (1.31 ± 0.3667 GPa in the flat-punch contact) assuming Hertzian contact [38] during the phase transition. Therefore, La-BFO softens by ≈6–10% during this crystalline phase transition, which contributes to its superior electromechanical response. Using this approach, phase transitions and coupled elasticity modulations can be detected and probed in a nanoscopic volume (≈10^{-24} m^3) directly beneath the AFM tip. Moreover, this approach is quite general and in principle is applicable to a wide range of electro-active systems and not just to piezoelectrics or ferroelectrics.

6.2.5 *Critical MPB*

BFO thin films with coexisting *T*- and *T′/R′*-like phases exhibit a high electromechanical response about a factor of 2–4 larger than the one for BFO films in the bulk-like rhombohedral phase and pure strained *T*-phase [17]. However, a critical degree of freedom that remains unexplored and analysed is the film thickness scaling of the response, which is known to control strain relaxation and the evolution of competing structural phases [40]. For undoped BFO thin films, the strain relaxation for the *T*-phase starts to occur at a thickness (t_c) of around 30 nm, while for La-doped BFO thin films, the chemical pressure exerted by La reduces this critical thickness to around 20 nm (Fig. 6.7a). A systematic investigation of quasi-static and high-frequency electromechanical response of morphotropic La-BFO as a

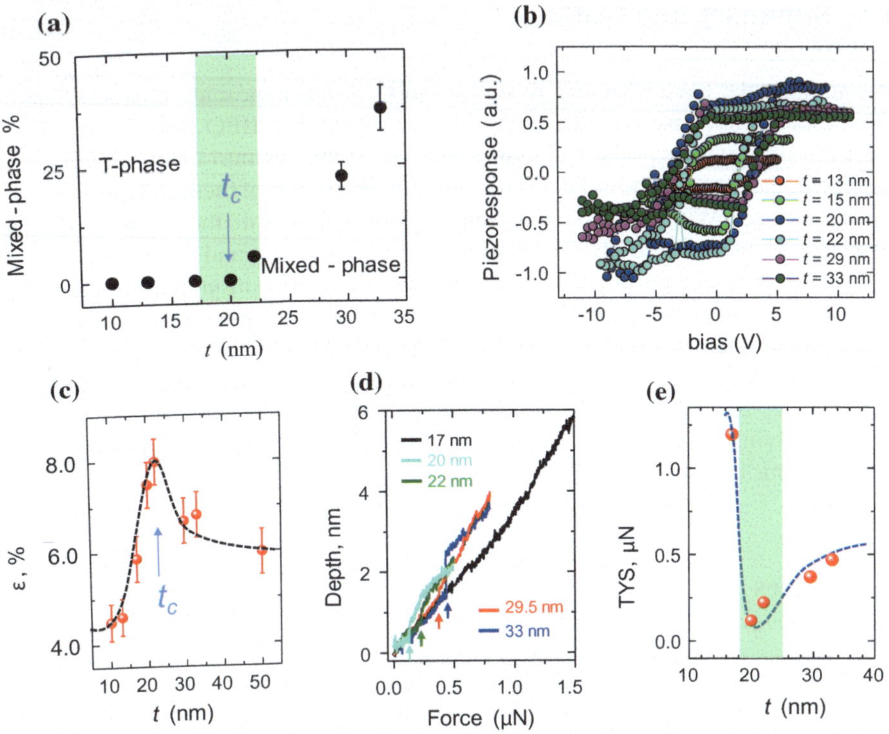

Fig. 6.7 Thickness-dependent critical MPB. **a** Areal fraction of the mixed-phase areas versus thickness of La-doped BFO films estimated from AFM topography images. **b** Dynamic piezoresponse measurements for La-BFO films as a function of thickness. **c** Electrically recoverable reversible stain (ε) as a function of thickness obtained from quasi-static piezoresponse measurements. **d** Depth versus force spectroscopic curves showing force-induced structural transition (marked by arrows from *T*-like phase to mixed-phase) for different thicknesses (17–33 nm). **e** Phase transition yield strength as a function of film thickness [40]

function of thickness (10–50 nm) reveals superior piezoelectricity at the crossover region between strained *T*-like phase and mixed-phase states (Fig. 6.7b, c). This crossover region exists in a small thickness window around t_c, at which nanoscale mixed-phase stripes first begin to spontaneously appear in the as-grown BFO. In this region, referred to as the critical MPB, the observed response is significantly enhanced compared to either of the surrounding regions (i.e. for *T*-like phase $t < t_c$; thick mixed-phase films $t > t_c$) and is approximately 40% higher than the previously reported results in mixed-phase morphotropic BFO ($t > t_c$). This enhancement is ascribed to an energetic degeneracy of the mixed-phase states and an increased mechanical softening (Fig. 6.7d, e) at the cross-over region [40].

6.3 Summary and Outlook

In summary, we have reviewed recent progress in the nanoscale characterization of morphotropic phase boundaries in strained crystalline BFO thin films using a spectrum of scanning probe techniques that allow insight into electromechanical, electronic, mechanical and coupled electroelastic properties of these responsive functional interfaces. With this approach, a microscopic insight into the factors underpinning superior piezoelectricity, correlations between elasticity and crystalline phases, electrically and mechanically activated crystalline phase transitions and coupled electroelastic and electronic anomalies is obtained. Detailed information of microstructural aspects, interactions and coupled electroelastic and electronic properties at the MPBs along with their rewritable control using electrical or mechanical activation is a critical step towards the realization of related applications, analogous to developments involving ferroelectric domain walls [41, 42] in the field of sensors [43, 44] and electronic devices [45–47].

References

1. A. Ohtomo, H.Y. Hwang, Nature **427**, 423 (2004)
2. Y.Y. Pai, A. Tylan-Tyler, P. Irvin, J. Levy, Rep. Prog. Phys. **81**, 036503 (2018)
3. E.Y. Tsymbal, E.R. Dagotto, C.-B. Eom, R. Ramesh, *Multifunctional Oxide Heterostructures* (Oxford University Press, Oxford, 2012)
4. H. Lu, C. W. Bark, D. Esque de los Ojos, J. Alcala, C. B. Eom, G. Catalan, A. Gruverman, Science **336**, 59 (2012)
5. P. Sharma, Z. Huang, M.S. Li, C.J. Li, S.B. Hu, H. Lee, J.W. Lee, C.B. Eom, S.J. Pennycook, J. Seidel, Ariando, and A. Gruverman. Adv. Funct. Mater. **28**, 1707159 (2018)
6. P. Sharma, S. Ryu, Z. Viskadourakis, T.R. Paudel, H. Lee, C. Panagopoulos, E.Y. Tsymbal, C.B. Eom, A. Gruverman, Adv. Funct. Mater. **25**, 6538 (2015)
7. P. Sharma, S. Ryu, J. D. Burton, T. R. Paudel, C. W. Bark, Z. Huang, Ariando, E. Y. Tsymbal, G. Catalan, C. B. Eom, A. Gruverman, Nano Lett. **15**, 3547 (2015)
8. M.E. Lines, A.M. Glass, *Principles and Applications of Ferroelectrics and Related Materials* (Oxford University Press, Oxford, 2001)
9. B. Jaffe, *Piezoelectric Ceramics*, vol. 3 (Elsevier, London, New York, 2012)
10. B. Noheda, J.A. Gonzalo, L.E. Cross, R. Guo, S.E. Park, D.E. Cox, G. Shirane, Phys. Rev. B **61**, 8687 (2000)
11. S.E. Park, T.R. Shrout, J. Appl. Phys. **82**, 1804 (1997)
12. Z. Wu, R.E. Cohen, Phys. Rev. Lett. **95**, 037601 (2005)
13. M. Ahart, M. Somayazulu, R.E. Cohen, P. Ganesh, P. Dera, H.K. Mao, R.J. Hemley, Y. Ren, P. Liermann, Z. Wu, Nature **451**, 545 (2008)
14. R.J. Zeches, M.D. Rossell, J.X. Zhang, A.J. Hatt, Q. He, C.H. Yang, A. Kumar, C.H. Wang, A. Melville, C. Adamo, G. Sheng, Y.H. Chu, J.F. Ihlefeld, R. Erni, C. Ederer, V. Gopalan, L.Q. Chen, D.G. Schlom, N.A. Spaldin, L.W. Martin, R. Ramesh, Science **326**, 977 (2009)
15. A.R. Damodaran, C.W. Liang, Q. He, C.Y. Peng, L. Chang, Y.H. Chu, L.W. Martin, Adv. Mater. **23**, 3170 (2011)
16. J.X. Zhang, R.J. Zeches, Q. He, Y.H. Chu, R. Ramesh, Nanoscale **4**, 6196 (2012)
17. J.X. Zhang, B. Xiang, Q. He, J. Seidel, R.J. Zeches, P. Yu, S.Y. Yang, C.H. Wang, Y.H. Chu, L.W. Martin, A.M. Minor, R. Ramesh, Nat. Nanotechnol. **6**, 98 (2011)

18. R.K. Vasudevan, M.B. Okatan, Y.Y. Liu, S. Jesse, J.C. Yang, W.I. Liang, Y.H. Chu, J.Y. Li, S.V. Kalinin, V. Nagarajan, Phys. Rev. B **88**, 020402 (2013)
19. J. Zhou, M. Trassin, Q. He, N. Tamura, M. Kunz, C. Cheng, J.X. Zhang, W.I. Liang, J. Seidel, C.L. Hsin, J.Q. Wu, J. Appl. Phys. **112**, 064102 (2012)
20. K.E. Kim, B.K. Jang, Y. Heo, J.H. Lee, M. Jeong, J.Y. Lee, J. Seidel, C.H. Yang, NPG Asia Mater. **6**, e81 (2014)
21. J. Seidel, M. Trassin, Y. Zhang, P. Maksymovych, T. Uhlig, P. Milde, D. Köhler, A.P. Baddorf, S.V. Kalinin, L.M. Eng, X. Pan, R. Ramesh, Adv. Mater. **26**, 4376 (2014)
22. R.K. Vasudevan, Y.Y. Liu, J.Y. Li, W.I. Liang, A. Kumar, S. Jesse, Y.C. Chen, Y.H. Chu, V. Nagarajan, S.V. Kalinin, Nano Lett. **11**, 3346 (2011)
23. Y.C. Chen, Q. He, F.N. Chu, Y.C. Huang, J.W. Chen, W.I. Liang, R.K. Vasudevan, V. Nagarajan, E. Arenholz, S.V. Kalinin, Y.-H. Chu, Adv. Mater. **24**, 3070 (2012)
24. P.S.S.R. Krishnan, J.A. Aguiar, Q.M. Ramasse, D.M. Kepaptsoglou, W.I. Liang, Y.H. Chu, N.D. Browning, P. Munroe, V. Nagarajan, J. Mater. Chem. C. **3**, 1835 (2015)
25. A.R. Damodaran, E. Breckenfeld, A.K. Choquette, L.W. Martin, Appl. Phys. Lett. **100**, 082904 (2012)
26. P. Sharma, Y. Heo, B.K. Jang, Y.Y. Liu, J.Y. Li, C.H. Yang, J. Seidel, Sci. Rep. **6**, 32347 (2016)
27. L. You, Z. Chen, X. Zou, H. Ding, W. Chen, L. Chen, G. Yuan, J. Wang, ACS Nano **6**, 5388 (2012)
28. P. Sharma, Y. Heo, B.K. Jang, Y.Y. Liu, V. Nagarajan, J.Y. Li, C.H. Yang, J. Seidel, Adv. Mater. Interfaces **3**, 1600033 (2016)
29. C.E. Cheng, H.J. Liu, F. Dinelli, Y.C. Chen, C.S. Chang, F.S. Chien, Y.H. Chu, Sci. Rep. **5**, 8091 (2015)
30. Y.J. Li, J.J. Wang, J.C. Ye, X.X. Ke, G.Y. Gou, Y. Wei, F. Xue, J. Wang, C.S. Wang, R.C. Peng, X.L. Deng, Y. Yang, X.B. Ren, L.Q. Chen, C.W. Nan, J.X. Zhang, Adv. Funct. Mater. **25**, 3405 (2015)
31. W. Siemons, M.D. Biegalski, J.H. Nam, H.M. Christen, Appl. Phys. Express **4**, 095801 (2011)
32. D. Mazumdar, V. Shelke, M. Iliev, S. Jesse, A. Kumar, S.V. Kalinin, A.P. Baddorf, A. Gupta, Nano Lett. **10**, 2555 (2010)
33. Y. Heo, B.K. Jang, S.J. Kim, C.H. Yang, J. Seidel, Adv. Mater. **26**, 7568 (2014)
34. Y. Heo, S. Hu, P. Sharma, K.E. Kim, B.K. Jang, C. Cazorla, C.H. Yang, J. Seidel, ACS Nano **11**, 2805 (2017)
35. B.J. Rodriguez, C. Callahan, S.V. Kalinin, R. Proksch, Nanotechnology **18**, 475504 (2007)
36. P. Sharma, K.R. Kang, B.K. Jang, C.H. Yang, J. Seidel, Adv. Electron. Mater. **2**, 1600283 (2016)
37. U. Rabe, K. Janser, W. Arnold, Rev. Sci. Instrum. **67**, 3281 (1996)
38. D. C. Hurley, in *Scanning Probe Microscopy of Functional Materials*, ed. by S. V. Kalinin, A. Gruverman (Springer, New York 2010)
39. S. Jesse, P. Maksymovych, S.V. Kalinin, Appl. Phys. Lett. **93**, 112903 (2008)
40. P. Sharma, K.R. Kang, Y.Y. Liu, B.K. Jang, J.Y. Li, C.H. Yang, J. Seidel, Nanotechnology **29**, 205703 (2018)
41. J. Seidel, G. Singh-Bhalla, Q. He, S.Y. Yang, Y.H. Chu, R. Ramesh, Phase Transit **86**, 53 (2013)
42. J. Seidel, J. Phys. Chem. Lett. **3**, 2905 (2012)
43. P. Zubko, J.C. Wojdel, M. Hadjimichael, S. Fernandez-Pena, A. Sene, I. Luk'yanchuk, J.M. Triscone, J. Iniguez, Nature **534**, 524 (2016)
44. Z. Gu, S. Pandya, A. Samanta, S. Liu, G. Xiao, C.J.G. Meyers, A.R. Damodaran, H. Barak, A. Dasgupta, S. Saremi, A. Polemi, L. Wu, A.A. Podpirka, A. Will-Cole, C.J. Hawley, P.K. Davies, R.A. York, I. Grinberg, L.W. Martin, J.E. Spanier, Nature **560**, 622 (2018)

45. P. Sharma, Q. Zhang, D. Sando, C.H. Lei, Y. Liu, J. Li, V. Nagarajan, J. Seidel, Sci. Adv. **3**, e1700512 (2017)
46. P. Sharma, D. Sando, Q. Zhang, X.X. Cheng, S. Prosandeev, R. Bulanadi, S. Prokhorenko, L. Bellaiche, L.Q. Chen, V. Nagarajan, J. Seidel, Adv. Funct. Mater. **29**, 1807523 (2019)
47. G. Sanchez-Santolino, J. Tornos, D. Hernandez-Martin, J.I. Beltran, C. Munuera, M. Cabero, A. Perez-Munoz, J. Ricote, F. Mompean, M. Garcia-Hernandez, Z. Sefrioui, C. Leon, S.J. Pennycook, M.C. Munoz, M. Varela, J. Santamaria, Nat. Nanotechnol. **12**, 655 (2017)

Chapter 7
Heat and Electro-Responsive Nanomaterials for Smart Windows

Jiadong Qin and Yu Lin Zhong

Abstract Energy-saving buildings have drawn increasing interest worldwide in the past 30 years, during which the growing population and expanding urbanization significantly increased the energy intensity of numerous cities. In the modern energy-saving buildings, smart windows are playing an important role in the efficient utilization of daylight and the intelligent control of heat exchange between indoor and outdoor, eventually reducing the energy waste associated with lighting and air-conditioning. The "intelligence" of smart windows originates from the responsive materials of which the optical properties are adaptive to temperature or applied voltage. Recently, the development of smart windows has been greatly motivated by the burgeoning nanomaterials. This chapter focuses on the development of heat and electro-responsive nanomaterials-based smart windows which outperform the conventional ones and, more importantly, likely to cost less for commercialization.

7.1 Introduction

The residential, public and commercial buildings consume considerable energy for heating, ventilation and air conditioning (HVAC) but more than 50% of the energy use is wasted due to the unwanted heat exchange through the conventional windows [1]. Additionally, due to the poor utilization of daylight, the maintenance of interior brightness still relies on the lightings even in the daytime. To reduce the unnecessary energy waste, smart windows are regarded as an efficient and promising energy conservation technology as they can dynamically adapt the heat flux and the light transmission to the changing exterior temperature and sunlight intensity. Furthermore, smart windows, through flexible switchability between a transparent state and an opaque state, can provide privacy protection and the control over indoor brightness

J. Qin · Y. L. Zhong (✉)
Centre for Clean Environment and Energy, School of Environment and Science, Griffith University, Gold Coast Campus, Gold Coast, QLD 4222, Australia
e-mail: y.zhong@griffith.edu.au

Z. Sun and T. Liao (eds.), *Responsive Nanomaterials for Sustainable Applications*, Springer Series in Materials Science 297,
https://doi.org/10.1007/978-3-030-39994-8_7

and temperature. Therefore, smart windows will eliminate the needs for curtains or blinds and provide additional controls.

The core of smart windows lies in the chromogenic materials whose transmittance is in response to the external stimuli, such as temperature (thermochromic) and applied voltage (electrochromic). Currently, thermochromic and electrochromic materials are playing the dominant roles in the smart windows. The temperature-responsive thermochromic windows are one of the most-studied smart windows as they can reduce the influx of heat in hot weather but welcome the heat from sunlight in cold weather. The principle is that the thermochromic materials can undergo a reversible phase transition or chemical reaction at a well-defined critical temperature (T_c). This will significantly change their optical properties to the near-infrared (NIR) light in the spectral range of 780–2500 nm, which is the predominant contributor of solar heat. They can block the NIR light at high temperature above T_c, whereas allowing the traverse of IR light at low temperature below T_c (as illustrated in Fig. 7.1a). Therefore, thermochromic windows with well-controlled heat exchange ability are helpful to reduce the energy consumption by HVAC. Electrochromic windows are another attractive type of smart windows. They can reversibly switch between coloured and bleached state, responsive to the applied voltage. A typical electrochromic window is based on an electrochemical cell structure where the active electrochromic components and the electrolyte layer are sandwiched between two transparent conducting electrodes (TCEs), as shown in Fig. 7.1b. When the device is switched on, the external voltage propels the migration of electric charges from the counter electrode, through the electrolyte, to the electrochromic electrode, giving rise to the change in its oxidation state and hence, the corresponding colouration [2]. When switched off, electric charges move back to the counter electrode for charge balance, and the electrochromic electrode concurrently reverts to its original bleached state. Electrochromic windows behave like an electrical curtain which facilitates the flexible broadband modulation of light transmittance [3].

Commercial smart windows have already emerged in the market. One of the best-known examples is the electrochromic windows employed in the Boeing 787 Dreamliner aeroplane, which can gradually darken under the external bias. Though advantageous in the improved energy conservation and dynamic energy management, smart windows fail to make a greater impact in the building market at the present stage because the conventional chromogenic materials are limited by high cost, inconsistent chromogenic ability as well as questionable working life. Fortunately, the recent promising researches in nanomaterials are likely to revive the commercialized smart windows in the near future. The functionality, reliability and stability of smart windows can be further improved by nanomaterials and nanostructuring via the facile and precise controls of morphology, nanocrystal size, surface chemistry and composition. Also, the employment of nanomaterials will benefit the scalable and cost-efficient manufacturing of smart windows. This is because a wide range of nanomaterials and their synthesis routes are solution-processable and free from high-temperature treatment [3]. Apart from the conventional methods (roll-to-roll printing and bar coating), many chromogenic nanomaterials can be deposited onto glass to form the "intelligently" responsive layers using the three-dimensional

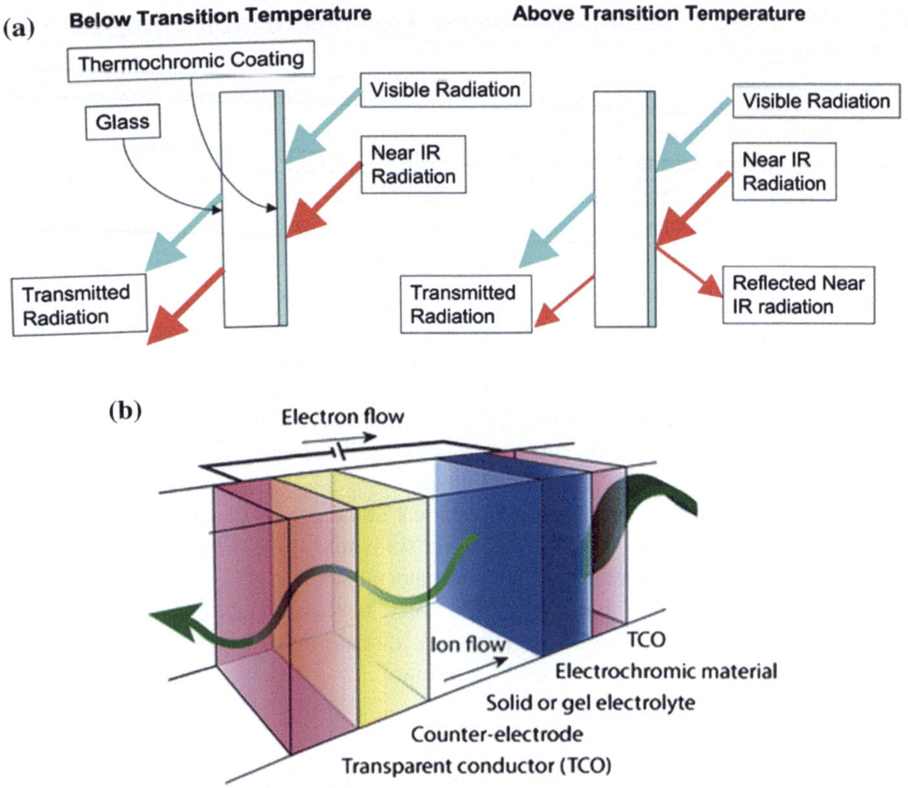

Fig. 7.1 Schematic mechanism of thermochromic and electrochromic windows. **a** For the thermochromic windows, the active thermochromic coating has a critical transition temperature. Below the transition temperature, most of the NIR light can pass through the window, thus transferring the solar heat into the building (keep the interior warm). However, the NIR light will be blocked outdoors above the transition temperature, which greatly reduces the transmitted heat radiation (keep the interior cool). Reproduction from [4] with permission from Royal Society of Chemistry, Copyright 2014. **b** When applying the voltage, the flowing ions will cause the reversible chromogenic reaction in the electrochromic layers, contributing the colour changes. Reproduction from [2] with permission from Royal Society of Chemistry, Copyright 2014

(3D) printing techniques [5–7]. With the advantages in the reproducible and customizable fabrication, the low-cost 3D printed smart windows show great promise in standardization and commercialization.

In this chapter, we review the recent progress in the responsive nanomaterials for smart windows. In Sects. 7.2 and 7.3, we focus on the nanomaterials-based thermochromic and electrochromic windows, respectively, from the perspective of working principles, light management ability and device engineering. Finally, to provide an outlook for the commercialization of smart windows, we discuss the importance of the utilization of nanomaterials and the challenges to be solved for the crucial laboratory-to-industries transition in Sect. 7.4.

7.2 Responsive Nanomaterials for Thermochromic Smart Windows

Temperature-responsive thermochromic smart windows can switch between the NIR-transparent and -blocking state at the transition temperature of T_c to realize the dynamic regulation of heat flux through the windows. For an ideal thermochromic smart window, it should simultaneously have high solar energy modulation ability ΔT_{solar} and high visible light transmittance (also known as luminous transmittance T_{lum} in the spectral range of 380–780 nm). In terms of ΔT_{solar}, it is the difference in the transmittance of solar energy, T_{solar}, between the two states in the range of 240–2500 nm, indicative of the energy conservation efficiency [8]. Apart from that, a suitable T_c in close proximity to room temperature (25–30 °C) is also needed for more effective thermal management. Besides the optical performance, thermochromic smart windows must be mechanically robust and chemically inert due to their long-time exposure to the weather.

Major conventional thermochromic windows make use of thermochromic polymers or inorganic solids to modulate the light transmission and the associated heat exchange [9, 10]. The advancement in nanomaterial science not only revitalizes the conventional materials but also introduces some new materials and novel device design with improved performance. Here, we discuss some of the representative thermochromic nanomaterials for smart windows.

7.2.1 Vanadium Dioxide-Based Thermochromic Nanomaterials

Among all of the conventional thermochromic materials, vanadium dioxide (VO_2) has attracted the most extensive studies due to its unique reversible metal-to-insulator transition (MIT) at a relatively low T_c of 68 °C (as shown in Fig. 7.2) and the accompanied abrupt change in the light transmittance in NIR region since the 1980s [12]. Upon exceeding T_c, the occurrence of MIT will immediately transform the crystalline structure of VO_2, from an insulating and NIR-transparent monoclinic phase (M phase, <68 °C) to a metallic and NIR-translucent rutile phase (R phase, >68 °C), within only 10^{-12} s [13]. In contrast, the phase transition of VO_2 does not apparently affect the transmittance of visible light (no obvious colour change) [8]. The T_c of VO_2 can be effectively reduced by elemental doping, such as H [14], W [15] and Mo [16], which makes VO_2 more applicable for practical use. This thermochromism behaviour, therefore, enables the temperature-adaptive regulation of solar heat. Nevertheless, the VO_2-based smart windows are limited by the poor ΔT_{solar} and T_{lum}, less than 20% and 50%, respectively [17]. The bigger challenge is that ΔT_{solar} and T_{lum} are seemingly mutually exclusive because it is difficult to enhance both parameters simultaneously [8]. In the traditional approach, T_{lum} can be improved by reducing the thickness of VO_2-based continuous film but at the expense

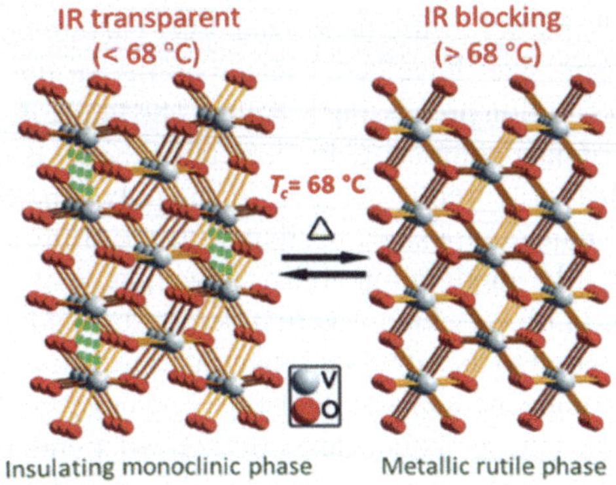

Fig. 7.2 Schematic of the change in the crystal structure of VO_2 at the T_c of 68 °C. When heated to 68 °C, the insulating monoclinic VO_2 (IR transparent) will transform into the metallic rutile phase with longer V-V distance (IR blocking). Conversely, the rutile VO_2 can recover to the monoclinic phase as cooled to below T_c. Reproduction from [11] with permission from Royal Society of Chemistry, Copyright 2011

of ΔT_{solar}, as illustrated in Fig. 7.3 [18]. Another approach to enhancing T_{lum} of VO_2 film is to apply the antireflection coatings and form the multilayered structures which utilize the light interference between the film interfaces [19]. However, the introduction of antireflection coatings normally requires the expensive and complex physical vapour deposition technology [20–22]. To fulfil the great potential of VO_2 for smart windows, a number of researches have attempted to achieve the optimal thermochromic properties, especially ΔT_{solar} and T_{lum} [8], via the forming of VO_2-based nanocomposites [23, 24] and the nanostructuring of VO_2 [25, 26].

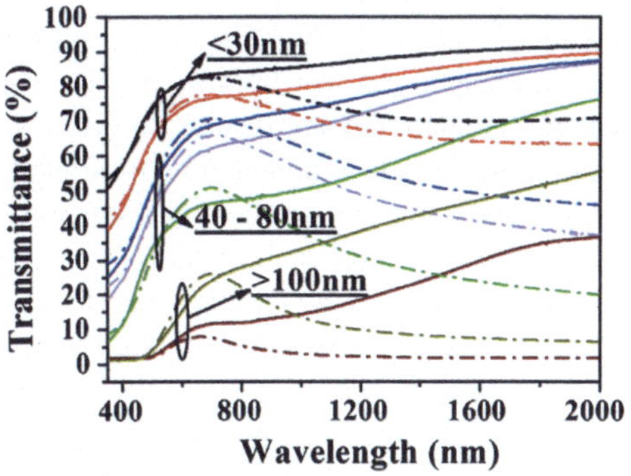

Fig. 7.3 UV-Vis-NIR transmittance spectra for VO_2 continuous films with different thickness. The solid and dashed lines represent the spectra measured at 20 and 90 °C, respectively. It is apparent that an ultrathin VO_2 film (<30 nm) can provide high T_{lum} (~80%) but low ΔT_{solar} (~20%). With the increased film thickness (>100 nm), ΔT_{solar} was noticeably enhanced (~50%) while T_{lum} was dropped sharply (~10%). Reproduction from [27] with permission from American Chemical Society, Copyright 2010

Relative to the bulk particles, the nanosized VO_2 particles (<40 nm) with high crystallinity and uniform sizes have better dispersibility in the aqueous solvent, in favour of the low-cost solution processing [28, 29]. The facile dispersion will greatly facilitate the forming of nanocomposites by dispersing the VO_2 nanocrystals into the dielectric matrix, such as polyurethane (PU) [30] and polydimethylsiloxane (PDMS) [31]. The VO_2-based nanocomposites, attributed to the minimized Mie scattering [32], can offer higher solar modulation ability as well as visible transmittance than the continuous VO_2 thin films, which was first demonstrated by the theoretical calculation [33]. As compared to the simulated spectral transmittance in Fig. 7.4a, b, the dispersive VO_2 nanocomposite can simultaneously achieve larger ΔT_{solar} and T_{lum} than the continuous film. On top of that, the particle size-dependent optical properties, which are obvious in Fig. 7.4a, imply the smaller VO_2 particles are more favourable for high ΔT_{solar} and T_{lum} [33]. Further, the advantages of the nanocomposite film have been confirmed by the practical experiments [8, 30, 31]. For example, Chen et al. reported a VO_2-based nanocomposite foil comprised of uniformly dispersed VO_2 nanoparticles in the PU matrix, as shown in Fig. 7.4c–e [30]. This type of nanocomposite film can deliver excellent optical properties ($\Delta T_{solar} = 22.3\%$ and $T_{lum} = 45.6\%$), which was superior to the continuous thin film [27], and VO_2-based multilayered film [34]. Because VO_2 is prone to be oxidized into V_2O_5 in the ambient air, VO_2 nanoparticles are usually coated with a stable oxide layer, such as SiO_2 [28] and ZnO [35], to form an oxidation-resistant core-shell structure. The model of dispersive VO_2-based nanocomposites also works well on the thin-walled VO_2 nanoparticles, providing an even higher ΔT_{solar} only at the cost of a slightly lesser T_{lum} [36]. Another advantage of scaling to nanometre-sized dimensions is that the thermochromic performance can also benefit from the localized surface plasmon resonance (LSPR) of the metallic VO_2 (R) nanoparticles [37]. The LSPR, initiated by the coupling of the resonance frequency of free electrons in VO_2 (R) with the wavelength of the incident light [38], will give rise to strong absorption in the NIR range and hence the enhanced ΔT_{solar} [39–42]. In addition to the improved optical properties, VO_2 nanoparticles are more resilient to the MIT-induced strain in the lattice, which guarantees the mechanical stability of thin films [43].

The optical properties of the VO_2 film can also be improved by the introduction of nanopores, of which the sizes are far smaller than the visible wavelength [25, 44–46]. Similar to the dispersive VO_2 nanocomposite, trapped air in the nanopores acts as a secondary phase. According to the simulated thermochromic performance (Fig. 7.5a), Zhang and his co-workers demonstrated that the increased porosity can enhance ΔT_{solar} of the nanoporous VO_2 films without the sacrifice of T_{lum} due to the higher ability to depress the reflection [25]. They also fabricated the VO_2 film with random porosity (pore size: 15–80 nm) by incorporating the removable polymer additives, and the resulting nanoporous film ($T_{lum} = 43.3\%$, $\Delta T_{solar} = 14.1\%$, Fig. 7.5b) had comparable optical performance to the multilayered VO_2 film with the optically optimized structure ($T_{lum} = 44\%$, $\Delta T_{solar} = 12\%$) [25]. In contrast to the random porous films, the highly ordered porosity can further improve the performance. For instance, Zhou et al. first reported the fabrication of periodic nanoporous VO_2 film, which featured a grid-like structure (Fig. 7.5c), using a feasible colloid lithography

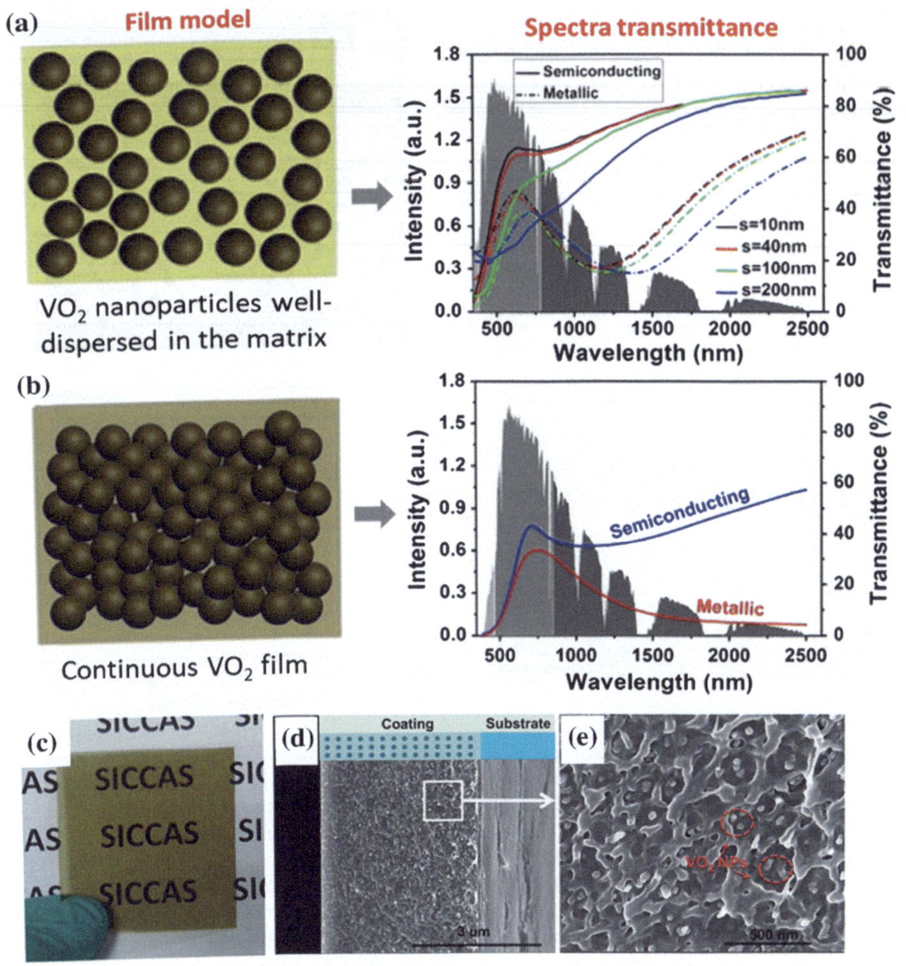

Fig. 7.4 Simulated and the experimental dispersive VO_2 nanocomposites. The ideal film models and their simulated light transmittance spectra for **a** a nanocomposite in which VO_2 nanoparticles (with different sizes: 10, 40, 100 and 200 nm, filling factor = 0.01) are uniformly embedded and **b** the continuous film in which the VO_2 nanoparticles cluster together. With the same film thickness of 10 μm, the dispersive VO_2 nanocomposite obviously has advantages in high ΔT_{solar} and T_{lum} over the continuous film. **a**, **b** Reproduction from [40] with permission from American Chemical Society, Copyright 2015. **c** The image of the nanocomposite-coated glass which shows a yellowish green colour. The composite coating is composed of the dispersed VO_2 nanoparticles in the PU matrix. **d** SEM images of the cross-sectional nanocomposite coating on the substrate and **e** the homogeneous dispersion of VO_2 nanoparticles in the PU matrix. **c–e** Reproduction from [30] with permission from Royal Society of Chemistry, Copyright 2014

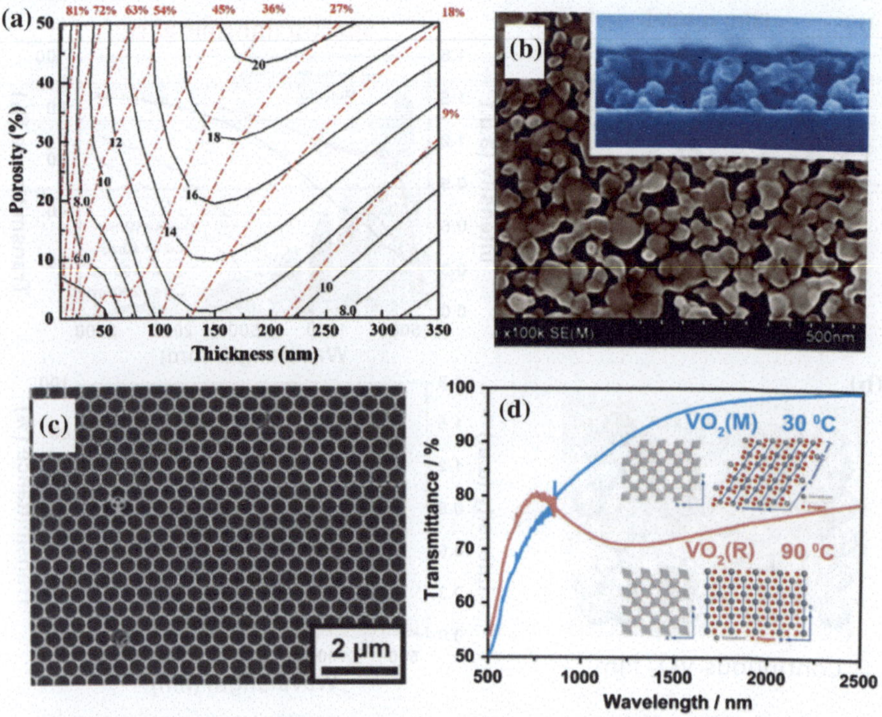

Fig. 7.5 Simulated and experimental optical performance of nanoporous VO_2 film. **a** The calculated T_{lum} of monoclinic VO_2 at low temperature (red dotted lines) and ΔT_{solar} of rutile VO_2 at high temperature(black solid lines) as a function of thickness and porosity of the VO_2 films. **b** SEM image of the VO_2 film with random porosity. The inset is the cross-sectional view of the nanoporous film. **a**, **b** Reproduction from [25] with permission from American Chemical Society, Copyright 2011. **c** SEM image of the single-layered periodic porous VO_2 film. (3) The spectral transmittance of the periodic porous film at low-temperature insulating (M phase, 30 °C) and high-temperature metallic state (R phase, 90 °C). It is apparent that the porous VO_2 highlights the ultrahigh visible transmittance regardless of the phase transition ($T_{lum} > 80\%$). **c**, **d** Reproduction from [45] with permission from Royal Society of Chemistry, Copyright 2014

method [45]. Owing to the optimal space occupancy of monoclinic VO_2 crystals, the periodic porous film possessed ultrahigh visible transmittance ($T_{lum} = 81\%$) and excellent solar modulation ability ($\Delta T_{solar} = 23\%$), as illustrated in Fig. 7.5d [45].

To fabricate the VO_2-based smart windows, the traditional approach is to employ the gas-phase deposition methods, including physical vapour deposition [47–49] and chemical vapour deposition [50–52], to form a uniform and high-quality VO_2 thermochromic film on the glass. However, the expensive and complex gas-phase deposition methods impede the mass production and commercialization of the smart windows. Due to the excellent solution processability, VO_2 nanoparticles are fully compatible with the scalable, low-cost and high-yield solution phase deposition methods, such as sol-gel coating [53, 54] and polymer-assisted deposition [25, 55].

Moreover, the dispersible VO_2 nanoparticles can be exploited as the active components in the thermochromic inks for 3D printed smart windows. In 2018, Ji et al. demonstrated the first inkjet printed VO_2 smart window with desirable performance (ΔT_{solar}= 15.31%) [7]. They carefully tuned the viscosity (8 mPa s) and surface tension (26.8 mN/m) of the ink based on VO_2 nanoparticles (30–50 nm). Therefore, the inks can be continuously deposited onto the glass and then solidify into the compact VO_2 thermochromic film via a simple thermal treatment at 80 °C after printing. This proof-of-concept 3D printed smart window is likely to open up new possibilities for the customizable fabrications of smart windows.

7.2.2 Polymer-Based Thermochromic Nanomaterials

Generally, thermochromism in the polymeric materials originates from the changes in light reflection or absorption induced by the thermally driven structural rearrangement or reversible reaction [3, 9]. In recent years, thermochromic hydrogels, which consist of the water-swollen networks of polymer chains, have emerged as a new class of temperature-responsive materials for smart window application [17, 56]. The solar heat can activate a hydrophilic-to-hydrophobic transition in thermochromic hydrogels, accompanied by the change in transparency, as long as the lower critical solution temperature (LCST) is surpassed (Fig. 7.6a). Below the LCST, intermolecular hydrogen bonds allow the polymer chains to be hydrated and swollen. When heated above the LCST, the breakage of intermolecular hydrogen bonds will cause dehydration of the polymer networks, leading to structural collapse and polymer aggregation [17, 57]. The aggregated polymer chains tend to invite more light scattering, which significantly reduces the transparency of the hydrogels (Fig. 7.6b) [58, 59]. The thermochromic hydrogels include but are not limited to hydroxypropyl cellulose (HPC) [60], polyampholyte hydrogel (PAH) [61] and poly(*N*-isopropylacrylamide) (PNIPAm) [62–64].

Due to the intrinsic cross-linked structure and temperature responsiveness, thermochromic hydrogels act as an ideal switchable matrix for photothermal nanoparticles to form a dual-responsive nanocomposite for smart windows [63]. Typically, the temperature-responsive materials only respond to the surrounding temperature but not the intensity of sunlight. As a result, they fail to provide dimming control in cold climates despite exposure to the intensive sunlight. In comparison, the dual-responsive nanocomposites can sense not only the environmental temperature but also the intensity of solar irradiation. The additional responsiveness to sunlight arises from the conversion of sunlight to heat by the embedded photothermal nanoparticles, such as graphene oxide (GO) and antimony tin oxide (ATO) [63, 66]. Once the converted heat transfers to the hydrogel matrix, the temperature increase will trigger the optical switching from transparent to translucent in the nanocomposite. For example, Lee et al. fabricated such dual-responsive nanocomposites by incorporating ATO nanoparticles into the PNIPAm hydrogel (Fig. 7.7a, b) [63]. When the solar irradiation is sufficiently intensive, the photothermal effect of ATO nanoparticles

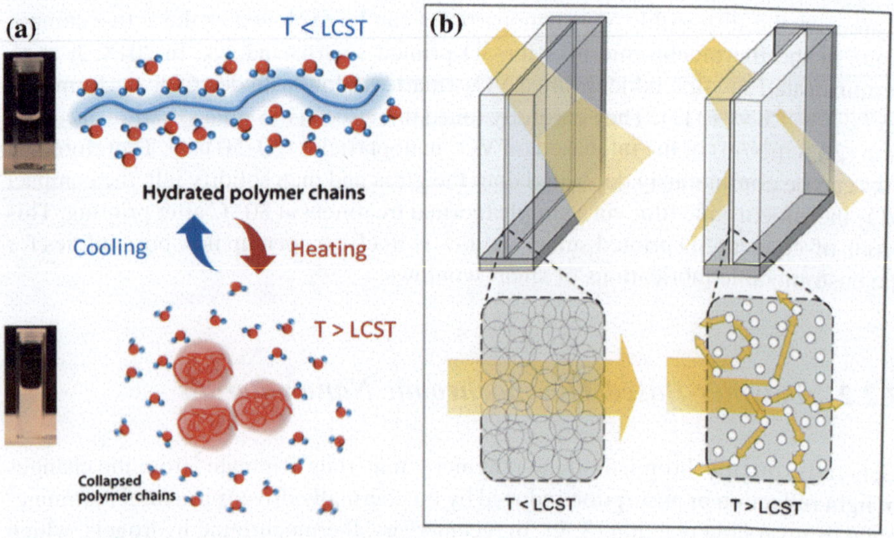

Fig. 7.6 Schematic of thermochromic hydrogels below and above the LCST. **a** The temperature-responsive arrangement of polymer chains. Below the LCST, the hydrophilic polymer chains are surrounded by water molecules and the hydrogel is transparent. Above the LCST, the polymer chains lose water molecules and collapse into polymer clusters which make the hydrogel opaque. Reproduction from [65] with permission from Elsevier, Copyright 2016. **b** The schematic of a smart window where the thermochromic hydrogel layer is sandwiched between two glass panels. The incident light can pass through the hydrogel layer below the LCST but is largely blocked above the LCST due to the light scattering. Reproduction from [42] with permission from Elsevier, Copyright 2019

will enable the immediate phase transition in the PNIPAm, making dimming control feasible even under the LCST of PNIPAm (~32 °C) in less than 5 min, as shown in Fig. 7.7c [63].

In addition to the photothermal nanoparticles, VO_2 nanoparticles can also be hybridized with the thermochromic hydrogel to form the aforementioned dispersive VO_2-based nanocomposite [60, 64]. In 2015, Zhou et al. reported a sandwich-type smart window based on a VO_2/PNIPAm hybrid nanocomposite (Fig. 7.8a, b) [64]. The highlight of their work is that the hybrid nanocomposite can afford excellent thermochromic performance ($T_{\mathrm{lum}} = 62.6\%$, $\Delta T_{\mathrm{solar}} = 34.7\%$), superior to pure VO_2 nanoparticles and pure PNIPAm hydrogel (Fig. 7.8c). The hybrid system exhibited dramatical contrast in the visible and NIR region, contributing to the prominent enhancement in $\Delta T_{\mathrm{solar}}$ [64].

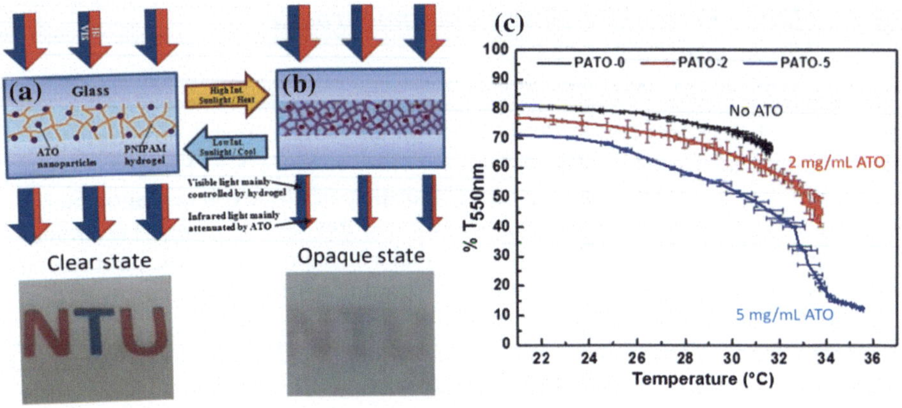

Fig. 7.7 Schematic of the dual-responsive PNIPAm/ATO nanocomposite in **a** clear state and **b** opaque state. The photothermal ATO nanoparticles are evenly distributed in the cross-linked hydrogel networks. Upon exposure to the strong sunlight, the ATO nanoparticles, acting as a nanoheater, convert sunlight to thermal energy, giving rise to temperature increase and phase transition in the PNIPAm. The hydrogel matrix is in charge of the visible light manipulation while the ATO nanoparticles can absorb the NIR light. **c** The optical switching of the nanocomposite is dependent on the content of ATO nanoparticles. With the addition of more ATO nanoparticles, the phase transition of PNIPAm tended to start at a lower temperature under the solar irradiation. Reproduction from [63] with permission from American Chemical Society, Copyright 2017

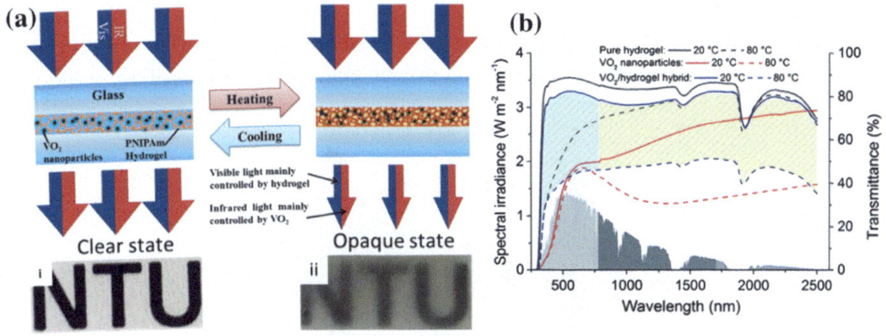

Fig. 7.8 **a** Schematic of the smart window based on the VO_2/PNIPAm nanocomposite. The PNIPAm matrix regulates the visible transmittance at ~30 °C while VO_2 nanoparticles regulate the NIR transmittance at ~68 °C. The insets (i) and (ii) show the clear state at room temperature (hydrated PNIPAm) and opaque state at 35 °C (dehydrated PNIPAm). **b** UV-Vis-NIR transmittance spectra of VO_2/PNIPAm nanocomposite, pure VO_2 nanoparticles and pure PNIPAm at 20 and 80 °C. It is obvious that VO_2/PNIPAm hybrid has higher T_{lum} than pure VO_2 due to the reduced light scattering. Further, VO_2/PNIPAm hybrid has greatly enhanced ΔT_{solar} relative to the other two. Reproduction from [64] with permission from Royal Society of Chemistry, Copyright 2015

7.2.3 *Halide Perovskite-Based Thermochromic Nanomaterials*

Over the past decades, halide perovskites of an ABX_3 crystal structure [A = $CH_3NH_3^+$ (MA), $HC(NH_2)_2^+$ (PA) or Cs^+; B = Pb^{2+}or Sn^{2+}; X = I^-, Br^- and Cl^-] as shown in Fig. 7.9 have been the research hotspots of photovoltaic nanomaterials due to their high-power conversion efficiency and low cost [67–69]. The tremendous research efforts on perovskites not only make huge progress in the photovoltaic devices but open up a new avenue for thermochromic smart windows as well.

In 2017, Bakr's group pioneered the work on the organic halide perovskites-based ($MAPbBr_{3-x}I_x$) thermochromic windows which exhibited yellow, orange, red and black colour at 25, 60, 90 and 120 °C, respectively (Fig. 7.10a, b) [71]. The thermochromism of the perovskite materials is associated with the inverse temperature crystallization (ITC) of halide perovskites, which means their solubility decreases at elevated temperature in certain solvents [72]. Another intriguing aspect is that both the crystallization temperature of ITC and the absorption of photons by the perovskites rely on the halogen constituent of the perovskites [73]. Therefore, with the increment of the temperature, orange $MAPbBr_{2.7}I_{0.3}$, red $MAPbBr_{2.4}I_{0.6}$ and black $MAPbBr_{1.8}I_{1.2}$ perovskite crystals successively precipitate out from the original yellow biconstituent perovskite solution at 60, 90 and 120 °C (Fig. 7.10c). The temperature-dependent precipitation and dissolution of perovskite crystals lead to the reversible chromatic variation.

Most importantly, the use of halide perovskites realizes the perfect combination of sunlight-regulating smart windows and power-generating photovoltaic windows

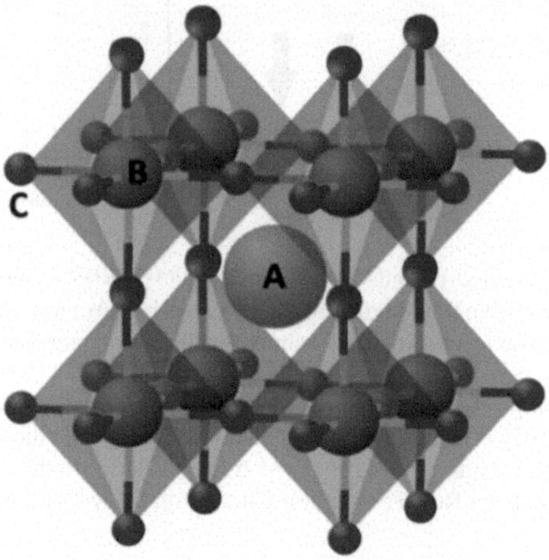

Fig. 7.9 Crystal structure of cubic halide perovskites with the general formula of ABX_3. The larger cation A (green) occupied the cuboctahedral voids formed by the 12 nearest-neighbouring X ions (purple). The smaller cation B is centred at the unit cell. Reproduction from [70] with permission from Nature Publishing Group, Copyright 2014

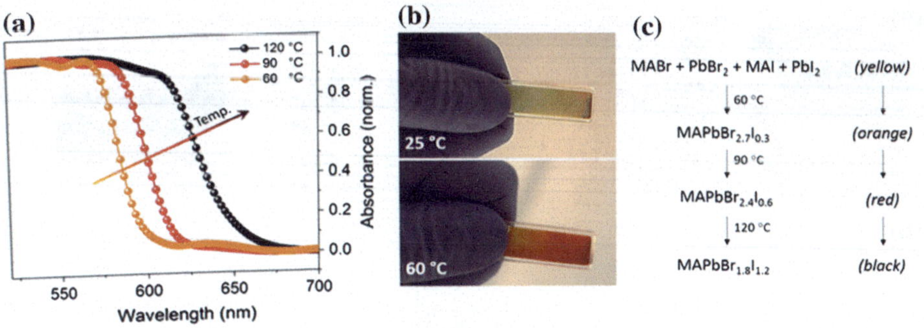

Fig. 7.10 **a** Absorption spectra of thermochromic perovskite crystals obtained at 60, 90 and 120 °C. The original yellow solution changed to orange at 60 °C, to bright red at 90 °C and finally to black at 120 °C. **b** Images of the thermochromic prototype annealed at 25 and 60 °C. **c** The formulation of the original perovskite solution and the major contributors of colour changes at 60, 90 and 120 °C. Reproduction from [71] with permission from American Chemical Society, Copyright 2017

at the same time (smart solar windows). One the one hand, perovskites can efficiently convert sunlight to electricity, as what they do in solar cells. On the other hand, their distinct chemistry and nanostructure render perovskites' optoelectronic properties responsive to the temperature change [74]. Consequently, the smart solar windows literally make full use of solar energy. In 2017, Wheeler et al. fabricated an inorganic–organic hybrid halide perovskite ($MAPbI_3$)-based smart solar window which can rapidly switch between transparent and opaque at 60 °C within a few minutes (Fig. 7.11a) [75]. At cooler temperature, the perovskite forms a visibly transparent complex with the methylamine (CH_3NH_2) in a sealed layer. When the temperature hits 60 °C, the window switches into an opaque state attributed to the dissociation of CH_3NH_2 from the complex and the strong absorption of sunlight by the perovskite (Fig. 7.11b). In the opaque state, the photovoltaic effect of the perovskite enables the transformation of absorbed solar energy into electricity, and the conversion efficiency reaches up to 11.3% (Fig. 7.11c). When cooled below 60 °C, CH_3NH_2 gas returns to the perovskite and reforms the complex, making the window return to the transparent state. Besides the chemical reaction-driven thermochromism, it has been found that the purely inorganic halide perovskites are responsive to temperature, varying a lot in their crystal structures and hence optical properties [76, 77]. In 2018, Yang's group reported a novel smart solar window utilizing the caesium-based perovskite ($CsPbI_{3-x}Br_x$) thin film [78]. Figure 7.11d shows the thermochromic phase transition of $CsPbI_{3-x}Br_x$. Once reaching a critical transition temperature (as low as 105 °C), the structural rearrangement occurs in the $CsPbI_{3-x}Br_x$ crystal: from a colourless room temperature non-perovskite phase (low-T phase, 81.7% transparency) to an orange-red-coloured high-temperature perovskite phase (high-T phase, 35.4% transparency), as shown in Fig. 7.11e. The exposure to moisture can initiate the back-conversion of the high-T phase to low-T phase. Therefore, this perovskite-based smart solar

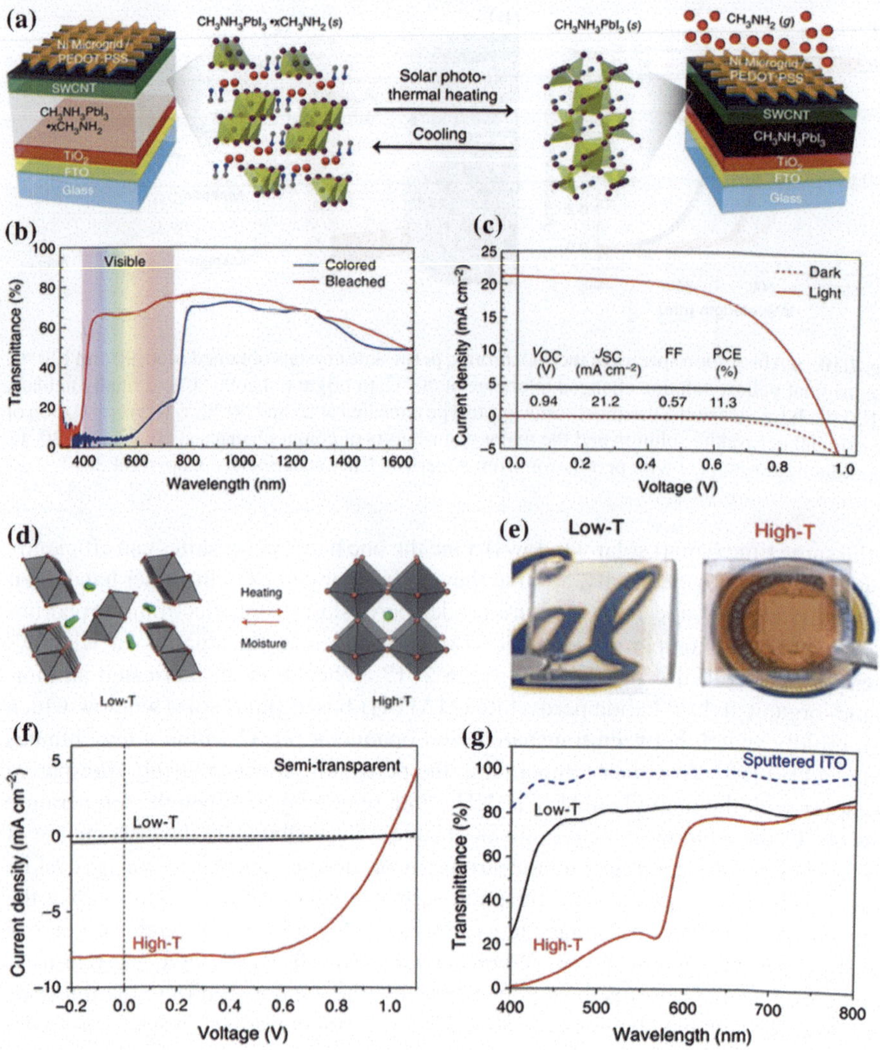

window can reversibly switch between a transparent low-temperature state with low-power output and a deeply coloured high-temperature state with high-energy output (conversion efficiency $\approx 7\%$) (Fig. 7.11f, g).

In spite of the fascinating features, the stability and efficiency of the perovskite-based smart solar windows are still far from satisfaction at the current stage. Wheeler's window, which is driven by chemical reaction, suffers from poor stability due to the loss of methylamine during the repetitive switching back and forth. Yang's structure transition-driven window, however, is limited by slow response (>10 h) and high transition temperature (>100 °C). In addition, both perovskite windows fail to

◂**Fig. 7.11** Halide perovskite smart solar windows based on the complexion–dissociation of methylamine (**a–c**) and the temperature-responsive structural change (**d–g**). **a** Schematic of the smart solar window architecture and the chemical reaction-induced thermochromism. In the architecture, TiO_2 is the electron transport layer, and fluorine-doped tin oxide (FTO) is the transparent conducting electrode. The single-walled carbon nanotube (SWCNT) is to provide hole extraction and improve lateral electrical transport. The topmost layer (microgrid/PEDOT:PSS) serves as an electric glue to package the whole device. **b** Transmittance spectra of the smart solar window in the transparent (red) and opaque (blue) states. It is obvious that the visible transmittance sharply drops after the dissociation of methylamine from the perovskite. **c** Current density versus voltage curve of the switchable smart solar window in the dark (dashed) and under illumination (solid). Reproduction from [75] with permission from Nature Publishing Group, Copyright 2017. **d** Schematic of the thermally driven, moisture-mediated reversible phase transition between the high-T and low-T phases. **e** The picture of the $CsPbIBr_2$ smart solar window in the colourless low-T state (left) and deeply coloured high-T state (right). The sputtered indium tin oxide (ITO) layer acts as the top transparent electrodes. **f** Current density versus voltage curve of the device. **g** The transmittance spectra of the smart solar windows at low-T (black) and high-T state (red). The spectrum for the sputtered ITO layer is as a reference. Reproduction from [78] with permission from Nature Publishing Group, Copyright 2018

regulate the light transmittance in the NIR range. Nevertheless, it is expected the smart solar windows will keep evolving with the persistent pursuit for new modified perovskites.

7.3 Responsive Nanomaterials for Electrochromic Smart Windows

Compared to the passive thermochromic windows with no energy input, electrochromic windows are powered by the electric supply, and they allow users to manage the light transmittance in a broad spectral range as needed by altering the applied voltage. Electrochromic materials are the core components in the whole device because they take control of the reversible redox switching between the bleached and coloured state as a response to electrical potential. To achieve the high optical performance of smart windows, electrochromic materials should provide high optical contrast, fast switching kinetics and high colouration efficiency [3]. Colouration efficiency is measured by the ratio of the optical absorbance change to the charge inserted into the electrochromic material per unit area [2]. In addition, the stability towards the long-term redox switching and sunlight exposure is of great importance. Various electrochromic materials, including transition metal oxides [79], conjugated polymers [80] and small organic and inorganic molecules [81, 82], have been employed in smart windows while they rarely reach a trade-off between optical performances, stability and production cost. These issues must be addressed, which is also the prerequisite for the commercialization of electrochromic smart windows. Numerous researches have proved that nanostructuring is an effective way to improve the conventional electrochromic materials [2, 3, 83]. Further, the introduction of nanomaterials endows smart windows with new functionalities, which will greatly broaden their application in the energy saving and light managing. In section, we will focus on the well-studied nanostructured metal oxides and the emerging two-dimensional (2D) nanomaterials for electrochromic smart windows.

7.3.1 Metal Oxides-Based Electrochromic Nanomaterials

The use of transition metal oxides (TMOs), such as tungsten oxide (WO_3) [79], molybdenum oxide (MoO_3) [84] and titanium oxide (TiO_2) [85], in electrochromic devices dates back to 1960s when Deb first demonstrated the electrochromic phenomenon in WO_3 [86]. In the 1980s, the successful introduction of electrochromic WO_3 to the dynamic light modulating windows flourished the researches on the electrochromic smart windows [79, 87]. The electrochromic nature of TMOs originates from the reversible change in the oxidation state of transition metal ions with the insertion/extraction of electrons (e^-) and ions (M^+) upon voltage bias [88]. For example, the reversible electrochromism of WO_3 between the transparent and blue state is shown by the following Faradaic reaction [89]:

$$WO_3(\text{transparent}) + x\left(M^+ + e^-\right) \leftrightarrow M_xW^{VI}_{1-x}W^{V}_{x}O_3(\text{blue}) \quad (7.1)$$

In general, the colouration efficiency, optical contrast and response time are closely related to the amount of reduced/oxidized metal ions (colouration centres) and the kinetics of redox reaction. The electrochromic redox reactions of TMOs involve the charge transfer at the interface between electrolyte and TMOs and the ion diffusion controlled redox reaction within the lattice of TMOs. Typically, electrochromic thin films based on bulk TMOs fall short of colouration efficiency (in the range of tens cm^2/s) and response time (>60 s) due to impeded charge transfer and long ion diffusion distance [90]. Therefore, the facilitation and promotion of the two processes are key to improving the electrochromic performance. Herein, nanostructuring of TMOs, including nanoparticles [91–93], nanorods [94–96], nanosheets [97–99], nanowires [100, 101] and quantum dots [102, 103], has been proven as a potent route to expedite the processes and enhance the performance. Owing to the large specific surface area, nanostructured TMOs greatly facilitate the charge transfer on more metal ions. Further, the diffusion path for ions is considerably shortened, from micrometre to a few hundred or tens of nanometres, within the lattice of nanosized TMOs. The junction of these effects not only increases the total amount of colouration centres but also accelerates the overall kinetics of redox reactions, resulting in the deeper optical contrast, shorter switching time and higher colouration efficiency [90, 104]. It is also worth noting that nanostructured TMOs with higher colouration efficiency tend to have better durability because less charge is needed to produce a given span of optical modulation [2, 92]. In terms of economic benefit, solution-processable TMOs nanomaterials have advantages in improved manufacturability, inexpensive production and good compatibility with 3D printing techniques [6, 105–107].

Among all electrochromic TMOs, tungsten oxides have attracted the most extensive studies since Deb's pioneering work and, accordingly, much progress has been made in the nanostructured WO_3. In 2011, Zhang et al. developed a hexagonal WO_3 nanowire array film using a low-cost solvothermal method [108]. In the previous works, thermal evaporation and physical vapour deposition were the prevailing methods to synthesize WO_3 nanowires on the conducting substrates but these techniques are costly and energy-intensive [109, 110]. In Zhang's solvothermal method, the ordered WO_3 nanowires directly grew from WO_3 seeds on the transparent conductive FTO-coated glass by controlling the reaction conditions, such as pH and reactants. Compared with the WO_3 micro-brick film, improved electrochromic performance was observed in the resulting WO_3 nanowire array film, with high colouration efficiency of 102.8 cm^2/s and fast response time of 7.6 and 4.2 s for colouring and bleaching, respectively (Fig. 7.12a, b) [108]. Further enhancement in the electrochromic performance can be achieved by downsizing the nanocrystalline tungsten oxide to quantum dots (QDs) with an average size of 1.6 nm via a facile colloid process, as reported by Cong and his co-workers [103]. In addition to the efficient charge

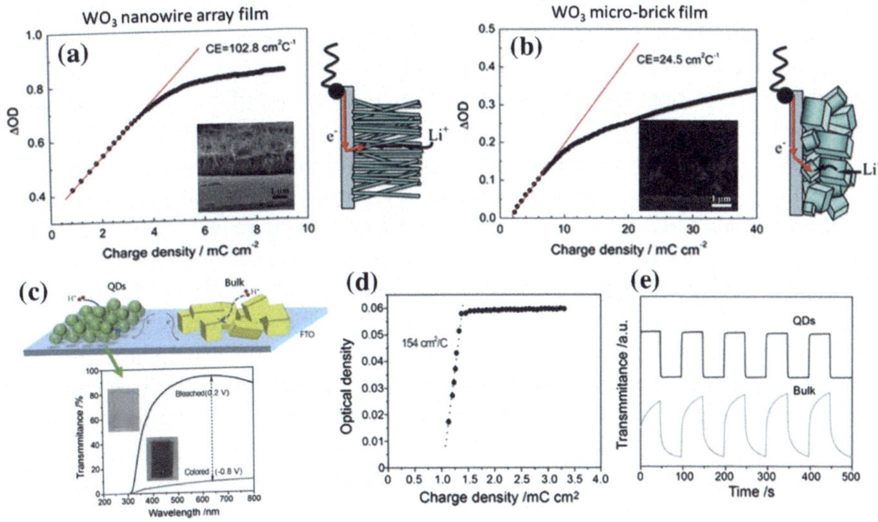

Fig. 7.12 Nanocrystalline WO_3 in the form of nanowires and quantum dots and the corresponding electrochromic performance. Colouration efficiency of WO_3 nanowire array film (**a**) and WO_3 micro-brick film (**b**). Reproduction from [108] with permission from Royal Society of Chemistry, Copyright 2011. **c** Tungsten oxide QDs provide more efficient ion diffusion and charge transfer than the bulk counterpart. Transmittance spectra for tungsten oxide QDs electrode deposited on FTO glass in the coloured (−0.8 V) and bleached (+0.2 V) states, respectively. The electrochromic window is highly transparent at bleached state (>90% of transmittance) and large modulation window (>80%). The insets are images of the electrode at bleached and coloured states, appearing almost transparent and dark blue colour, respectively. **d** The colouration efficiency of the QDs-based electrochromic window. **e** Alternate transmittance switching for bulk (green) and QDs (black) tungsten oxide showing significantly faster response for QDs. Reproduction from [103] with permission from Wiley-VCH, Copyright 2014

transfer and ion diffusion, tungsten oxide quantum dots are even more active in electrochromic reactions due to the ultrahigh proportion of surface atoms (Fig. 7.12c). Consequently, the QDs-based electrochromic window showed a remarkably high colouration efficiency (154 cm^2/s) and super-fast switching time (<1 s for both colouring and bleaching), as shown in Fig. 7.12d, e. The significant performance improvements can also be achieved in other TMOs at the nanoscale, such as TiO_2, [111] MoO_3 [84] and NiO [112].

In addition to the improved electrochromic performance, some nanostructured metal oxides with strong LSPR absorption can deliver a new functionality of selective NIR shielding to smart windows [113]. The LSPR effect can be flexibly tuned across a wide range of wavelengths by electrochemical charging or doping of plasmonic nanocrystals [2]. With the combination of intrinsic Faradaic reaction-based electrochromism and NIR-selective LSPR, the use of plasmonic nanocrystals allows the dynamic and independent control of NIR and visible light transmittance by varying the applied voltage, namely the dual-band electrochromism. In 2013, Milliron's group fabricated such a dual-band electrochromic window, for the first time, via incorporating the tin-doped indium oxide (ITO) nanocrystals into niobium oxide glass (NbO_x) [114]. Figure 7.13a illustrates the architecture and working principle of this smart window. The nanocrystal-in-glass composite electrode is coupled with a lithium counter electrode, both of which are immersed in the Li^+ electrolyte. Under the external bias, the amorphous NbO_x matrix modulates mostly the visible light while the ITO nanocrystals mainly block the NIR light through LSPR. As a whole, the nanocrystal-in-glass composite film progressively switches between three distinct operating modes: bright mode at the open-circuit voltage, cool mode at an intermediate voltage and dark mode at a low voltage (Fig. 7.13b). At bright mode, neither the insertion/extraction of Li^+ and electrons nor the LSPR happens, allowing the traverse of both visible and NIR light. At cool mode, the LSPR effect takes place in the charged ITO nanocrystals which have a sufficiently high concentration of free electrons, whereupon they only block most of NIR light while still welcome the visible light [115]. At dark mode, the electrochromic NbO_x comes into effect which results in the broadband blocking of NIR and visible light. Moreover, relative to the pure NbO_x film, the ITO-NbO_x composite is more durable and has better visible light modulation ability. This dual-band electrochromic smart window is more energy efficient than conventional smart windows owing to its controllable and selective regulation of solar heat exchange and interior lighting. Inspired by Milliron's work, the recent five years have witnessed the spring up of various metal oxide nanocomposite-based dual-band electrochromic smart windows [116–119]. Apart from the nanocomposite system, a single-component metal oxide nanocrystals with defects or dopants can also be utilized as an active material for this smart window [120–122]. For example, Lee's group reported a dual-band electrochromic smart window which exhibited bright mode at 1 V, cool mode at −2.8 V and dark mode at −4 V (as shown in Fig. 7.13c, d) [123]. Designed based on the monoclinic oxygen-deficient WO_3 (m-WO_{3-x}) nanowires, this smart window makes use of the LSPR property of m-WO_{3-x} nanowires and the phase transition (from dielectric and metallic) to modulate the NIR transmittance and the bandgap transitions (intraband

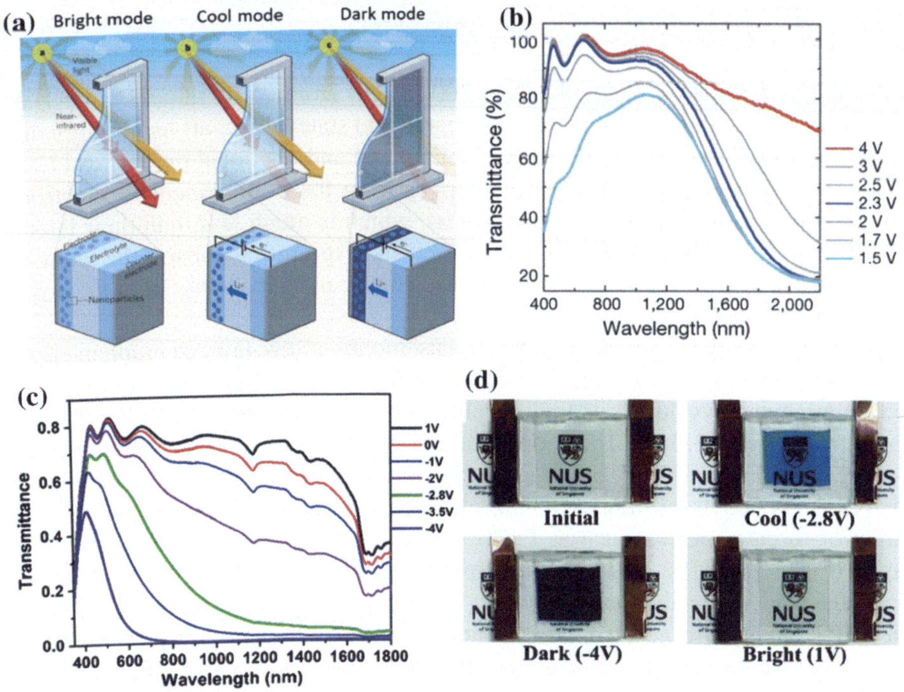

Fig. 7.13 Dual-band smart windows which can dynamically and independently manage the transmission of visible and NIR light. **a** The smart window based on plasmonic nanocrystal-in-glass exhibits three distinct modes. Reproduction from [125] with permission from Nature Publishing Group, Copyright 2013. **b** Changes in the transmittance spectra of the ITO-in-NbO_x film with the varying voltages versus Li/Li^+. The composite shows bright mode at 4 V, cool mode at 2.3 V and dark mode at 1.5 V. Reproduction from [114] with permission from Nature Publishing Group, Copyright 2013. **c** The transmittance spectra of the m-WO_{3-x} nanowires-based dual-band smart window under different external biases. **d** The three modes under different voltages. Reproduction from [123] with permission from Royal Society of Chemistry, Copyright 2018

and interband transition) to the visible transmittance. In addition, they replaced the commonly used monovalent Li^+ with the smaller multivalent Al^{3+} as the insertion ions and thus the smart window can deliver excellent solar light modulation ability, high colouration efficiencies (254 and 121 cm^2/C at NIR and visible region, respectively), fast response time (within 2 s for colouration/bleaching at NIR and visible region) and good stability and cyclability. The dual-band electrochromic smart windows are especially advantageous in the selective modulation of solar irradiation without sacrificing the visible transmittance for daylighting, helpful to reduce the energy consumption on HVAC and interior lighting [3, 124].

7.3.2 2D Electrochromic Nanomaterials

Since the first isolation of graphene in 2004 [126] atomically, thin 2D materials have been the spotlight of fundamental research and state-of-the-art technology due to their extraordinary electrical, mechanical, optical and chemical properties [127–131]. The uniqueness of 2D materials derives from the confinement of electrons and photons within their nanoscale thickness, which is much less than their lateral dimensions [132]. The recent in-depth research and exploration unearth the interesting optoelectronic properties of 2D materials, which can be made use of in the novel electrochromic applications.

As the best-known example of 2D materials, mono- or few-layered graphene is an ideal alternative to the traditional ITO transparent conductive electrode in the electrochromic windows due to its high mechanical strength, optical transparency and carrier mobility [133, 134]. The optical transparency tends to decrease with the number of graphene layers [135]. However, multilayered graphene will reveal a significant increase in optical transmittance and/or electrical conductivity towards the intercalation of foreign species, such as $FeCl_3$, Li^+ and Na^+, between the neighbouring layers (Fig. 7.14a) [136–139]. The principle behind this exceptional phenomenon is that intercalation heavily dopes graphene, which fills up the conduction band (higher electrical conductivity) and suppresses interband optical transitions by Pauli blocking (higher visible transmittance) [137]. The controllable and reversible intercalation of graphene can be achieved by electrochemical methods [132, 140]. For example, Bao et al. fabricated an encapsulated electrochemical cell in which the multilayered CVD graphene electrode received Li^+ from the Li counter electrode, and hence, Li^+ intercalation was enabled. As shown in Fig. 7.14b, c, the transparency of 40 and 80 nm thick graphene increases pronouncedly upon the external voltage is applied. Therefore, electrochemical intercalation of graphene potentially provides an emerging and promising strategy for making a simply structured electrochromic device with tunable transparency. In this device, graphene acts not only as a transparent conductive electrode but also as an active electrochromic contributor.

In 2015, Wan and his co-workers demonstrated such a reversible electrochromic device based on the electrochemical Na^+ intercalation of reduced graphene oxide (rGO) network (Fig. 7.14d, e) [139]. Its optical transmittance increases from 36 to 79% within 10 min when the Na^+ intercalation reaches its maximum (Fig. 7.14f). In comparison with Bao's work, they adopted the low-cost rGO as the dual-functional component instead of the costly CVD graphene. In addition, rGO has expanded interlayer spacing due to the presence of some oxygenated function groups which greatly facilitates the insertion of Na^+. Moreover, although both Na^+ and Li^+ are both highly reactive, this Na^+ intercalation electrochromic system is much more stable in air than Bao's Li^+ intercalation system. This is possibly related to the poorer diffusivity of Na^+ in the barrier layer at the edge of rGO sheets formed by the reaction between Na^+ and water/O_2/CO_2, thus well protecting the intercalated Na^+ inside the sheets. This electrochromic Na^+-intercalated rGO system provides a good example for the commercial intercalated graphene electrochromic windows in

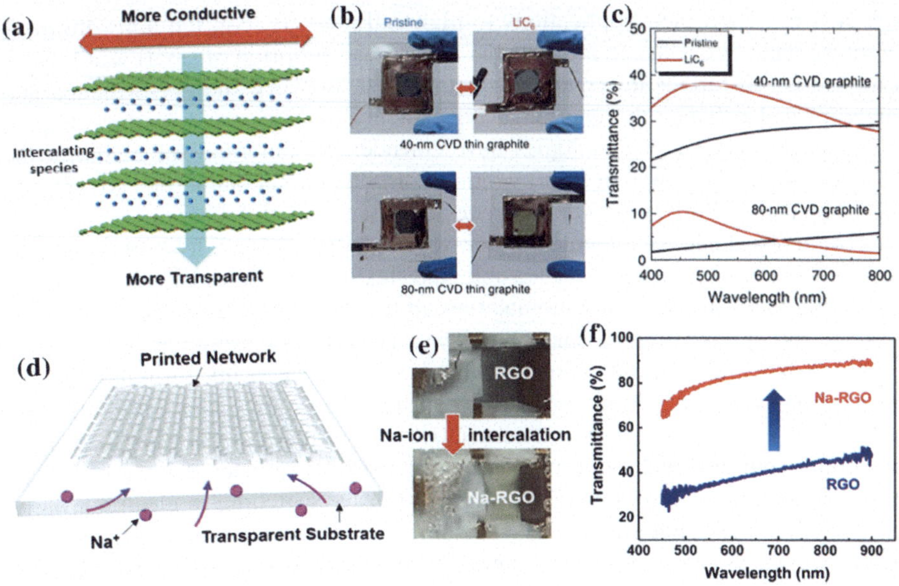

Fig. 7.14 Intercalated graphene-based electrochromic device. **a** Schematic of the multilayered graphene intercalated by metal ions. Reproduction from [132] with permission from Royal Society of Chemistry, Copyright 2016. **b** Images of the encapsulated CVD graphene electrochromic device with different thickness of graphene electrode. **c** The optical transmittance before and after the Li^+ intercalation. Both CVD graphene electrodes showed increased transparency after applying the external voltage to the devices. Reproduction from [137] with permission from Nature Publishing Group, Copyright 2014. **d** Schematic of Na^+ intercalation in the rGO film on the transparent substrate. **e** The images show the transparency increase in the Na^+-intercalated rGO electrochromic device. **f** The optical transmittance of rGO film before and after Na^+ intercalation. Reproduction from [139] with permission from American Chemical Society, Copyright 2015

the future. In addition to graphene, the intercalation-induced reversible change in the light transmittance also occurs in 2D molybdenum disulphide (MoS_2) and titanium carbide MXene flakes [141–143]. This type of 2D materials-based electrochromic devices is still at the very initial stage but holds great promise for low-cost and high-efficiency smart windows.

7.4 Conclusion and Outlook

Smart windows are appealing to green energy-saving buildings and smart home systems because of their dynamic management of light transmittance and solar heat exchange controlled by temperature (thermochromic) or users' preferences (electrochromic). After decades of development, thermochromic and electrochromic smart windows have obtained the preliminary success in commercialization, but their high cost restricted their applications to high-end markets, e.g. the auto-dimming aeroplane windows. Thus, to promote the widespread adoption of smart windows, it is imperative to further lower the production cost, improve the performance and durability and enhance the energy-saving benefits. As discussed in this chapter, nanomaterials and nanotechnologies have considerable potential for enabling the broader commercialization of smart windows, ascribed to the numerous advantages offered by them. Firstly, the solution-processable nanomaterials are compatible with low cost and scalable wet chemical processes, including sol-gel, hydrothermal and electrochemical approaches. The feasible and flexible control of the composition, morphology and structure of nanomaterials in the wet chemical processes greatly facilitate the optimization of smart windows. The excellent solution processability also makes the as-prepared nanomaterials amendable to ink formulations for 3D printings and applications on various surfaces. Secondly, nanocrystallization and nanostructuring of responsive materials can contribute to the improved thermochromic and electrochromic properties, for example, greater ΔT_{solar} and T_{lum} in thermochromic windows and the deeper and faster switching in electrochromic windows, and the better stability and durability. In particular, nanomaterials with unique properties can even introduce new functionality to smart windows, such as the electrical power generation and the independent control of visible and NIR light. Lastly, nanomaterials-based smart windows are expected to be much more efficient in saving energy consumed for HVAC than the static glazed windows or conventional blinds, consequently compensating the relatively higher priced smart windows.

Despite the advantages of the utilization of nanomaterials, some fundamental issues must be addressed before further development and large-scale commercialization of smart windows can be realised. In terms of the wet chemical synthesis of nanomaterials, one of the major challenges is the level of defects which is more difficult to control than the more expensive chemical vapour deposition methods. The presence of unwanted defects might degrade the performance and durability of smart windows. In addition, it is still problematic to scale up the synthesis of nanocrystals because the increased thermal input in the industrial-level production will drastically influence the nanocrystal growth, different from the laboratory-scale synthesis, which is undesirable for smart windows [2]. Although multiple challenges are still awaiting solutions, the recent encouraging advances in the nanomaterials-powering smart windows point to a promising future in this field, which provides the possibility of achieving greater market penetration and commercial success in the near future.

References

1. Y. Gao, H. Luo, Z. Zhang, L. Kang, Z. Chen, J. Du, M. Kanehira, C. Cao, Nanoceramic VO_2 thermochromic smart glass: a review on progress in solution processing. Nano Energy **1**(2), 221–246 (2012)

2. E.L. Runnerstrom, A. Llordés, S.D. Lounis, D.J. Milliron, Nanostructured electrochromic smart windows: traditional materials and NIR-selective plasmonic nanocrystals. Chem. Commun. **50**(73), 10555–10572 (2014)
3. Y. Wang, E.L. Runnerstrom, D.J. Milliron, Switchable materials for smart windows. Annu. Rev. Chem. Biomol. Eng. **7**(1), 283–304 (2016)
4. M.E.A. Warwick, R. Binions, Advances in thermochromic vanadium dioxide films. J. Mater. Chem. A **2**(10), 3275–3292 (2014)
5. C. Costa, C. Pinheiro, I. Henriques, C.A.T. Laia, Inkjet printing of sol-gel synthesized hydrated tungsten oxide nanoparticles for flexible electrochromic devices. ACS Appl. Mater. Interfaces **4**(3), 1330–1340 (2012)
6. M. Layani, P. Darmawan, W.L. Foo, L. Liu, A. Kamyshny, D. Mandler, S. Magdassi, P.S. Lee, Nanostructured electrochromic films by inkjet printing on large area and flexible transparent silver electrodes. Nanoscale **6**(9), 4572–4576 (2014)
7. H. Ji, D. Liu, H. Cheng, C. Zhang, Inkjet printing of vanadium dioxide nanoparticles for smart windows. J. Mater. Chem. C **6**(10), 2424–2429 (2018)
8. Y. Cui, Y. Ke, C. Liu, Z. Chen, N. Wang, L. Zhang, Y. Zhou, S. Wang, Y. Gao, Y. Long, Thermochromic VO_2 for energy-efficient smart windows. Joule **2**(9), 1707–1746 (2018)
9. A. Seeboth, D. Lötzsch, R. Ruhmann, O. Muehling, Thermochromic polymers—function by design. Chem. Rev. **114**(5), 3037–3068 (2014)
10. J.H. Day, Thermochromism of inorganic compounds. Chem. Rev. **68**(6), 649–657 (1968)
11. Z. Lu, C. Li, Y. Yin, Synthesis and thermochromic properties of vanadium dioxide colloidal particles. J. Mater. Chem. **21**(38), 14776–14782 (2011)
12. S.M. Babulanam, T.S. Eriksson, G.A. Niklasson, C.G. Granqvist, Thermochromic VO_2 films for energy-efficient windows. Solar Energy Mater. **16**(5), 347–363 (1987)
13. P. Baum, D.-S. Yang, A.H. Zewail, 4D visualization of transitional structures in phase transformations by electron diffraction. Science **318**(5851), 788–792 (2007)
14. K.H. Warnick, B. Wang, S.T. Pantelides, Hydrogen dynamics and metallic phase stabilization in VO_2. Appl. Phys. Lett. **104**(10), 101913 (2014)
15. T.D. Manning, I.P. Parkin, M.E. Pemble, D. Sheel, D. Vernardou, Intelligent window coatings: atmospheric pressure chemical vapor deposition of tungsten-doped vanadium dioxide. Chem. Mater. **16**(4), 744–749 (2004)
16. L.Q. Mai, B. Hu, T. Hu, W. Chen, E.D. Gu, Electrical property of Mo-doped VO_2 nanowire array film by melting–quenching sol–gel method. J. Phys. Chem. B **110**(39), 19083–19086 (2006)
17. Y. Ke, C. Zhou, Y. Zhou, S. Wang, S.H. Chan, Y. Long, Emerging thermal-responsive materials and integrated techniques targeting the energy-efficient smart window application. Adv. Funct. Mater. **28**(22), 1800113 (2018)
18. G. Xu, P. Jin, M. Tazawa, K. Yoshimura, Tailoring of luminous transmittance upon switching for thermochromic VO_2 films by thickness control. Jpn. J. Appl. Phys. **43**(1), 186–187 (2004)
19. H.K. Raut, V.A. Ganesh, A.S. Nair, S. Ramakrishna, Anti-reflective coatings: a critical, in-depth review. Energy Environ. Sci. **4**(10), 3779–3804 (2011)
20. J. Zheng, S. Bao, P. Jin, TiO_2(R)/VO_2(M)/TiO_2(A) multilayer film as smart window: combination of energy-saving, antifogging and self-cleaning functions. Nano Energy **11**, 136–145 (2015)
21. G. Xu, P. Jin, M. Tazawa, K. Yoshimura, Optimization of antireflection coating for VO_2-based energy efficient window. Sol. Energy Mater. Sol. Cells **83**(1), 29–37 (2004)
22. H. Koo, D. Shin, S.-H. Bae, K.-E. Ko, S.-H. Chang, C. Park, The effect of CeO_2 antireflection layer on the optical properties of thermochromic VO_2 film for smart window system. J. Mater. Eng. Perform. **23**(2), 402–407 (2014)
23. Y. Gao, S. Wang, H. Luo, L. Dai, C. Cao, Y. Liu, Z. Chen, M. Kanehira, Enhanced chemical stability of VO_2 nanoparticles by the formation of SiO_2/VO_2 core/shell structures and the application to transparent and flexible VO_2-based composite foils with excellent thermochromic properties for solar heat control. Energy Environ. Sci. **5**(3), 6104–6110 (2012)

24. Z. Chen, C. Cao, S. Chen, H. Luo, Y. Gao, Crystallised mesoporous TiO_2(A)–VO_2(M/R) nanocomposite films with self-cleaning and excellent thermochromic properties. J. Mater. Chem. A **2**(30), 11874–11884 (2014)
25. L. Kang, Y. Gao, H. Luo, Z. Chen, J. Du, Z. Zhang, Nanoporous thermochromic VO_2 films with low optical constants, enhanced luminous transmittance and thermochromic properties. ACS Appl. Mater. Interfaces **3**(2), 135–138 (2011)
26. Y. Choi, D.M. Sim, Y.H. Hur, H.J. Han, Y.S. Jung, Synthesis of colloidal VO_2 nanoparticles for thermochromic applications. Sol. Energy Mater. Sol. Cells **176**, 266–272 (2018)
27. Z. Zhang, Y. Gao, Z. Chen, J. Du, C. Cao, L. Kang, H. Luo, Thermochromic VO_2 thin films: solution-based processing, improved optical properties, and lowered phase transformation temperature. Langmuir **26**(13), 10738–10744 (2010)
28. Y. Gao, C. Cao, L. Dai, H. Luo, M. Kanehira, Y. Ding, Z.L. Wang, Phase and shape controlled VO_2 nanostructures by antimony doping. Energy Environ. Sci. **5**(9), 8708–8715 (2012)
29. K. Laaksonen, S.Y. Li, S.R. Puisto, N.K.J. Rostedt, T. Ala-Nissila, C.G. Granqvist, R.M. Nieminen, G.A. Niklasson, Nanoparticles of TiO_2 and VO_2 in dielectric media: conditions for low optical scattering, and comparison between effective medium and four-flux theories. Sol. Energy Mater. Sol. Cells **130**, 132–137 (2014)
30. Z. Chen, Y. Gao, L. Kang, C. Cao, S. Chen, H. Luo, Fine crystalline VO_2 nanoparticles: synthesis, abnormal phase transition temperatures and excellent optical properties of a derived VO_2 nanocomposite foil. J. Mater. Chem. A **2**(8), 2718–2727 (2014)
31. T. Moot, C. Palin, S. Mitran, J.F. Cahoon, R. Lopez, Designing plasmon-enhanced thermochromic films using a vanadium dioxide nanoparticle elastomeric composite. Adv. Opt. Mater. **4**(4), 578–583 (2016)
32. Y. Naoi, J. Amano, Optimization of VO_2 nanowire polymer composite thermochromic films by optical simulation. J. Appl. Phys. **120**(23), 235301 (2016)
33. S.-Y. Li, G.A. Niklasson, C.G. Granqvist, Nanothermochromics: calculations for VO_2 nanoparticles in dielectric hosts show much improved luminous transmittance and solar energy transmittance modulation. J. Appl. Phys. **108**(6), 063525 (2010)
34. C.G. Granqvist, P.C. Lansåker, N.R. Mlyuka, G.A. Niklasson, E. Avendaño, Progress in chromogenics: new results for electrochromic and thermochromic materials and devices. Sol. Energy Mater. Sol. Cells **93**(12), 2032–2039 (2009)
35. Y. Chen, X. Zeng, J. Zhu, R. Li, H. Yao, X. Cao, S. Ji, P. Jin, High performance and enhanced durability of thermochromic films using VO_2@ZnO core-shell nanoparticles. ACS Appl. Mater. Interfaces **9**(33), 27784–27791 (2017)
36. S.-Y. Li, G.A. Niklasson, C.G. Granqvist, Nanothermochromics with VO_2-based core-shell structures: calculated luminous and solar optical properties. J. Appl. Phys. **109**(11), 113515 (2011)
37. Y. Ke, X. Wen, D. Zhao, R. Che, Q. Xiong, Y. Long, Controllable fabrication of two-dimensional patterned VO_2 nanoparticle, nanodome, and nanonet arrays with tunable temperature-dependent localized surface plasmon resonance. ACS Nano **11**(7), 7542–7551 (2017)
38. K. Manthiram, A.P. Alivisatos, Tunable localized surface plasmon resonances in tungsten oxide nanocrystals. J. Am. Chem. Soc. **134**(9), 3995–3998 (2012)
39. N. Wang, M. Duchamp, C. Xue, R.E. Dunin-Borkowski, G. Liu, Y. Long, Single-crystalline W-doped VO_2 nanobeams with highly reversible electrical and plasmonic responses near room temperature. Adv. Mater. Interfaces **3**(15), 1600164 (2016)
40. J. Zhu, Y. Zhou, B. Wang, J. Zheng, S. Ji, H. Yao, H. Luo, P. Jin, Vanadium dioxide nanoparticle-based thermochromic smart coating: high luminous transmittance, excellent solar regulation efficiency, and near room temperature phase transition. ACS Appl. Mater. Interfaces **7**(50), 27796–27803 (2015)
41. Y. Zhou, A. Huang, Y. Li, S. Ji, Y. Gao, P. Jin, Surface plasmon resonance induced excellent solar control for VO_2@SiO_2 nanorods-based thermochromic foils. Nanoscale **5**(19), 9208–9213 (2013)

42. Y. Ke, Y. Yin, Q. Zhang, Y. Tan, P. Hu, S. Wang, Y. Tang, Y. Zhou, X. Wen, S. Wu, T.J. White, J. Yin, J. Peng, Q. Xiong, D. Zhao, Y. Long, Adaptive thermochromic windows from active plasmonic elastomers. Joule **3**(3), 858–871 (2019)
43. J. Cao, E. Ertekin, V. Srinivasan, W. Fan, S. Huang, H. Zheng, J.W.L. Yim, D.R. Khanal, D.F. Ogletree, J.C. Grossman, J. Wu, Strain engineering and one-dimensional organization of metal–insulator domains in single-crystal vanadium dioxide beams. Nat. Nanotechnol. **4**, 732 (2009)
44. X. Cao, N. Wang, J.Y. Law, S.C.J. Loo, S. Magdassi, Y. Long, Nanoporous thermochromic VO_2 (M) thin films: controlled porosity, largely enhanced luminous transmittance and solar modulating ability. Langmuir **30**(6), 1710–1715 (2014)
45. M. Zhou, J. Bao, M. Tao, R. Zhu, Y. Lin, X. Zhang, Y. Xie, Periodic porous thermochromic VO_2(M) films with enhanced visible transmittance. Chem. Commun. **49**(54), 6021–6023 (2013)
46. M. Liu, B. Su, Y.V. Kaneti, Z. Chen, Y. Tang, Y. Yuan, Y. Gao, L. Jiang, X. Jiang, A. Yu, Dual-phase transformation: spontaneous self-template surface-patterning strategy for ultra-transparent VO_2 solar modulating coatings. ACS Nano **11**(1), 407–415 (2017)
47. C. Ba, S.T. Bah, M. D'Auteuil, P.V. Ashrit, R. Vallée, Fabrication of high-quality VO_2 thin films by ion-assisted dual ac magnetron sputtering. ACS Appl. Mater. Interfaces **5**(23), 12520–12525 (2013)
48. J.P. Fortier, B. Baloukas, O. Zabeida, J.E. Klemberg-Sapieha, L. Martinu, Thermochromic VO_2 thin films deposited by HiPIMS. Sol. Energy Mater. Sol. Cells **125**, 291–296 (2014)
49. J.-C. Orlianges, J. Leroy, A. Crunteanu, R. Mayet, P. Carles, C. Champeaux, Electrical and optical properties of vanadium dioxide containing gold nanoparticles deposited by pulsed laser deposition. Appl. Phys. Lett. **101**(13), 133102 (2012)
50. R. Binions, G. Hyett, C. Piccirillo, I.P. Parkin, Doped and un-doped vanadium dioxide thin films prepared by atmospheric pressure chemical vapour deposition from vanadyl acetylacetonate and tungsten hexachloride: the effects of thickness and crystallographic orientation on thermochromic properties. J. Mater. Chem. **17**(44), 4652–4660 (2007)
51. M.E.A. Warwick, C.W. Dunnill, J. Goodall, J.A. Darr, R. Binions, Hybrid chemical vapour and nanoceramic aerosol assisted deposition for multifunctional nanocomposite thin films. Thin Solid Films **519**(18), 5942–5948 (2011)
52. H. Yin, K. Yu, C. Song, Z. Wang, Z. Zhu, Low-temperature CVD synthesis of patterned core–shell VO_2@ZnO nanotetrapods and enhanced temperature-dependent field-emission properties. Nanoscale **6**(20), 11820–11827 (2014)
53. D. Li, W. Huang, L. Song, Q. Shi, Thermal stability of VO_2 thin films deposited by sol–gel method. J. Sol-Gel. Sci. Technol. **75**(1), 189–197 (2015)
54. J. Wu, W. Huang, Q. Shi, J. Cai, D. Zhao, Y. Zhang, J. Yan, Effect of annealing temperature on thermochromic properties of vanadium dioxide thin films deposited by organic sol–gel method. Appl. Surf. Sci. **268**, 556–560 (2013)
55. L. Kang, Y. Gao, H. Luo, A novel solution process for the synthesis of VO_2 thin films with excellent thermochromic properties. ACS Appl. Mater. Interfaces **1**(10), 2211–2218 (2009)
56. H. Watanabe, Intelligent window using a hydrogel layer for energy efficiency. Sol. Energy Mater. Sol. Cells **54**(1), 203–211 (1998)
57. H.G. Schild, D.A. Tirrell, Microcalorimetric detection of lower critical solution temperatures in aqueous polymer solutions. J. Phys. Chem. **94**(10), 4352–4356 (1990)
58. K. Jain, R. Vedarajan, M. Watanabe, M. Ishikiriyama, N. Matsumi, Tunable LCST behavior of poly(N-isopropylacrylamide/ionic liquid) copolymers. Polym. Chem. **6**(38), 6819–6825 (2015)
59. C. Wu, S. Zhou, Laser light scattering study of the phase transition of poly(N-isopropylacrylamide) in Water. 1. Single Chain. Macromolecules **28**(24), 8381–8387 (1995)
60. Y.-S. Yang, Y. Zhou, F.B.Y. Chiang, Y. Long, Tungsten doped VO_2/microgels hybrid thermochromic material and its smart window application. RSC Adv. **7**(13), 7758–7762 (2017)
61. T.-G. La, X. Li, A. Kumar, Y. Fu, S. Yang, H.-J. Chung, Highly flexible, multipixelated thermosensitive smart windows made of tough hydrogels. ACS Appl. Mater. Interfaces **9**(38), 33100–33106 (2017)

62. Y. Zhou, Y. Cai, X. Hu, Y. Long, Temperature-responsive hydrogel with ultra-large solar modulation and high luminous transmission for "smart window" applications. J. Mater. Chem. A **2**(33), 13550–13555 (2014)
63. H.Y. Lee, Y. Cai, S. Bi, Y.N. Liang, Y. Song, X.M. Hu, A dual-responsive nanocomposite toward climate-adaptable solar modulation for energy-saving smart windows. ACS Appl. Mater. Interfaces **9**(7), 6054–6063 (2017)
64. Y. Zhou, Y. Cai, X. Hu, Y. Long, VO_2/hydrogel hybrid nanothermochromic material with ultra-high solar modulation and luminous transmission. J. Mater. Chem. A **3**(3), 1121–1126 (2015)
65. V.R. de la Rosa, P. Woisel, R. Hoogenboom, Supramolecular control over thermoresponsive polymers. Mater. Today **19**(1), 44–55 (2016)
66. D. Kim, E. Lee, H.S. Lee, J. Yoon, Energy efficient glazing for adaptive solar control fabricated with photothermotropic hydrogels containing graphene oxide. Sci. Rep. **5**, 7646 (2015)
67. X. Li, D. Bi, C. Yi, J.-D. Décoppet, J. Luo, S.M. Zakeeruddin, A. Hagfeldt, M. Grätzel, A vacuum flash–assisted solution process for high-efficiency large-area perovskite solar cells. Science **353**(6294), 58–62 (2016)
68. M.M. Lee, J. Teuscher, T. Miyasaka, T.N. Murakami, H.J. Snaith, Efficient hybrid solar cells based on meso-superstructured organometal halide perovskites. Science **338**(6107), 643–647 (2012)
69. M. Liu, M.B. Johnston, H.J. Snaith, Efficient planar heterojunction perovskite solar cells by vapour deposition. Nature **501**, 395 (2013)
70. M. Grätzel, The light and shade of perovskite solar cells. Nat. Mater. **13**, 838 (2014)
71. M. De Bastiani, M.I. Saidaminov, I. Dursun, L. Sinatra, W. Peng, U. Buttner, O.F. Mohammed, O.M. Bakr, Thermochromic perovskite inks for reversible smart window applications. Chem. Mater. **29**(8), 3367–3370 (2017)
72. M.I. Saidaminov, A.L. Abdelhady, B. Murali, E. Alarousu, V.M. Burlakov, W. Peng, I. Dursun, L. Wang, Y. He, G. Maculan, A. Goriely, T. Wu, O.F. Mohammed, O.M. Bakr, High-quality bulk hybrid perovskite single crystals within minutes by inverse temperature crystallization. Nat. Commun. **6**, 7586 (2015)
73. G. Maculan, A.D. Sheikh, A.L. Abdelhady, M.I. Saidaminov, M.A. Haque, B. Murali, E. Alarousu, O.F. Mohammed, T. Wu, O.M. Bakr, $CH_3NH_3PbCl_3$ single crystals: inverse temperature crystallization and visible-blind UV-photodetector. J. Phys. Chem. Lett. **6**(19), 3781–3786 (2015)
74. C.C. Stoumpos, C.D. Malliakas, M.G. Kanatzidis, Semiconducting tin and lead iodide perovskites with organic cations: phase transitions, high mobilities, and near-infrared photoluminescent properties. Inorg. Chem. **52**(15), 9019–9038 (2013)
75. L.M. Wheeler, D.T. Moore, R. Ihly, N.J. Stanton, E.M. Miller, R.C. Tenent, J.L. Blackburn, N.R. Neale, Switchable photovoltaic windows enabled by reversible photothermal complex dissociation from methylammonium lead iodide. Nat. Commun. **8**(1), 1722 (2017)
76. C.K. MØller, Crystal structure and photoconductivity of cæsium plumbohalides. Nature **182**(4647), 1436 (1958)
77. S. Sharma, N. Weiden, A. Weiss, Phase diagrams of quasibinary systems of the type: ABX3—A′BX3; ABX3—AB′X3, and ABX3—ABX′3; X = Halogen. Z. Phys. Chem. (Muenchen, Ger.) **175**, 63 (1992)
78. J. Lin, M. Lai, L. Dou, C.S. Kley, H. Chen, F. Peng, J. Sun, D. Lu, S.A. Hawks, C. Xie, F. Cui, A.P. Alivisatos, D.T. Limmer, P. Yang, Thermochromic halide perovskite solar cells. Nat. Mater. **17**(3), 261–267 (2018)
79. G.A. Niklasson, C.G. Granqvist, Electrochromics for smart windows: thin films of tungsten oxide and nickel oxide, and devices based on these. J. Mater. Chem. **17**(2), 127–156 (2007)
80. C.M. Amb, A.L. Dyer, J.R. Reynolds, Navigating the color palette of solution-processable electrochromic polymers. Chem. Mater. **23**(3), 397–415 (2011)
81. H.C. Moon, T.P. Lodge, C.D. Frisbie, Solution processable, electrochromic ion gels for Sub-1 V, flexible displays on plastic. Chem. Mater. **27**(4), 1420–1425 (2015)

82. M. Stolar, J. Borau-Garcia, M. Toonen, T. Baumgartner, Synthesis and tunability of highly electron-accepting, N-benzylated "phosphaviologens". J. Am. Chem. Soc. **137**(9), 3366–3371 (2015)
83. Y. Li, H. Qu, Z. Tong, J. Zhao, Nanostructured electrochromic materials, in *Electrochromic Smart Materials: Fabrication and Applications* (The Royal Society of Chemistry, 2019), pp. 430–474
84. L. Zheng, Y. Xu, D. Jin, Y. Xie, Novel metastable hexagonal MoO_3 nanobelts: synthesis, photochromic, and electrochromic properties. Chem. Mater. **21**(23), 5681–5690 (2009)
85. R. Cinnsealach, G. Boschloo, S. Nagaraja Rao, D. Fitzmaurice, Coloured electrochromic windows based on nanostructured TiO_2 films modified by adsorbed redox chromophores. Sol. Energy Mater. Sol. Cells **57**(2), 107–125 (1999)
86. S.K. Deb, A novel electrophotographic system. Appl. Opt. **8**(S1), 192–195 (1969)
87. J.S.E.M. Svensson, C.G. Granqvist, Electrochromic coatings for "smart windows". Solar Energy Mater. **12**(6), 391–402 (1985)
88. W. Wu, M. Wang, J. Ma, Y. Cao, Y. Deng, Electrochromic metal oxides: recent progress and prospect. Adv. Electron. Mater. **4**(8), 1800185 (2018)
89. R.J. Mortimer, Electrochromic materials. Annu. Rev. Mater. Res. **41**(1), 241–268 (2011)
90. V.K. Thakur, G. Ding, J. Ma, P.S. Lee, X. Lu, Hybrid materials and polymer electrolytes for electrochromic device applications. Adv. Mater. **24**(30), 4071–4096 (2012)
91. Y.-Y. Song, Z.-D. Gao, J.-H. Wang, X.-H. Xia, R. Lynch, Multistage coloring electrochromic device based on TiO_2 nanotube arrays modified with WO_3 nanoparticles. Adv. Funct. Mater. **21**(10), 1941–1946 (2011)
92. S.-H. Lee, R. Deshpande, P.A. Parilla, K.M. Jones, B. To, A.H. Mahan, A.C. Dillon, Crystalline WO_3 nanoparticles for highly improved electrochromic applications. Adv. Mater. **18**(6), 763–766 (2006)
93. G. Cai, X. Wang, M. Cui, P. Darmawan, J. Wang, A.L.-S. Eh, P.S. Lee, Electrochromo-supercapacitor based on direct growth of NiO nanoparticles. Nano Energy **12**, 258–267 (2015)
94. R.S. Devan, S.-Y. Gao, W.-D. Ho, J.-H. Lin, Y.-R. Ma, P.S. Patil, Y. Liou, Electrochromic properties of large-area and high-density arrays of transparent one-dimensional β-Ta_2O_5 nanorods on indium-tin-oxide thin-films. Appl. Phys. Lett. **98**(13), 133117 (2011)
95. G. Yuan, C. Hua, S. Khan, S. Jiang, Z. Wu, Y. Liu, J. Wang, C. Song, G. Han, Improved electrochromic performance of WO_3 films with size controlled nanorods. Electrochim. Acta **260**, 274–280 (2018)
96. R.A. Patil, R.S. Devan, J.-H. Lin, Y.-R. Ma, P.S. Patil, Y. Liou, Efficient electrochromic properties of high-density and large-area arrays of one-dimensional NiO nanorods. Sol. Energy Mater. Sol. Cells **112**, 91–96 (2013)
97. A. Azam, J. Kim, J. Park, T.G. Novak, A.P. Tiwari, S.H. Song, B. Kim, S. Jeon, Two-dimensional WO_3 nanosheets chemically converted from layered WS_2 for high-performance electrochromic devices. Nano Lett. **18**(9), 5646–5651 (2018)
98. Z. Bi, X. Li, Y. Chen, X. He, X. Xu, X. Gao, Large-scale multifunctional electrochromic-energy storage device based on tungsten trioxide monohydrate nanosheets and prussian white. ACS Appl. Mater. Interfaces **9**(35), 29872–29880 (2017)
99. D. Ma, G. Shi, H. Wang, Q. Zhang, Y. Li, Hierarchical NiO microflake films with high coloration efficiency, cyclic stability and low power consumption for applications in a complementary electrochromic device. Nanoscale **5**(11), 4808–4815 (2013)
100. C. Xiong, A.E. Aliev, B. Gnade, K.J. Balkus, Fabrication of silver vanadium oxide and V_2O_5 nanowires for electrochromics. ACS Nano **2**(2), 293–301 (2008)
101. J.-L. Wang, Y.-R. Lu, H.-H. Li, J.-W. Liu, S.-H. Yu, Large area co-assembly of nanowires for flexible transparent smart windows. J. Chem. Soc. **139**(29), 9921–9926 (2017)
102. A. Hasani, Q.V. Le, M. Tekalgne, W. Guo, S.H. Hong, K.S. Choi, T.H. Lee, H.W. Jang, S.Y. Kim, Tungsten trioxide doped with CdSe quantum dots for smart windows. ACS Appl. Mater. Interfaces **10**(50), 43785–43791 (2018)
103. S. Cong, Y. Tian, Q. Li, Z. Zhao, F. Geng, Single-crystalline tungsten oxide quantum dots for fast pseudocapacitor and electrochromic applications. Adv. Mater. **26**(25), 4260–4267 (2014)

104. C.-P. Li, C.A. Wolden, A.C. Dillon, R.C. Tenent, Electrochromic films produced by ultrasonic spray deposition of tungsten oxide nanoparticles. Sol. Energy Mater. Sol. Cells **99**, 50–55 (2012)
105. G. Cai, P. Darmawan, X. Cheng, P.S. Lee, Inkjet printed large area multifunctional smart windows. Adv. Energy Mater. **7**(14), 1602598 (2017)
106. C.-W. Chang-Jian, E.-C. Cho, S.-C. Yen, B.-C. Ho, K.-C. Lee, J.-H. Huang, Y.-S. Hsiao, Facile preparation of WO_3/PEDOT:PSS composite for inkjet printed electrochromic window and its performance for heat shielding. Dyes Pigm. **148**, 465–473 (2018)
107. G. Cai, P. Darmawan, M. Cui, J. Chen, X. Wang, A.L.-S. Eh, S. Magdassi, P.S. Lee, Inkjet-printed all solid-state electrochromic devices based on NiO/WO_3 nanoparticle complementary electrodes. Nanoscale **8**(1), 348–357 (2016)
108. J. Zhang, J.-P. Tu, X.-H. Xia, X.-L. Wang, C.-D. Gu, Hydrothermally synthesized WO_3 nanowire arrays with highly improved electrochromic performance. J. Mater. Chem. **21**(14), 5492–5498 (2011)
109. J. Zhou, Y. Ding, S.Z. Deng, L. Gong, N.S. Xu, Z.L. Wang, Three-dimensional tungsten oxide nanowire networks. Adv. Mater. **17**(17), 2107–2110 (2005)
110. Y. Li, Y. Bando, D. Golberg, Quasi-aligned single-crystalline W18O49 nanotubes and nanowires. Adv. Mater. **15**(15), 1294–1296 (2003)
111. R.A. Patil, R.S. Devan, Y. Liou, Y.-R. Ma, Efficient electrochromic smart windows of one-dimensional pure brookite TiO_2 nanoneedles. Sol. Energy Mater. Sol. Cells **147**, 240–245 (2016)
112. S.-H. Lee, C.E. Tracy, Y. Yan, J.R. Pitts, S.K. Deb, Solid-State nanocomposite electrochromic pseudocapacitors. Electrochem. Solid-State Lett. **8**(4), A188–A190 (2005)
113. A. Agrawal, S.H. Cho, O. Zandi, S. Ghosh, R.W. Johns, D.J. Milliron, Localized surface plasmon resonance in semiconductor nanocrystals. Chem. Rev. **118**(6), 3121–3207 (2018)
114. A. Llordés, G. Garcia, J. Gazquez, D.J. Milliron, Tunable near-infrared and visible-light transmittance in nanocrystal-in-glass composites. Nature **500**, 323 (2013)
115. G. Garcia, R. Buonsanti, E.L. Runnerstrom, R.J. Mendelsberg, A. Llordes, A. Anders, T.J. Richardson, D.J. Milliron, Dynamically modulating the surface plasmon resonance of doped semiconductor nanocrystals. Nano Lett. **11**(10), 4415–4420 (2011)
116. A. Llordés, Y. Wang, A. Fernandez-Martinez, P. Xiao, T. Lee, A. Poulain, O. Zandi, C.A. Saez Cabezas, G. Henkelman, D.J. Milliron, Linear topology in amorphous metal oxide electrochromic networks obtained via low-temperature solution processing. Nat. Mater. **15**, 1267 (2016)
117. C.J. Barile, D.J. Slotcavage, M.D. McGehee, Polymer-nanoparticle electrochromic materials that selectively modulate visible and near-infrared light. Chem. Mater. **28**(5), 1439–1445 (2016)
118. J. Kim, G.K. Ong, Y. Wang, G. LeBlanc, T.E. Williams, T.M. Mattox, B.A. Helms, D.J. Milliron, Nanocomposite architecture for rapid, spectrally-selective electrochromic modulation of solar transmittance. Nano Lett. **15**(8), 5574–5579 (2015)
119. H. Gu, C. Guo, S. Zhang, L. Bi, T. Li, T. Sun, S. Liu, Highly efficient, near-infrared and visible light modulated electrochromic devices based on polyoxometalates and W18O49 nanowires. ACS Nano **12**(1), 559–567 (2018)
120. C.J. Dahlman, Y. Tan, M.A. Marcus, D.J. Milliron, Spectroelectrochemical signatures of capacitive charging and ion insertion in doped anatase titania nanocrystals. J. Am. Chem. Soc. **137**(28), 9160–9166 (2015)
121. S. Cao, S. Zhang, T. Zhang, Q. Yao, J.Y. Lee, A visible light-near-infrared dual-band smart window with internal energy storage. Joule **3**(4), 1152–1162 (2019)
122. S. Zhang, S. Cao, T. Zhang, Q. Yao, A. Fisher, J.Y. Lee, Monoclinic oxygen-deficient tungsten oxide nanowires for dynamic and independent control of near-infrared and visible light transmittance. Mater. Horiz. **5**(2), 291–297 (2018)
123. S. Zhang, S. Cao, T. Zhang, A. Fisher, J.Y. Lee, Al^{3+} intercalation/de-intercalation-enabled dual-band electrochromic smart windows with a high optical modulation, quick response and long cycle life. Energy Environ. Sci. **11**(10), 2884–2892 (2018)

124. N. DeForest, A. Shehabi, J. O'Donnell, G. Garcia, J. Greenblatt, E.S. Lee, S. Selkowitz, D.J. Milliron, United States energy and CO_2 savings potential from deployment of near-infrared electrochromic window glazings. Build. Environ. **89**, 107–117 (2015)
125. B.A. Korgel, Composite for smarter windows. Nature **500**, 278 (2013)
126. K.S. Novoselov, A.K. Geim, S.V. Morozov, D. Jiang, Y. Zhang, S.V. Dubonos, I.V. Grigorieva, A.A. Firsov, Electric field effect in atomically thin carbon films. Science **306**(5696), 666–669 (2004)
127. K. Novoselov, Mind the gap. Nat. Mater. **6**, 720 (2007)
128. M. Chhowalla, H.S. Shin, G. Eda, L.-J. Li, K.P. Loh, H. Zhang, The chemistry of two-dimensional layered transition metal dichalcogenide nanosheets. Nat. Chem. **5**, 263 (2013)
129. C. Lee, X. Wei, J.W. Kysar, J. Hone, Measurement of the elastic properties and intrinsic strength of monolayer graphene. Science **321**(5887), 385–388 (2008)
130. K.S. Kim, Y. Zhao, H. Jang, S.Y. Lee, J.M. Kim, K.S. Kim, J.-H. Ahn, P. Kim, J.-Y. Choi, B.H. Hong, Large-scale pattern growth of graphene films for stretchable transparent electrodes. Nature **457**, 706 (2009)
131. L. Gomez De Arco, Y. Zhang, C.W. Schlenker, K. Ryu, M.E. Thompson, C. Zhou, Continuous, highly flexible, and transparent graphene films by chemical vapor deposition for organic photovoltaics. ACS Nano **4**(5), 2865–2873 (2010)
132. J. Wan, S.D. Lacey, J. Dai, W. Bao, M.S. Fuhrer, L. Hu, Tuning two-dimensional nanomaterials by intercalation: materials, properties and applications. Chem. Soc. Rev. **45**(24), 6742–6765 (2016)
133. K. Mallikarjuna, H. Kim, Highly transparent conductive reduced graphene oxide/silver nanowires/silver grid electrodes for low-voltage electrochromic smart windows. ACS Appl. Mater. Interfaces **11**(2), 1969–1978 (2019)
134. J. Palenzuela, A. Viñuales, I. Odriozola, G. Cabañero, H.J. Grande, V. Ruiz, Flexible viologen electrochromic devices with low operational voltages using reduced graphene oxide electrodes. ACS Appl. Mater. Interfaces **6**(16), 14562–14567 (2014)
135. S. Bae, H. Kim, Y. Lee, X. Xu, J.-S. Park, Y. Zheng, J. Balakrishnan, T. Lei, H. Ri Kim, Y.I. Song, Y.-J. Kim, K.S. Kim, B. Özyilmaz, J.-H. Ahn, B.H. Hong, S. Iijima, Roll-to-roll production of 30-inch graphene films for transparent electrodes. Nat. Nanotechnol. **5**, 574 (2010)
136. I. Khrapach, F. Withers, T.H. Bointon, D.K. Polyushkin, W.L. Barnes, S. Russo, M.F. Craciun, Novel highly conductive and transparent graphene-based conductors. Adv. Mater. **24**(21), 2844–2849 (2012)
137. W. Bao, J. Wan, X. Han, X. Cai, H. Zhu, D. Kim, D. Ma, Y. Xu, J.N. Munday, H.D. Drew, M.S. Fuhrer, L. Hu, Approaching the limits of transparency and conductivity in graphitic materials through lithium intercalation. Nat. Commun. **5**, 4224 (2014)
138. K. Kanetani, K. Sugawara, T. Sato, R. Shimizu, K. Iwaya, T. Hitosugi, T. Takahashi, Ca intercalated bilayer graphene as a thinnest limit of superconducting C_6Ca. Proc. Natl. Acad. Sci. U. S. A. **109**(48), 19610–19613 (2012)
139. J. Wan, F. Gu, W. Bao, J. Dai, F. Shen, W. Luo, X. Han, D. Urban, L. Hu, Sodium-ion intercalated transparent conductors with printed reduced graphene oxide networks. Nano Lett. **15**(6), 3763–3769 (2015)
140. J. Wang, K.K. Manga, Q. Bao, K.P. Loh, High-yield synthesis of few-layer graphene flakes through electrochemical expansion of graphite in propylene carbonate electrolyte. J. Am. Chem. Soc. **133**(23), 8888–8891 (2011)
141. J. Wan, W. Bao, Y. Liu, J. Dai, F. Shen, L. Zhou, X. Cai, D. Urban, Y. Li, K. Jungjohann, M.S. Fuhrer, L. Hu, In situ investigations of Li-MoS_2 with planar batteries. Adv. Energy Mater. **5**(5), 1401742 (2015)
142. F. Xiong, H. Wang, X. Liu, J. Sun, M. Brongersma, E. Pop, Y. Cui, Li Intercalation in MoS_2: in situ observation of its dynamics and tuning optical and electrical properties. Nano Lett. **15**(10), 6777–6784 (2015)
143. K. Hantanasirisakul, M.-Q. Zhao, P. Urbankowski, J. Halim, B. Anasori, S. Kota, C.E. Ren, M.W. Barsoum, Y. Gogotsi, Fabrication of $Ti_3C_2T_x$ MXene transparent thin films with tunable optoelectronic properties. Adv. Electron. Mater. **2**(6), 1600050 (2016)

Chapter 8
Proton-Responsive Nanomaterials for Fuel Cells

Xi Xu and Lei Bi

Abstract Proton-conducting oxides receive more and more attention in the current world because of its wide applications in renewable and sustainable devices. Among all these applications, the use of proton-conducting oxide for fuel cells is becoming quite a hot topic as it avoids the problems for traditional fuel cells using oxide electrolyte and also lowers the working temperature of fuel cells, making them possible for practical applications. With the framework of proton-conducting solid oxide fuel cells, the utilization of nanomaterials now playing an essential part in the whole community, and this chapter will briefly summarize the nanomaterials for protonic oxide fuel cells that are also known as proton-conducting solid oxide fuel cells (SOFCs).

8.1 Proton Conduction in Oxides

Ionic conduction has been discovered in oxides long back, and for a long period, people can only find the migration of oxygen ions in the oxide at high temperatures, and there is no evidence suggesting proton could migrate in the oxide. However, Iwahara et al. [1] discovered that some oxides exhibit proton conductions under a wet atmosphere which opens a new door in the solid-state ionic's field. Although the proton migration mechanism in oxides is not completely revealed, most of the evidence suggests that H_2O incorporates into the lattice and fills the oxygen vacancies to form the proton defects [2]:

$$H_2O + V_o^{\bullet\bullet} + O_o^{\times} \Leftrightarrow 2OH^{\bullet}$$

Compared with the traditional oxygen-ion conductors, proton-conducting oxides have smaller activation energies that are helpful for their low-temperature operations [3]. It should be noted that the low-temperature operation of SOFCs is important

X. Xu · L. Bi (✉)
Institute of Materials for Energy and Environment, College of Materials Science and Engineering, Qingdao University, Ningxia Road No. 308, Qingdao 266071, China
e-mail: bilei81@gmail.com; bilei@qdu.edu.cn

Z. Sun and T. Liao (eds.), *Responsive Nanomaterials for Sustainable Applications*, Springer Series in Materials Science 297,
https://doi.org/10.1007/978-3-030-39994-8_8

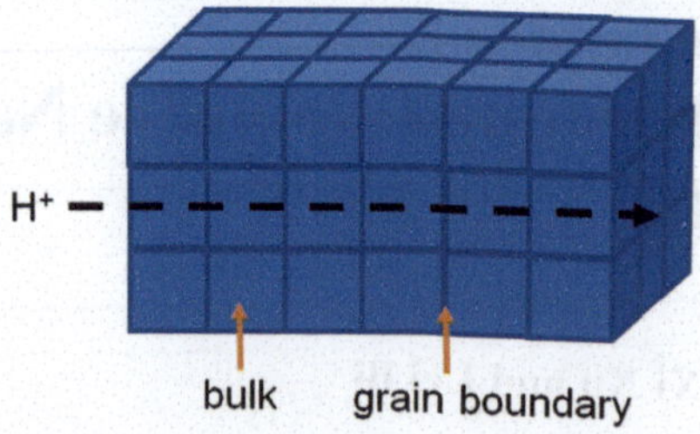

Fig. 8.1 Scheme for the proton transportation in polycrystalline electrolyte

as it could extend the lifetime of the fuel cells and also provide the possibility for the quick start-up of fuel cell devices [4]. In addition, the fuel cells using proton-conducting electrolyte offer another advantage of producing H_2O at the cathode side that will not dilute the fuels that always happens for the oxygen-ion conducting SOFCs, making the development of proton-conducting SOFCs becomes a hot topic in the community [5]. As a fuel cell device, it usually contains the components including electrolyte, anode, and cathode. In the following sections, we will briefly introduce the state-of-the-art as well as the recent development of nanomaterials for protonic SOFCs.

8.2 Proton-Conducting Electrolytes

Proton conduction was firstly found in doped $SrCeO_3$ materials [1] and then different oxides are reported to show proton conduction in the past four decades [6–8]. However, the most classical proton-conducting oxides are still the ABO_3-based perovskite oxides, and they are usually used as electrolytes for proton-conducting SOFCs [9]. As the fuel cell electrolyte, it is desirable to have a low resistance which could avoid the high ohmic loss in the fuel cell operations. This requires high conductivity for the electrolyte materials. According to the classical "brick-layer model," the solid-state electrolyte consisted of grain interiors (also called as "bulk") and the grain boundaries, and protons have to pass through the grains and the grain boundaries for the transportation [10], as schemed in Fig. 8.1. It has been demonstrated that the resistance of the grain boundary is a few orders of magnitude of larger than that of the grains, suggesting the reduction of the grain boundary resistance is the key for the improvement of the overall electrolyte conductivity [10]. Generally, there are two routes to improve the grain boundary conductivity. One is to decrease the thickness of the grain boundary so that protons can pass through less distance with the high resistance grain boundaries. In this case, the specific grain boundary resistance could be decreased. The other approach is to increase grain size which decreases the volume of the grain boundaries. Although the specific grain boundary resistance is not changed with this strategy, the total grain boundary resistance is reduced because of the reduced grain boundary volume that allows the protons to meet fewer grain boundaries during their transportation. Nanotechnologies and nanomaterials play an important role in both strategies and let us introduce them in detail.

8.2.1 Reducing the Specific Grain Boundary Resistance

Comparing with the grain which has typically a size of a few micrometers, the grain boundary usually has a thickness of a few to tens of nanometers. Therefore, tailoring the thickness at the nanoscale is quite challenging in early studies but now some effective strategies have been used to reduce the specific grain boundary resistance by tailoring the grain boundaries. Y doped $BaZrO_3$ (BZY) is one of the most investigated proton-conducting oxides in the field but its conductivity is highly hindered by its grain boundary resistance [11]. To overcome this drawback, Sun et al. [12] have used Li_2O as the sintering aid to improve the sinterability of BZY. Different to other sintering aid-modified BZY materials that have large grains, the Li_2O modified BZY pellet shows a small grain size of a few hundred nanometers. However, electrochemical measurements indicate that the conductivity of Li_2O modified BZY shows high conductivity compared with other sintering aids-modified BZY. Further, material analysis reveals that there are two factors helping the conductivity for the Li_2O modified BZY. One is that Li_2O participates in the sintering procedure but it evaporates leaving the sample without Li element. In this case, the electrochemical properties of the sintered BZY electrolyte are free of the influence of the Li element. It has been demonstrated that although sintering aids could improve the sinterability of the samples, they impair the electrical properties, so the dense sample free of the sintering aid could maintain the high conductivity of BZY. Second, the microstructure analysis reveals that the formed BZY pellet has a very clear and thin grain boundary, as shown in Fig. 8.2. The thickness of the grain boundary is about 4 nm, which narrows the distance that protons must pass through for each grain boundary and thus improves the conductivity.

One may think how about the complete removal of the grain boundaries; although this task is difficult for traditional protonic SOFCs, it becomes feasible in the filed thin-film SOFC. Pergolesi et al. [13] have fabricated a grain boundary-free BZY thin films by the pulsed laser deposition (PLD) method on the MgO substrate and found that the conductivity of the BZY film is much larger than that of the polycrystalline BZY pellet and the conductivity value is also higher than most of the oxygen-ion conductors, implying the advantage in electrochemistry by using proton conductors. Furthermore, the measured conductivity of the BZY films matches well with the theoretical bulk conductivity of BZY, suggesting the grain boundary is indeed removed and the total conductivity is equal to the bulk conductivity. This study demonstrates for the first time that the grain boundary of proton conductor could be removed from an experimental way, achieving high conductivity for the sample. It should be noted that this BZY thin film is fabricated on the MgO substrate that is not the configuration for fuel cell devices, so the fabrication of nanoscale electrolyte on the conventional NiO-based electrolyte is necessary to explore the possibility of using BZY thin films

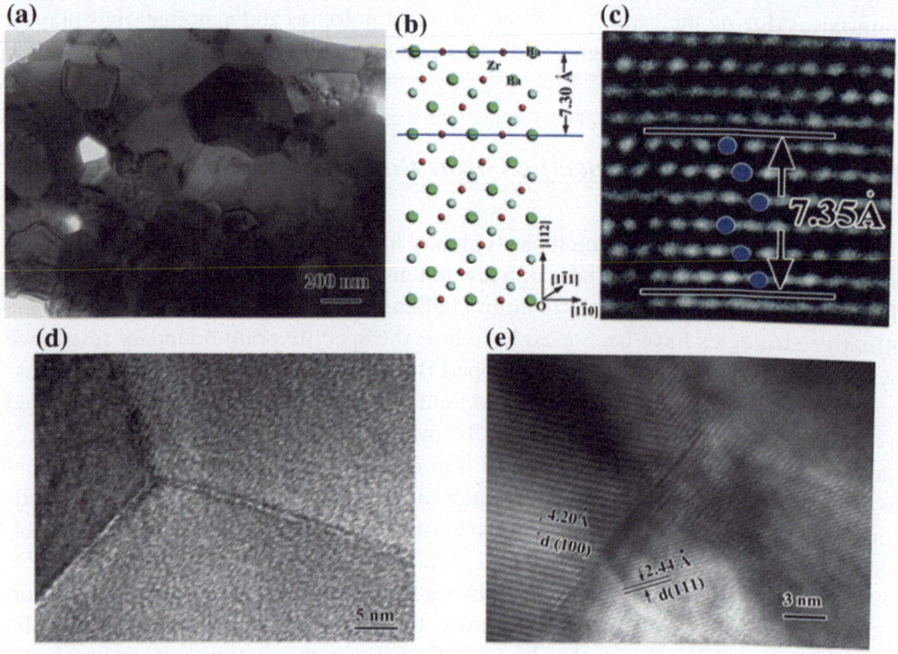

Fig. 8.2 Microstructure of BZY electrolyte sintered with the addition of Li_2O, showing clean and thin grain boundaries [12]. Reproduced by permission of the PCCP Owner Societies

for fuel cell applications. Fabbri et al. have employed to PLD method to fabricate a thin BZY film on the NiO-based anode and tested in the fuel cell condition. However, the fuel cell performance of this fuel cell is not high, due to the use of the single-chamber mode. In addition, the BZY electrolyte film was fabricated on the NiO-based anode. Unlike the single-crystal MgO cathode used in the previous study, the NiO-based anode did not have a specific orientation that made the growth of single-crystal BZY film difficult. Therefore, although the thin-film BZY has been fabricated by the PLD method, the grain boundary still existed for this electrolyte and decreased the conductivity of the electrolyte and thus decreased the overall fuel cell performance [14]. Similarly, the BZY thin film was also fabricated on the sintered BCY electrolyte, forming the bilayer electrolyte structure [15]. The main purpose of this structure is to shield the unstable BCY electrolyte with the stable BZY thin film, making this electrolyte has better chemical stability. Although the conductivity of the bilayer electrolyte is obviously larger than that of the pure BZY electrolyte, the value is still apparently lower than that of the BCY electrolyte from the protective layer. In spite of the thin BZY electrolyte used, the polycrystalline BZY electrolyte still influences the overall conductivity of the electrolyte. Recently, Bae et al. [16] have tried to solve this problem by fabricating the BZY column-like crystalline electrolyte. As seen in Fig. 8.3, different from the conventional polycrystalline BZY electrolyte, the column-like grain allows the transportation of protons from the anode side to the

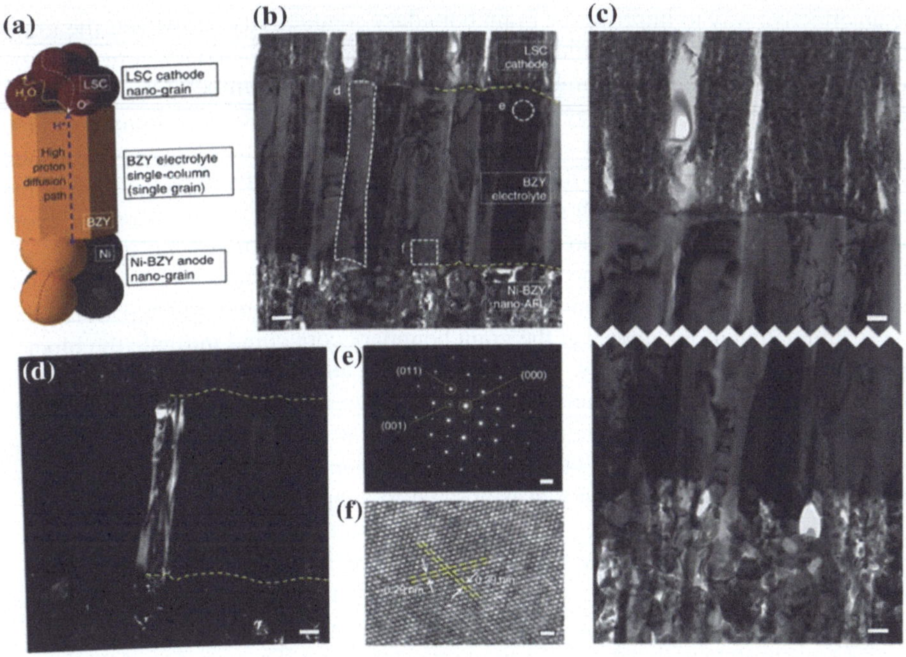

Fig. 8.3 Microstrure for the BZY thin-film electrolyte with columnar structure grain boundaries, facilitating the migration of protons. Reproduced with permission from [16]

cathode side without passing through the grain boundaries, facilitating the migration procedure and thus improves the electrolyte conductivity. Although this work and that work presented in [14] both use PLD fabricated thin-film BZY electrolyte, the current work shows much higher fuel cell performance, suggesting the removal of the grain boundary during the proton transport is essential for its high performance. However, it should be noted that all these works have to rely on the PLD method and PLD is an advanced physical deposition method that is more suitable for small-scale applications. The tailoring of the grain boundary at the nanoscale with the traditional processing method could be more significant for fuel cell applications.

The above strategies have been proposed to reduce the grain boundary thickness or eliminate the high resistance grain boundaries, but these methods did not aim to improve the conductivity of the specific grain boundary conductivity at a given thickness, while it is reasonable to assume that the conductivity of the grain boundary could be enhanced if the grain boundaries becomes conductive. Although the high resistance of grain boundary is recognized, the reason for this high resistance is still debating, and the most popular opinion is that the form of the space charge layer is the main reason for the high resistive grain boundaries [17, 18]. The formation of the space charge layer is due to the accumulation of positive charge of the core of the grain boundary that blocks the transportation of protons. Therefore, neutralizing the positive charge at the grain boundary core to suppress the space charge layer could be an effective way to improve the grain boundary conductivity. However, the grain boundary is only a few nm, while the thickness of the grain boundary core is only a few Å that makes the modification difficult. Recently, high-temperature treatments were found to be a possible way to tailor the properties of the grain boundary [19, 20]. After sintering the pellet by spark plasma sintering (SPS), the sample was further treatment at high temperatures that allowed the diffusion of the dopant to the grain boundary region. The SPS method was used here to guarantee the high densification of the pellet as well as almost no grain growth in further heat treatment. In this case, the difference in grain size (the grain boundary volume) contribution to the grain boundary conductivity is negligible. Due to the negative charge of the dopant (M'_{Zr}), the accumulation of the dopant at the grain boundary core could mitigate the blocking effect of the grain boundary. The microstructure analysis has confirmed that the dopant indeed diffuses into the grain boundary region, while the electrochemical studies indicate that the conductivity of the sample is dramatically improved compared with the sample without tailoring the grain boundary. This evidence suggests that tailoring the grain boundary at nanoscale to suppress the space charge layer effect is the key to improve the grain boundary properties.

Besides tailoring the grain boundary itself, the introduction of noble metal nanoparticles is reported to be an effective method to improve the grain boundary conductivity. Tong et al. [21] have reported the conductivity enhancement for the BZY electrolyte with the use of nano-ionic composite. They have used Pd metal nanoparticles as the dispersed phase and the Pd has little effect on the conductivity when it incorporates into the BZY lattice. After reducing in H_2 atmosphere, Pd nanoparticles exsolve from the BZY lattice, forming the nanocomposite, and the exsolved Pd nanoparticles are found to accumulate at the BZY grain boundaries, as shown in Fig. 8.4. Although all the BZY-Pd nanoparticle composites show higher conductivity than that of BZY without exsolved Pd, the sample with the small Pd nanoparticles (5–10 nm) shows the best conductivity which is 2.7 times higher than

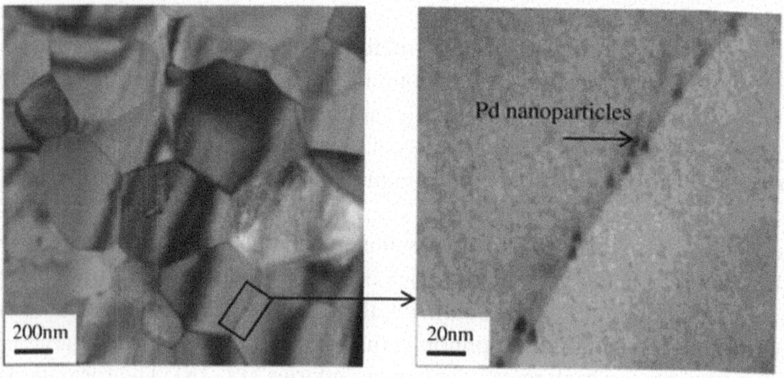

Fig. 8.4 The use of Pd nanoparticles at the grain boundary of BZY could enhance the conductivity. Reproduced from [21] with permission from Elsevier B.V.

that of the pure BZY sample, suggesting the smaller Pd particle size could better promote the conductivity of the whole sample.

8.2.2 *Reducing the Overall Grain Boundary Resistance*

The other strategy to improve the electrolyte conductivity is to increase the grain size of the electrolyte and thus decreases the volume of the grain boundaries and thus reduces the overall grain boundary resistance. According to the ceramic sintering principle, the oxide powder with small particle size having high surface energy could be easily sintered and the better growth of grain size could be expected [22]. Proton-conducting oxides are traditionally synthesized by a solid-state reaction, in which metal oxides powders are mixed together and then co-fired at a high temperature to achieve the desired phase for the target proton-conducting oxides. The solid-state reaction method provides a simple and straightway to prepare proton-conducting oxides, but high calcination temperatures have to be used in this method to accelerate the inter-diffusion between the starting oxide materials, and the high calcination temperature inevitably leads to the formation of coarse powders [23]. It is recognized that the coarse powder is not good for the later material sintering procedure, and ultra-high sintering temperature has to be used to density the pellet and sometimes causes the material evaporation because of the too high sintering temperature used. In contrast, the nano-sized powder is more desirable for the electrolyte fabrication and wet chemical routes are used to prepare nano-sized proton-conducting oxides. It has been reported that the solid-state reaction prepared BZY powder has to be sintered at 1800–2200 °C, while the sintering temperature of 1600 °C is sufficient for sintering the nano-sized proton-conducting powders [23]. Yamazaki et al. [24] have used the sol–gel method to prepare nanometric BZY powder and the dense BZY electrolyte pellet could be obtained after sintering at 1600 °C. In addition, the grain size of the BZY pellet is around 1 μm that dramatically reduces the volume of the grain boundaries and thus leads to high total conductivity of 0.01 S cm^{-1} at 450 °C.

Similar results are also obtained by Fabbri et al., who use Pr and Y co-doped $BaZrO_3$ (BZPY) as the electrolyte for protonic SOFCs [25]. The combustion method was employed to prepare nano-sized BZPY powders. The sinterability tests indicate that the incorporation of Pr into BZY can significantly promote the sinterability of the material, leading to not only the higher shrinkage but also the large grains. As a result, the BZYP pellet shows higher conductivity than that of the BZY pellet without using Pr, leading a conductivity value of 0.01 S cm^{-1} at 600 °C, without impairing the chemical stability. Furthermore, the introduction of 10 mol% Pr into BZY does not bring obvious electronic conduction at intermediate temperatures. Recent work reveals that even with a small addition of Pr (5 mol%) forming the nano-sized particles, the sinterability and conductivity could be significantly improved [26].

NiO also plays a similar role in the sinterability and conductivity improvements for BZY. Shafi et al. [27] use NiO as the dopant instead of external sintering aid for

BZY. It is found that the doping concentration of NiO in BZY is less than 10 mol% and the sinterability of the material greatly improves with an only doping amount of 4 mol%. As a result, the electrolyte can be easily sintered at 1400 °C and the anode supported fuel cell generates a peak power density of 428 mW cm^{-2} at 700 °C which is one of the largest power outputs for BZY-based cells at the time reported.

Apart from the use of wet chemical routes, the utilization of proper calcination method is recently found to be feasible for further reducing the initial particle size of the powder that helps the electrolyte densification and grain growth in the later sintering procedure. Microwave sintering is a sintering method that has been employed in ceramic sintering and it offers the advantage of fast sintering, eco-friendly, and low sintering temperatures [28]. Although the phrase "sintering" usually refers to the heat treatment to densify the material, the basic working principle of microwave sintering makes it also feasible for the calcination of the powder, in spite it has not been used for proton-conducting oxides before. However, a recent study indicates that the use of microwave sintering to calcine the $BaZr_{0.1}Ce_{0.7}Y_{0.2}O_{3-\delta}$ (BCZY) proton-conducting oxide powder could lower the phase formation temperature and also reduce the dwell time, which leads to small particle size for the initial powder [29]. It has been reported that the BCZY powder forms a pure phase at 900 °C with a dwell time of 1 h when the microwave sintering method was used. In contrast, the BCZY powder calcined in the conventional condition needs a thermal treatment condition of 1000 °C and 5 h for getting a pure phase. Both the calcination temperature and the dwell time are larger than those used in the microwave condition. As a result, the particle size of the microwave sintering method prepared BCZY powder is about 25 nm, while the corresponding value for the conventionally calcined BCZY powder is 100 nm, as shown in Fig. 8.5. The small initial powder size for the microwave method prepared BCZY allows the densification of the BCZY electrolyte membrane at 1300 °C. Although the electrolyte membrane with the conventionally calcined BCZY powder appears to be dense after firing at 1300 °C, the grain size of this electrolyte is much smaller than that of the electrolyte with the microwave prepared BCZY powder. This is because the small initial powder with small particle size could be better sintered in the later sintering step, leading to improved grain growth of the electrolyte. The large grain is desirable for the electrolyte which could reduce the volume of grain boundaries and thus improve the fuel cell performance. As a result, the single cell with the BCZY electrolyte started from the microwave prepared powder generated a peak power density of 791 mW cm^{-2}, which is at the high level for protonic SOFCs. All these evidences suggest that nanomaterials used in the electrolyte materials are beneficial for the overall fuel cell performance of protonic fuel cells.

8.3 Anode Nanomaterials for Protonic SOFCs

In fuel cells, the anode should be catalytic active for H_2 that splits H_2 into protons. Then protons pass through the electrolyte to react with O_2 at the cathode side. For

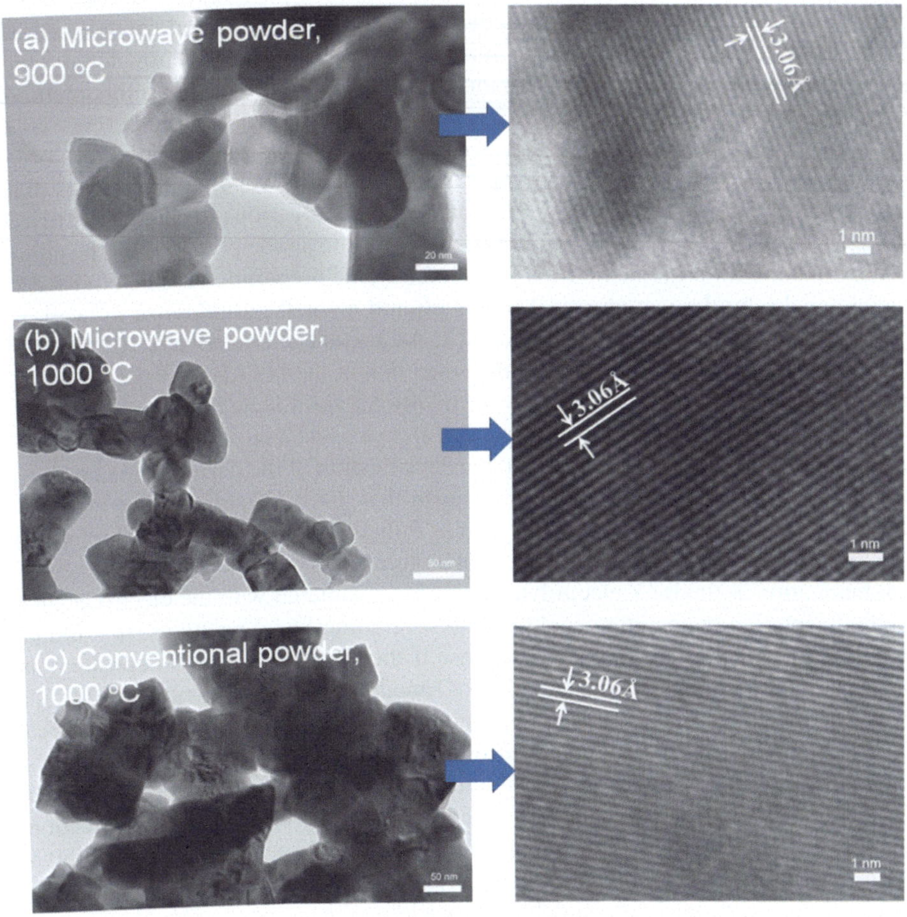

Fig. 8.5 The particle of BCZY material prepared under different thermal treatment conductions, showing the microwave method could reduce the initial size of the BCZY powder. Reproduced from [29] with permission from Elsevier B.V.

SOFCs, the anode is not only used as the electrode for catalytic purposes but also serves as the support for the complete cell. Therefore, the anode is normally thicker than the electrolyte layer and the cathode layer to guarantee the mechanical strength for the complete cell. Up to now, the most used anode is the NiO-based anode in which NiO powder is mixed with the electrolyte powder to assure first the thermo-compatibility with the electrolyte and second improved the triple-phase boundaries (TPBs). In protonic SOFCs, the anode is usually the composite anode consisted of NiO and proton-conducting oxides. As the anode research is relatively mature and there are no many other materials that could compete with the NiO-based anode now, the research on anode for protonic SOFCs is much less than that for electrolytes and cathodes. However, it is that the use of nanomaterials in the anode could be beneficial for the anode performance as well as for the cell fabrications. The anode powder is traditionally prepared by the mechanical mixing method, in which NiO powder and the proton-conducting oxide powder were mixed by ball milling. Although this method is straightforward and cost-effective, the mixing of NiO and proton-conducting oxide might not be homogenous. In addition, the size of the NiO is relatively large. Some attempts have been carried out to prepare anode powder in wet chemical routes that could make the powder in a nanometric scale with homogenous distribution of two phases. Bi et al. [30] have used the NiO-BZY composite anode prepared by the combustion method for protonic SOFCs. In their study, the preparation of the NiO-BZY anode powder by the wet chemical routes offered a small particle size as well as a homogenous distribution of NiO and BZY. As a result, the nano-sized NiO-BZY anode powder showed better performance than that of the NiO-BZY anode powder with the same composition prepared by the traditional solid-state reaction method. In addition to the improved electrochemical performance, the nano-sized anode powder showed better sinterability that allowed the densification of the deposited electrolyte during the co-sintering procedure. It is known that the BZY suffers from its very poor sinterability and the sintering for BZY is difficulty, usually requiring high temperatures. However, the high sintering temperature is not feasible for use in the fuel cell fabrication as the BZY electrolyte is usually co-sintered with the electrode, and the high sintering temperature could lead to severe interfacial reactions. It is found that the BZY electrolyte cannot reach dense if the electrolyte film is deposited on the traditional mechanically mixed BZY-NiO anode, even after sintering at 1400 °C. However, the situation was changed by employing the wet chemical route prepared BZY-NiO anode. As the wet chemical route prepared BZY-NiO anode powder is nano-sized powder, it shows better sinterability than that of mechanically mixed BZY-NiO anode powder, leading to a higher shrinkage. During the cell fabrication procedure, the electrolyte and anode are co-sintered and shrink together. The high shrinkage of the anode could promote the deposited BZY film, allowing the densification of the electrolyte without deliberately adding the sintering aids [31]. Similar work was also carried out for NiO-BCY anode for protonic SOFCs, and the composite NiO-BCY also showed excellent electrochemical performance [32].

However, it is also noted that the anode nanomaterials have to be prepared by wet chemical routes, which are more complicated and less straightforward than the traditional mechanical mixing routes. The mass production of this kind of powder is challenging and its application in the practical application seems limited. Therefore, the use of these nano-sized anode materials as the functional layer instead of the whole anode could be more proper for practical applications. Bi et al. [33] have investigated the effect of the anode functional layer on the performance of protonic SOFCs. In this study, the nano-sized anode functional layer power was prepared by a combustion method and then fabricated on the anode substrate as a transition layer for the cell. One can see that the use of the anode functional layer significantly improves the microstructure for the cell. Obvious cracks appear for the cell without the anode functional layer, which is probably due to the inhomogeneous distribution of proton-conducting phase and NiO phase as well as the coarse powder used. When the anode functional layer was used, the interface of the anode/electrolyte, where the electrochemical reaction happens, becomes the interface of the anode functional layer and the electrolyte. The nano-sized anode functional layer extends the TPBs compared with the anode due to its small particle size. Furthermore, the anode functional layer makes an improved contact between the anode part with the electrolyte layer by eliminating macro-pores at the interface. As a result, both ohmic resistance and polarization resistance of the cell with the anode functional layer have been greatly reduced, leading to higher fuel cell performance compared with the traditional cell without the anode functional layer. It is also true that many high-performance protonic SOFCs now tend to use an anode functional layer to improve the electrolyte/anode interfacial contact and thus to enhance the fuel cell performance [34].

8.4 Cathode Nanomaterials for Protonic SOFCs

Compared with the relatively few researches on anode for protonic SOFCs, the research on cathode materials for protonic SOFCs is now becoming the major research topic in the field. Different materials and microstructures are proposed, aiming to improve cathode performance. For protonic SOFCs in the early years, people just take the cathode materials that work well in oxygen-ion SOFCs for protonic SOFCs [35]. Although these cathodes can generate reasonable fuel cell performance, it is not a rational way. In oxygen-ion SOFCs, oxygen ions and electrons participate the cathode reactions, and the electronic conduction in the cathode is usually a few orders of magnitude larger than that of the oxygen-ion conduction. Therefore, the design and discovery of cathode materials with high oxygen-ion conduction are critical for cathodes for oxygen-ion SOFCs [36]. However, the cathode reaction for protonic SOFCs is different from that of oxygen-ion SOFCs, making the design of the cathode different. Peng et al. [37] have investigated the cathode process for protonic SOFCs and found that both oxygen-ion migration and proton migration are involved at triple-phase boundaries (TPBs). Therefore, the development of cathode materials with both oxygen-ion conduction and proton conduction is essential for high-performance protonic SOFCs. Although most traditional cathode materials have oxygen-ion conduction and electronic conduction, the proton conduction is not revealed before and the search of cathode materials with proton conduction is a hot topic in the community but also quite challenging. Figure 8.6 shows the scheme for the cathode reaction for protonic SOFCs using oxygen-ion and electron mixed conductor as the cathode; one can see that the TPBs are restricted at the cathode and electrolyte interface, which limits the reaction active area and thus restricts the cathode performance. In order to improve the cathode performance, one of the most feasible approaches is to couple the cathode material with the electrolyte material, which could extend the TPBs to the connections where electrolyte and cathode meet. In this case, the TPBs are extended compared with the single-phase cathode. This method is widely used in the past decade and even now for protonic SOFCs and the use of nanomaterials is critical for the enhancement of cathode performance as

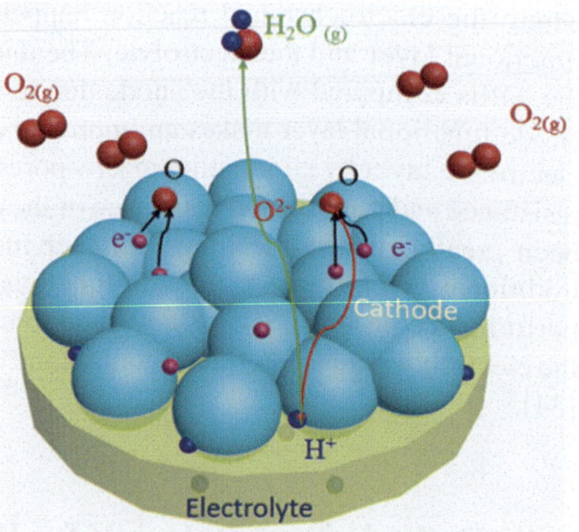

Fig. 8.6 Scheme for the cathode reaction using traditional oxygen-ion and electron mixed conducting cathodes. Reproduced from [60] with permission from The Royal Society of Chemistry

nanomaterials could enlarge the TPBs compared with the large size cathodes. In the past two decades, many studies have been carried out to find the suitability of the cathode for protonic SOFCs and most of the studies use the nano-sized composite cathode [38]. Usually, wet chemical routes are employed to prepare the cathode nanomaterials as well as the proton-conducting electrolyte materials, and then, these two nanomaterials are mixed together to form the composite cathode [39, 40]. Many different cathode materials were evaluated for protonic SOFCs and they generally produced decent fuel cell performance [9]. In order to further maximize the TPBs, the impregnation method was employed to coat cathode nanoparticles on the electrolyte backbone, which allows a further improvement in TPBs and thus higher electrochemical performance [41]. This method receives more attention these days due to its demonstrated suitability of improving the fuel cell performance for protonic SOFCs. Da'as et al. [42] have used the impregnated Sr-doped $LaMnO_3$ nanoparticles as the cathode for BZY-based fuel cells and the performance is impressive. In their study, LSM precursor was deposited on the BZY backbone forming the LSM-BZY composite cathode by an ink-jet printing method (Fig. 8.7). LSM is the first-generation cathode for SOFCs and it is regarded not appropriate for intermediate temperature SOFCs due to its pure electronic conducting feature [43]. It is true that most LSM-based SOFCs have to work at high temperatures (above 700 °C) and the performance of LSM cathode for BZY-based fuel cells is reported to be low [44]. However, in the study of Da'as et al., the peak power density of the BZY cell using the LSM nanoparticles reaches 200 mW cm^{-2} at 600 °C. The power output value is one order of magnitude larger than that of the same cell using the mechanically mixed LSM cathode, suggesting the extension of the TPBs by tailoring the particle size of LSM to nanoscale is the key for the much-improved fuel cell performance. This study somehow turnovers the traditional thinking that LSM is not proper to use at intermediate temperatures, extending the application of the LSM material. It also should be highlighted that although impregnation method for preparing cathode nanoparticles is widely reported in the community, this study presents an interesting

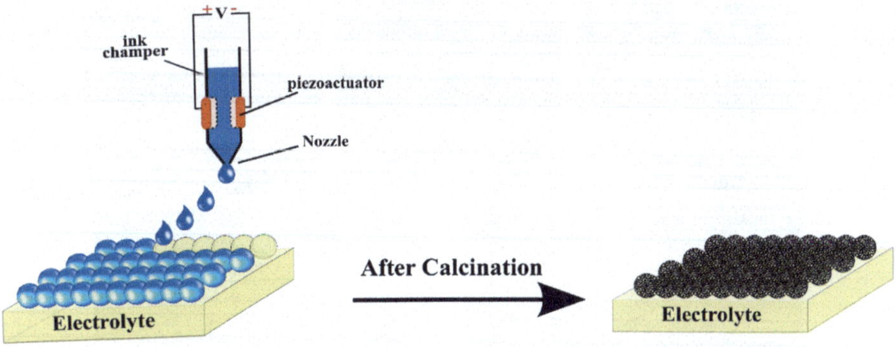

Fig. 8.7 Scheme for the preparation of nanoparticles cathodes for proton-conducting SOFCs with the impregnation method integrated with the ink-jet printing technique. Reproduced from [42] with permission from Science China Press and Springer-Verlag GmbH Germany, part of Springer Nature

strategy of using ink-jet printing for impregnation [42]. Different from the traditional manual impregnation method that confronts difficulties in controllability and reproducibility, the impregnation with ink-jet printing technique allows the fabrication of nanostructured cathodes in a controllable manner in which the whole impregnation process is carried by the machine. In this case, the cathode fabrication is independent of the people once the parameters for the ink-jet printer are fixed. Furthermore, the traditional impregnation method is only suitable for small-scale applications and the impregnation, and firing procedure has to be repeated several times until the required loading amount of LSM has been achieved. In the ink-jet printing procedure, only one step of impregnation and firing procedure is used, saving much time and effort. Also, the ink-jet printing can precisely control the volume of the drops, and the use of very tiny drops in the impregnation procedure is feasible. In this case, the backbone could be coated with a large number of tiny drops that could guarantee the homogeneity of cathodes. Although the performance of the fuel cell with the impregnated LSM cathode has been dramatically improved with the traditional study, the power output is still not at the top level of the reported protonic SOFCs which is mainly due to the pure electronic conduction nature of LSM. By using LSM nanoparticles covering the BZY backbone, indeed the TPBs are extended and the performance is enhanced. However, it is also noticed that the oxygen-ion can only transport via the LSM surface while the transportation of oxygen-ion through the bulk is unlikely for LSM. Therefore, it is reasonable to assume that the fuel cell performance could be further improved by using the oxygen-ion and electron mixed conductors. Bi et al. [45] have attempted to use the impregnated $Sm_{0.5}Sr_{0.5}CoO_3$(SSC) cathode for BZY-based protonic SOFCs. Compared with the almost pure electronic conductor LSM, SSC is a well-known mixed conductor that allows the diffusion of oxygen ions and the cathode reaction activity could be much improved with the use of SSC cathode. By using the SSC nanostructured cathode, the BZY cell generates a high peak power density of 602 mW cm^{-2} at 600 °C and this value is one of the largest ever reported for BZY cells and it is also at the high level for protonic SOFCs (Fig. 8.8). To further

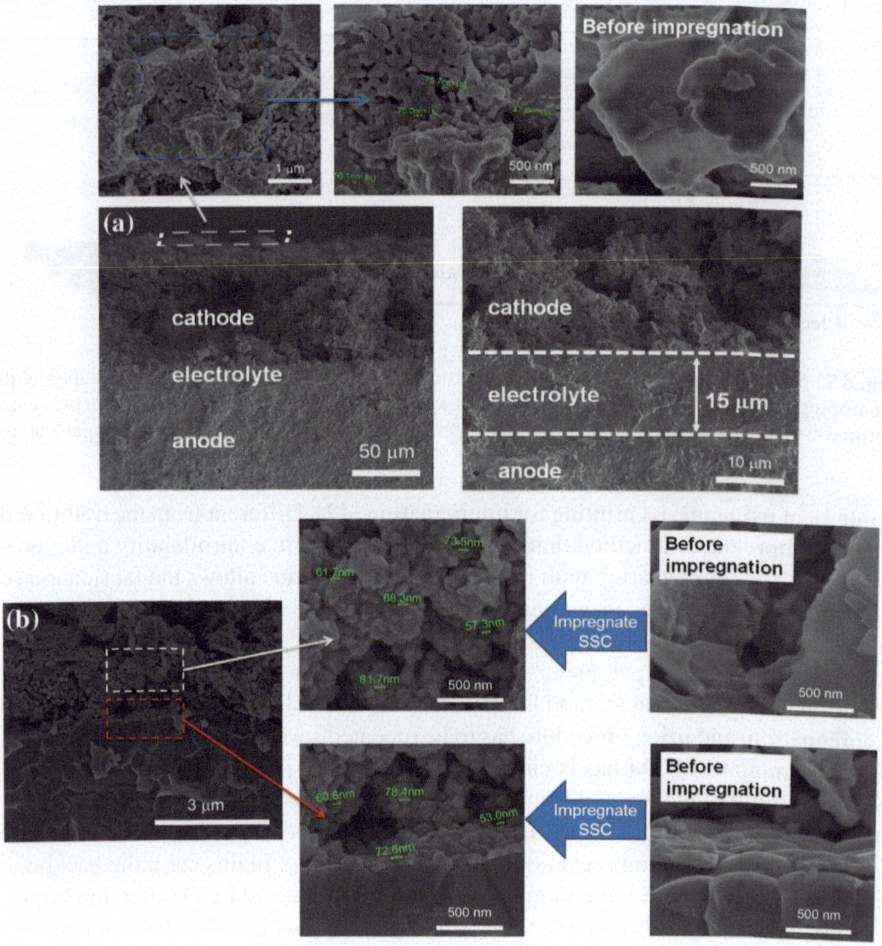

Fig. 8.8 BZY cells with impregnated SSC as the cathode: **a** surface and **b** cross-sectional view. Reproduced from [45] with permission from WILEY-VCH Verlag GmbH & Co. KGaA, Weinheim

demonstrate the feasibility of using nanostructured mixed conductors for protonic SOFCs, the impregnated $PrBaCo_2O_{5+x}$(PBCO) material is also used for BZY-based cells. It is interesting to find that the PBCO without obtaining the pure phase shows better performance than that of the cell with pure phase PBCO cathode. Unlike the SSC material that can form the pure phase at 800 °C, the formation of the PBCO phase requires a high firing temperature of 1200 °C. However, with the increase of the calcination temperature, the size of cathode powder inevitably increases. Although the calcination temperature of 1200 °C could allow the formation of pure phase PBCO impregnated particles, such a high firing temperature also leads to the loss of the nanostructure for the cathode. In contrast, the PBCO precursor fired at 800°C cannot achieve the pure PBCO double perovskite phase but form the $BaCoO_3$ and

$PrCoO_3$-based intermediates, while the nanostructured of the cathode is maintained. The BZY cell with the 800 °C fired PBCO cathode shows a record-high performance of 650 mW cm^{-2} at 600 °C. This study suggests that the formation of the nanostructured for the cathode sometimes is more important than the formation of the pure phase of the target cathode material, provided the intermediates still have the catalytic activity for protonic SOFCs.

However, it should be noted that although nanomaterials were used in these studies, the extension of TPBs is still limited at the interface of the proton-conducting phase (normally the electrolyte material in the composite cathode) and electron-oxygen mixed conducting phase. The use of impregnated nanoparticles as cathode for protonic SOFCs just further improves the TPB area compared with the mechanically mixed two-phase composite cathode and it is a microstructure optimization without using the new materials. According to the reaction mechanism of protonic SOFC cathodes, the ideal cathode for protonic SOFCs should have the proton, oxygen-ion, and electron conductions which could extend the reaction active area to the whole electrode surface instead of only the connection between the cathode phase and the proton-conducting electrolyte phase, as schemed in Fig. 8.9. Although the advantage of using materials with proton, oxygen-ion, and electron conductions is evidently recognized, the development of materials with proton, oxygen-ion, and electron conductions is quite challenging. Unlike the oxygen–electron mixed conducting materials that are widely available [36], few materials are reported to have proton, oxygen-ion, and electron conductions and the situation is changed only in recent years.

The mixed proton-electron conducting materials were firstly designed by tailoring the traditional proton conductor with proper dopants. $BaCeO_3$, which is a well-known proton conductor, is doped with Bi-element, allowing the material to show both protonic and electron conduction. $Ba(Ce_{1-x}Bi_x)O_3$ ($x = 0.0$–0.5) are reported to show both electronic conduction and protonic conduction [46], although the total

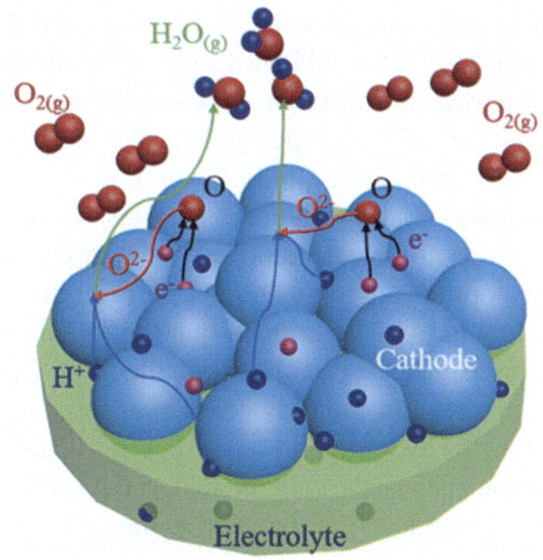

Fig. 8.9 Scheme for the use of triple-conducting material (proton, oxygen-ion, and electron) as the cathode for proton-conducting SOFCs, extending the reaction active area to the whole cathode surface. Reproduced from [60] with permission from The Royal Society of Chemistry

conductivity of these materials is relatively low, reaching only about 0.1 S cm^{-1} at 600 °C. The following work has demonstrated that the use of $BaCe_{0.5}Bi_{0.5}O_3$ material as cathode could allow reasonable performance for protonic SOFCs, reaching 321 mW cm^{-2} at 700 °C with a polarization of 0.28 Ω cm^2 at the same temperature [47]. This power output value is at the high level for protonic SOFCs at the time reported and the polarization resistance is similar to many composite cathodes in the literature. It should be noted that only single-phase $BaCe_{0.5}Bi_{0.5}O_3$ material was used as the cathode instead of the conventional composite cathode, implying protonic conduction in this material. Otherwise, the polarization resistance could be huge due to the absence of proton conduction in the cathode [48]. These studies suggest that the modification of existing proton conductors could be a feasible route to design cathode materials with proton and electron mixed conductivity. Following this idea, other materials have been designed and prepared. Compared with the mild electronic conduction potential of Bi-doping materials, the modification with transition metal oxides could be a more straightforward way due to the valence change of the metal elements. Similar to $BaCe_{0.5}Bi_{0.5}O_3$ material, Fe was used as a dopant to modify the $BeCeO_3$ proton conductor forming the Fe doped $BaCeO_3$ materials as cathodes for protonic SOFCs [49]. Different phase structures have been obtained with the different Fe-doping level and the $BaCe_{0.5}Fe_{0.5}O_3$ material shows the best electrochemical performance. It is also noted that doped $BaCeO_3$ is a well-known proton conductor but also gets criticized for its poor chemical stability. Therefore, the modification of stable proton conductors could be a more rational way of designing the cathodes with proton condition. Rao et al. [50] have used Co-doped $BaZrO_3$ as the cathode materials for protonic SOFCs. In their research, they have demonstrated the $BaZrO_3$ material with 40 mol% Co-doping shows superior performance than that of $Sm_{0.5}Sr_{0.5}CoO_3$-based composite cathode, even though $Sm_{0.5}Sr_{0.5}CoO_3$ is regarded as one of the most promising cathode materials for intermediate temperature SOFCs, implying the proton conduction in the material. A similar approach has been used to dope Pr in $BaZrO_3$, finding that the high concentration of Pr-doping allows the appearance of electronic conduction for $BaZrO_3$. Moreover, the ionic conduction is still maintained for these materials, making them attractive as cathodes for protonic SOFCs [51]. However, it should be noted that the Pr-doping strategy does not introduce sufficient electronic conduction to make Pr-doped $BaZrO_3$ to be used independently, and they have to be used together with other electronic conductors as the composite cathode. Even in this case, the 30 mol% Pr-doped $BaZrO_3$ used in the cathode allows an encouraging electrochemical performance [52], making the use of proton-electron mixed conductor as one of the most interesting directions for developing cathode materials for protonic SOFCs.

It should be mentioned that the direct measurement of proton conduction in the cathode material is still quite challenging. The direct measurement of the conductivity of the material can only give the value of the total conductivity, including electronic conductivity, oxygen-ion conductivity, and protonic conductivity. In a mixed conductor, such as the cathode material, the electronic conduction is usually a few orders of magnitude larger than that of the ionic (both oxygen-ion and proton) conductivity, which makes the value of the total conductivity almost equals to that of the electronic conduction. The oxygen-ion conductivity in the mixed conductor could be further separated by using some strategies, such as oxygen permeation tests or concentration cell tests in oxygen-containing atmospheres. However, such strategies cannot be applied for the separation of the protonic conductivity from the total conductivity. The cathode material tends to decomposes in the reducing atmosphere, so the tests, such as hydrogen permeation tests or hydrogen concentration cell tests, are not feasible. As a result, it is difficult to use conventional experimental methods to measure the specific proton conduction in the cathodes. In order to overcome this technical difficulty, theoretical approaches have been employed to investigate the proton mobility in the cathode materials from an atomic point of view.

The density functional theory (DFT) method has been recently employed in the field of protonic SOFCs to calculate the proton migration ability in the cathode, which could avoid the technical problems that appeared in experimental approaches [53, 54]. Wang et al. [55] have evaluated the performance of $Sr_3Fe_2O_{7-\delta}$ as the cathode for protonic SOFCs. Although $Sr_3Fe_2O_{7-\delta}$ shows similar oxygen-ion conduction with the traditional oxygen–electron mixed conductor, its application in protonic SOFCs shows superior performance suggesting the possible proton migration in this material. DFT method is employed to investigate the proton migration energies in this material, suggesting feasibility for proton migration in this material. In addition, the $Sr_3Fe_2O_{7-\delta}$ shows a decreased performance when coupling it with proton-conducting oxide as the composite cathode, indicating the occupancy of the $Sr_3Fe_2O_{7-\delta}$ by proton-conducting oxide impairs the fuel cell performance. Generally, the composite cathode using a certain amount of proton-conducting oxide shows improved electrochemical performance than that of the single-phase cathode, and this study shows an opposite trend which suggests the different electrochemical reaction for the $Sr_3Fe_2O_{7-\delta}$ cathode.

To introduce proton conduction in the cathode materials, the evaluation of hydration ability is another rational approach to screen the possible cathodes with proton and electron conduction. Poetzsch et al. [56] have measured the proton uptake of $Ba_{0.5}Sr_{0.5}Fe_{0.8}Zn_{0.2}O_{3-\delta}$ from the hydration ability tests, suggesting $Ba_{0.5}Sr_{0.5}Fe_{0.8}Zn_{0.2}O_{3-\delta}$ is a mixed conductor with protonic conduction and the hydration ability is the key for the protonic conduction. Based on this concept, some new materials have been proposed for protonic SOFCs, aiming to improve the hydration ability and thus the proton migration ability. $Ba_{0.5}Sr_{0.5}Co_{0.8}Fe_{0.2}O_{3-\delta}$, which is a high-performance cathode for oxygen-ion SOFCs [57], does not show similar high performance in protonic SOFCs [58]. It is well known that $Ba_{0.5}Sr_{0.5}Co_{0.8}Fe_{0.2}O_{3-\delta}$ has good oxygen-ion migration ability and the lack of sufficient proton mobility could be the reason for the low performance for $Ba_{0.5}Sr_{0.5}Co_{0.8}Fe_{0.2}O_{3-\delta}$ in protonic SOFCs. To enhance the proton mobility, one straightforward strategy could be the improvement of hydration ability in the material. Therefore, K was used as the dopant for $Ba_{0.5}Sr_{0.5}Co_{0.8}Fe_{0.2}O_{3-\delta}$ and K partially replaced Ba in the material, aiming to enhance the hydration ability of the material [59]. As K has stronger basicity than that of Ba, the K-doped $Ba_{0.5}Sr_{0.5}Co_{0.8}Fe_{0.2}O_{3-\delta}$ material could possess higher hydration ability than that of the K-free sample. The hydration ability tests indicate the $Ba_{0.4}K_{0.1}Sr_{0.5}Co_{0.8}Fe_{0.2}O_{3-\delta}$ has higher hydration ability than that of $Ba_{0.5}Sr_{0.5}Co_{0.8}Fe_{0.2}O_{3-\delta}$ material. The improved hydration is also demonstrated from theoretical point of view, in which $Ba_{0.4}K_{0.1}Sr_{0.5}Co_{0.8}Fe_{0.2}O_{3-\delta}$ has lower hydration energy than that of $Ba_{0.5}Sr_{0.5}Co_{0.8}Fe_{0.2}O_{3-\delta}$. In addition, the proton mobility investigation carried out by DFT calculations also suggests that the K-doping strategy changes the proton hopping and rotation energies, leading to an easier proton migration in $Ba_{0.4}K_{0.1}Sr_{0.5}Co_{0.8}Fe_{0.2}O_{3-\delta}$. As a result, the protonic SOFC using the $Ba_{0.4}K_{0.1}Sr_{0.5}Co_{0.8}Fe_{0.2}O_{3-\delta}$ cathode shows an encouraging fuel cell performance of 1275 mW cm^{-2} at 700 °C.

Sn and Bi co-doping strategy also shows a similar impact on the cathode. Xia et al. [60] have used Sn and Bi as the co-dopant for traditional $BaFeO_3$-based cathode materials. The material shows higher hydration ability with an increased Bi-doping level until 30 mol%. After that doping concentration, the material cannot maintain a single phase. Although the doping of Bi could reduce the total conductivity of the sample, the conductivity value is still higher than 1 S cm^{-1} at the SOFC testing temperature range, making it suitable as the electrode for fuel cells. Moreover, the oxygen-ion and proton migration ability improve with the Bi-doping, allowing the triple-conduction in the target material. As a result, the cell using the $BaFe_{0.5}Sn_{0.2}Bi_{0.3}O_3$ cathode generates a high peak power density of 1277 mW cm^{-2} at 700 °C, which is the highest ever reported for protonic SOFCs using cobalt-free cathodes.

Besides the design of new materials as cathodes for protonic SOFCs, interfacial engineering is demonstrated to be important for cell performance. Choi et al. have used $PrBa_{0.5}Sr_{0.5}Co_{1.5}Fe_{0.5}O_{5+\delta}$ as the cathode for protonic SOFCs [61]. Due to the proton and electron conduction nature of $PrBa_{0.5}Sr_{0.5}Co_{1.5}Fe_{0.5}O_{5+\delta}$, the cell exhibits high fuel cell performance reaching 800 mW cm^{-2} at 600 °C. Further, tailoring the cathode/electrolyte interface with a thin layer of dense $PrBa_{0.5}Sr_{0.5}Co_{1.5}Fe_{0.5}O_{5+\delta}$ (around 100 nm) could boost the fuel cell performance, leading to a peak power density of 1098 mW cm^{-2} at the same testing temperature. The fuel cell performance is record-high for protonic SOFCs until now. The addition of the thin $PrBa_{0.5}Sr_{0.5}Co_{1.5}Fe_{0.5}O_{5+\delta}$ layer could improve the contact at the cathode/electrolyte interface, reducing the ohmic resistance of the cell and thus improving the overall fuel cell performance. In addition, the improvement in fuel cell performance by adding the dense $PrBa_{0.5}Sr_{0.5}Co_{1.5}Fe_{0.5}O_{5+\delta}$ layer also demonstrates the protonic conduction in the $PrBa_{0.5}Sr_{0.5}Co_{1.5}Fe_{0.5}O_{5+\delta}$. If the material has no proton conduction, the dense layer could block the migration of proton from the anode to the cathode side and then dramatically reduces the fuel cell performance. All the above evidence suggests that the development of new materials with proton conduction and the optimization of microstructure are now two hot directions for the cathode designs for protonic SOFCs

8.5 Conclusions

The discovery of proton conduction in some oxides opens a new door in the solid-state ionic field and many devices based on proton-conducting oxides have been developed. Among all these devices, protonic SOFCs become a hot topic recently due to their unique advantages over the traditional SOFCs. Nanomaterials now play more and more important roles in protonic SOFCs, including the utilization of nano materials in electrolyte, anode, and cathode. The electrolyte powder is preferred to be restricted at the nanoscale that will be beneficial for later sintering procedures to produce high-quality electrolyte membranes. Different methods have been employed to produce nano-electrolyte powders and wet chemical routes are more favorable than the traditional solid-state reaction methods. In addition, the microwave method has been recently used for the synthesis of proton-conducting oxide powders. It is found that the microwave sintering method could allow the formation of pure phase at relatively low temperatures with a shorter dwell time compared with the traditional calcination method. As a result, the size of the electrolyte powder could be further reduced. Although microwave sintering is not widely used in the community of protonic SOFCs, its advantages over traditional sintering make it appealing not only in nano-sized powder preparation but also in electrolyte membrane preparations. The low preparation temperature not only reduces the particle size for the initial powders but also depresses metal element evaporations during the co-sintering procedure, leading to high membrane conductivity and thus high fuel cell performance. Also, the tailoring of the grain boundary core at sub-nanoscale has been demonstrated to improve the specific grain boundary conductivity, although now the high-temperature thermal treatment method is the only way to tailor the grain boundary core. The development of proper strategies to design the core of grain boundary could be interesting but also challenging for the improvement of proton-conducting electrolyte materials.

Anode materials are less studied due to the maturity of NiO-based anode. NiO-based anodes are widely used in protonic SOFCs due to their good catalytic activity and chemical compatibility. Rather than the investigations on the materials as anode for protonic SOFCs, the optimization of the microstructure of the anode is critical for the overall fuel cell performance. The use of the composite nanomaterials as the anode functional layer is found to be beneficial for the cell performance. The nano-sized composite anode provides a more active reaction area compared with the traditional mechanically mixed anode. In addition, the anode functional layer serves as a transition layer which improves the interfacial contacts between the anode and the electrolyte layer, reducing the overall ohmic resistance and thus improving the fuel cell performance.

Among all the key components for protonic SOFCs, cathodes are mostly investigated. Nanomaterials are widely used in this aspect, including the fabrication of nano-sized cathode particles and the design of novel cathode materials with protonic conduction. The use of the impregnation method could deposit cathode nanoparticles on the electrolyte backbone, allowing the extension of TPBs compared with the traditional mechanically mixed composite cathode. Moreover, one recent study demonstrates that the maintenance of nanoparticles for the cathode is somehow more important than the formation of targeting structure, provided the intermediates still have sufficient catalytic activities. Due to the different reaction mechanisms of cathodes for protonic SOFCs compared with that for traditional oxygen-ion conducting

SOFCs, the introduction of proton conduction is the key to the development of high-performance cathodes for protonic SOFCs. Different strategies have been proposed, aiming to enhance the proton migration ability and also the hydration ability of the materials, and it is believed that this area will be still the hot topic for protonic SOFCs in the following years.

References

1. H. Iwahara, T. Esaka, H. Uchida, N. Maeda, Proton conduction in sintered oxides and its application to steam electrolysis for hydrogen-production. Solid State Ionics **3–4**, 359–363 (1981)
2. T. Norby, M. Wideroe, R. Glockner, Y. Larring, Hydrogen in oxides. Dalton Trans. **19**, 3012–3018 (2004)
3. K.D. Kreuer, Proton-conducting oxides. Annu. Rev. Mater. Res. **33**, 333–359 (2003)
4. E.D. Wachsman, K.T. Lee, Lowering the temperature of solid oxide fuel cells. Science **334**(6058), 935–939 (2011)
5. H.K. Dai, H.N. Kou, H.Q. Wang, L. Bi, Electrochemical performance of protonic ceramic fuel cells with stable $BaZrO_3$-based electrolyte: a mini-review. Electrochem. Commun. **96**, 11–15 (2018)
6. H. Iwahara, H. Uchida, K. Ono, K. Ogaki, Proton conduction in sintered oxides based on $BaCeO_3$. J. Electrochem. Soc. **135**(2), 529–533 (1988)
7. R. Haugsrud, T. Norby, Proton conduction in rare-earth ortho-niobates and ortho-tantalates. Nat. Mater. **5**(3), 193–196 (2006)
8. E. Fabbri, D. Pergolesi, E. Traversa, Materials challenges toward proton-conducting oxide fuel cells: a critical review. Chem. Soc. Rev. **39**(11), 4355–4369 (2010)
9. E. Fabbri, L. Bi, D. Pergolesi, E. Traversa, Towards the next generation of solid oxide fuel cells operating below 600 °C with chemically stable proton-conducting electrolytes. Adv. Mater. **24**(2), 195–208 (2012)
10. S.M. Haile, G. Staneff, K.H. Ryu, Non-stoichiometry, grain boundary transport and chemical stability of proton conducting perovskites. J. Mater. Sci. **36**(5), 1149–1160 (2001)
11. A. D'Epifanio, E. Fabbri, E. Di Bartolomeo, S. Licoccia, E. Traversa, Design of $BaZr_{0.8}Y_{0.2}O_3$ protonic conductor to improve the electrochemical performance in intermediate temperature solid oxide fuel cells (IT-SOFCs). Fuel Cells **8**(1), 69–76 (2008)
12. Z.Q. Sun, E. Fabbri, L. Bi, E. Traversa, Lowering grain boundary resistance of $BaZr_{0.8}Y_{0.2}O_3$ with $LiNO_3$ sintering-aid improves proton conductivity for fuel cell operation. Phys. Chem. Chem. Phys. **13**(17), 7692–7700 (2011)
13. D. Pergolesi, E. Fabbri, A. D'Epifanio, E. Di Bartolomeo, A. Tebano, S. Sanna, S. Licoccia, G. Balestrino, E. Traversa, High proton conduction in grain-boundary-free yttrium-doped barium zirconate films grown by pulsed laser deposition. Nat. Mater. **9**(10), 846–852 (2010)
14. D. Pergolesi, E. Fabbri, E. Traversa, Chemically stable anode-supported solid oxide fuel cells based on Y-doped barium zirconate thin films having improved performance. Electrochem. Commun. **12**(7), 977–980 (2010)
15. E. Fabbri, D. Pergolesi, A. D'Epifanio, E. Di Bartolomeo, G. Balestrino, S. Licoccia, E. Traversa, Design and fabrication of a chemically-stable proton conductor bilayer electrolyte for intermediate temperature solid oxide fuel cells (IT-SOFCs). Energy Environ. Sci. **1**(3), 355–359 (2008)
16. K. Bae, D.Y. Jang, H.J. Choi, D. Kim, J. Hong, B.K. Kim, J.H. Lee, J.W. Son, J.H. Shim, Demonstrating the potential of yttrium-doped barium zirconate electrolyte for high-performance fuel cells. Nat. Commun. **8**, 14553 (2017)

17. E.E. Helgee, A. Lindman, G. Wahnstrom, Origin of space charge in grain boundaries of proton-conducting $BaZrO_3$. Fuel Cells **13**(1), 19–28 (2013)
18. M. Shirpour, R. Merkle, J. Maier, Space charge depletion in grain boundaries of $BaZrO_3$ proton conductors. Solid State Ionics **225**, 304–307 (2012)
19. M. Shirpour, B. Rahmati, W. Sigle, P.A. van Aken, R. Merkle, J. Maier, Dopant segregation and space charge effects in proton-conducting $BaZrO_3$ perovskites. J. Phys. Chem. C **116**(3), 2453–2461 (2012)
20. M. Shirpour, R. Merkle, J. Maier, Evidence for space charge effects in Y-doped $BaZrO_3$ from reduction experiments. Solid State Ionics **216**, 1–5 (2012)
21. J.H. Tong, A. Subramaniyan, H. Guthrey, D. Clark, B.P. Gorman, R. O'Hayre, Electrical conductivities of nano ionic composite based on yttrium-doped barium zirconate and palladium metal. Solid State Ionics **211**, 26–33 (2012)
22. V. Esposito, E. Traversa, Design of electroceramics for solid oxides fuel cell applications: playing with ceria. J. Am. Ceram. Soc. **91**(4), 1037–1051 (2008)
23. L. Bi, E. Traversa, Synthesis strategies for improving the performance of doped-$BaZrO_3$ materials in solid oxide fuel cell applications. J. Mater. Res. **29**(1), 1–15 (2014)
24. Y. Yamazaki, R. Hernandez-Sanchez, S.M. Haile, High total proton conductivity in large-grained yttrium-doped barium zirconate. Chem. Mater. **21**(13), 2755–2762 (2009)
25. E. Fabbri, L. Bi, H. Tanaka, D. Pergolesi, E. Traversa, Chemically stable Pr and Y co-doped barium zirconate electrolytes with high proton conductivity for intermediate-temperature solid oxide fuel cells. Adv. Funct. Mater. **21**(1), 158–166 (2011)
26. H.L. Dai, Proton conducting solid oxide fuel cells with chemically stable $BaZr_{0.75}Y_{0.2}Pr_{0.05}O_3$ electrolyte. Ceramics International **43**(9), 7362–7365 (2017)
27. S.P. Shafi, L. Bi, S. Boulfrad, E. Traversa, Y and Ni co-doped $BaZrO_3$ as a proton-conducting solid oxide fuel cell electrolyte exhibiting superior power performance. J. Electrochem. Soc. **162**(14), F1498–F1503 (2015)
28. C.K. Ng, S. Ramesh, C.Y. Tan, A. Muchtar, M.R. Somalu, Microwave sintering of ceria-doped scandia stabilized zirconia as electrolyte for solid oxide fuel cell. Int. J. Hydrogen Energy **41**(32), 14184–14190 (2016)
29. B. Wang, X.H. Liu, L. Bi, X.S. Zhao, Fabrication of high-performance proton-conducting electrolytes from microwave prepared ultrafine powders for solid oxide fuel cells. J. Power Sources **412**, 664–669 (2019)
30. L. Bi, E. Fabbri, Z.Q. Sun, E. Traversa, $BaZr_{0.8}Y_{0.2}O_3$-NiO composite anodic powders for proton-conducting SOFCs prepared by a combustion method. J. Electrochem. Soc. **158**(7), B797–B803 (2011)
31. L. Bi, E. Fabbri, Z.Q. Sun, E. Traversa, Sinteractive anodic powders improve densification and electrochemical properties of $BaZr_{0.8}Y_{0.2}O_3$ electrolyte films for anode-supported solid oxide fuel cells. Energy Environ. Sci. **4**(4), 1352–1357 (2011)
32. L. Chevallier, M. Zunic, V. Esposito, E. Di Bartolomeo, E. Traversa, A wet-chemical route for the preparation of Ni-$BaCe_{0.9}Y_{0.1}O_3$ cermet anodes for IT-SOFCs. Solid State Ionics **180**(9–10), 715–720 (2009)
33. L. Bi, E. Fabbri, E. Traversa, Effect of anode functional layer on the performance of proton-conducting solid oxide fuel cells (SOFCs). Electrochem. Commun. **16**(1), 37–40 (2012)
34. X. Xu, H.Q. Wang, J.M. Ma, W.Y. Liu, X.F. Wang, M. Fronzi, L. Bi, Impressive performance of proton-conducting solid oxide fuel cells using a first-generation cathode with tailored cations. J. Mater. Chem. A **7**(32), 18792–18798 (2019)
35. E. Fabbri, L. Bi, J.L.M. Rupp, D. Pergolesi, E. Traversa, Electrode tailoring improves the intermediate temperature performance of solid oxide fuel cells based on a Y and Pr co-doped barium zirconate proton conducting electrolyte. RSC Adv. **1**(7), 1183–1186 (2011)
36. J.A. Kilner, M. Burriel, Materials for intermediate-temperature solid-oxide fuel cells. Annu. Rev. Mater. Res. **44**, 365–393 (2014)
37. R.R. Peng, T.Z. Wu, W. Liu, X.Q. Liu, G.Y. Meng, Cathode processes and materials for solid oxide fuel cells with proton conductors as electrolytes. J. Mater. Chem. **20**(30), 6218–6225 (2010)

38. Y. Chen, S. Yoo, K. Pei, D.C. Chen, L. Zhang, B. deGlee, R. Murphy, B.T. Zhao, Y.X. Zhang, Y. Chen, M.L. Liu, An in situ formed, dual-phase cathode with a highly active catalyst coating for protonic ceramic fuel cells. Adv. Funct. Mater. **28**(5), 1704907 (2018)
39. W. He, R.H. Yuan, F.F. Dong, X.L. Wu, M. Ni, High performance of protonic solid oxide fuel cell with $BaCo_{0.7}Fe_{0.22}Sc_{0.08}O_3$ electrode. Int. J. Hydrogen Energy **42**(39), 25021–25025 (2017)
40. N.A. Danilov, A.P. Tarutin, J.G. Lyagaeva, E.Y. Pikalova, A.A. Murashkina, D.A. Medvedev, M.V. Patrakeev, A.K. Demin, Affinity of $YBaCo_4O_7$-based layered cobaltites with protonic conductors of cerate-zirconate family. Ceram. Int. **43**(17), 15418–15423 (2017)
41. S.P. Jiang, Nanoscale and nano-structured electrodes of solid oxide fuel cells by infiltration: advances and challenges. Int. J. Hydrogen Energy **37**(1), 449–470 (2012)
42. E.H. Da'as, L. Bi, S. Boulfrad, E. Traversa, Nanostructuring the electronic conducting $La_{0.8}Sr_{0.2}MnO_3$ cathode for high-performance in proton-conducting solid oxide fuel cells below 600 °C. Sci. China Mater. **61**(1), 57–64 (2018)
43. S.P. Jiang, Development of lanthanum strontium manganite perovskite cathode materials of solid oxide fuel cells: a review. J. Mater. Sci. **43**(21), 6799–6833 (2008)
44. C. Peng, J. Melnik, J.X. Li, J.L. Luo, A.R. Sanger, K.T. Chuang, ZnO-doped $BaZr_{0.85}Y_{0.15}O_3$-proton-conducting electrolytes: Characterization and fabrication of thin films. J. Power Sources **190**(2), 447–452 (2009)
45. L. Bi, S.P. Shafi, E.H. Da'as, E. Traversa, Tailoring the cathode-electrolyte interface with nanoparticles for boosting the solid oxide fuel cell performance of chemically stable proton-conducting electrolytes. Small **14**(32), 1801231 (2018)
46. Z. Hui, P. Michele, Preparation, chemical stability, and electrical properties of $Ba(Ce_{1-x}Bi_x)O_3$ ($x = 0.0–0.5$). J. Mater. Chem. **12**(12), 3787–3791 (2002)
47. Z.T. Tao, L. Bi, L.T. Yan, W.P. Sun, Z.W. Zhu, R.R. Peng, W. Liu, A novel single phase cathode material for a proton-conducting SOFC. Electrochem. Commun. **11**(3), 688–690 (2009)
48. E. Fabbri, S. Licoccia, E. Traversa, E.D. Wachsman, Composite cathodes for proton conducting electrolytes. Fuel Cells **9**(2), 128–138 (2009)
49. Z.T. Tao, L. Bi, Z.W. Zhu, W. Liu, Novel cobalt-free cathode materials $BaCe_xFe_{1-x}O_3$-delta for proton-conducting solid oxide fuel cells. J. Power Sources **194**(2), 801–804 (2009)
50. Y.Y. Rao, S.H. Zhong, F. He, Z.B. Wang, R.R. Peng, Y.L. Lu, Cobalt-doped $BaZrO_3$: a single phase air electrode material for reversible solid oxide cells. Int. J. Hydrogen Energy **37**(17), 12522–12527 (2012)
51. E. Fabbri, I. Markus, L. Bi, D. Pergolesi, E. Traversa, Tailoring mixed proton-electronic conductivity of $BaZrO_3$ by Y and Pr co-doping for cathode application in protonic SOFCs. Solid State Ionics **202**(1), 30–35 (2011)
52. E. Fabbri, L. Bi, D. Pergolesi, E. Traversa, High-performance composite cathodes with tailored mixed conductivity for intermediate temperature solid oxide fuel cells using proton conducting electrolytes. Energy Environ. Sci. **4**(12), 4984–4993 (2011)
53. A.B. Munoz-Garcia, M. Tuccillo, M. Pavone, Computational design of cobalt-free mixed proton-electron conductors for solid oxide electrochemical cells. J. Mater. Chem. A **5**(23), 11825–11833 (2017)
54. A.B. Munoz-Garcia, M. Pavone, First-principles design of new electrodes for proton-conducting solid-oxide electrochemical cells: a-site doped $Sr_2Fe_{1.5}Mo_{0.5}O_6$ perovskite. Chem. Mater. **28**(2), 490–500 (2016)
55. Z.Q. Wang, W.Q. Yang, S.P. Shafi, L. Bi, Z.B. Wang, R.R. Peng, C.R. Xia, W. Liu, Y.L. Lu, A high performance cathode for proton conducting solid oxide fuel cells. J. Mater. Chem. A **3**(16), 8405–8412 (2015)
56. D. Poetzsch, R. Merkle, J. Maier, Proton uptake in the H^+-SOFC cathode material $Ba_{0.5}Sr_{0.5}Fe_{0.8}Zn_{0.2}O_3$: transition from hydration to hydrogenation with increasing oxygen partial pressure. Faraday Discuss. **182**, 129–143 (2015)
57. W. Zhou, R. Ran, Z.P. Shao, Progress in understanding and development of $Ba_{0.5}Sr_{0.5}Co_{0.8}Fe_{0.2}O_3$-based cathodes for intermediate-temperature solid-oxide fuel cells: a review. J. Power Sources **192**(2), 231–246 (2009)

58. Y.M. Guo, Y. Lin, R. Ran, Z.P. Shao, Zirconium doping effect on the performance of proton-conducting $BaZr_yCe_{0.8-y}Y_{0.2}O_3$ ($0.0 \leq y \leq 0.8$) for fuel cell applications. J. Power Sources **193**(2), 400–407 (2009)
59. X. Xu, H.Q. Wang, M. Fronzi, X.F. Wang, L. Bi, E. Traversa, Tailoring cations in a perovskite cathode for proton-conducting solid oxide fuel cells with high performance. J. Mater. Chem. A **7**(36), 20624–20632 (2019)
60. Y.P. Xia, Z.Z. Jin, H.Q. Wang, Z. Gong, H.L. Lv, R.R. Peng, W. Liu, L. Bi, A novel cobalt-free cathode with triple-conduction for proton-conducting solid oxide fuel cells with unprecedented performance. J. Mater. Chem. A **7**(27), 16136–16148 (2019)
61. S. Choi, C.J. Kucharczyk, Y.G. Liang, X.H. Zhang, I. Takeuchi, H.I. Ji, S.M. Haile, Exceptional power density and stability at intermediate temperatures in protonic ceramic fuel cells. Nat. Energy **3**(3), 202–210 (2018)

Chapter 9
Thermo-Responsive Nanomaterials for Thermoelectric Generation

Wei-Di Liu, Jin Zou and Zhi-Gang Chen

Abstract In the past decades, as an emission-free technique capable of realizing direct energy conversion between heat and electricity, thermoelectric materials/applications have attracted extensive attention. The efficiency of thermoelectric modules/devices is dominated by the material dimensionless figure of merit, zT. zT of thermoelectric materials can be enhanced through both optimizing carrier transportation properties and refraining the lattice thermal conductivity. Module design can also influence the energy conversion efficiency. Proper module design, such as segmented or cascade design, can effectively utilize the potential of composing materials. Furthermore, through proper device design, the thermoelectric modules can be designed as both flexible and rigid types and applied in both niche and macro-fields.

9.1 Introduction

Since the discovery of the Seebeck effect more than two hundred years ago, thermoelectric materials, capable of direct energy conversion between heat and electricity without direct environment pollution, have attracted extensive research interest [1, 2]. The direct energy conversion between heat and electricity has enabled thermoelectric materials promising substitution/supplementation for fossil fuels. For example, thermoelectric generator can be used for the waste heat recovery in various sectors, such as vehicle engines and power plants, which may boost the energy utilization

W.-D. Liu · J. Zou
Materials Engineering, The University of Queensland, Brisbane, QLD 4072, Australia

J. Zou (✉)
Centre for Microscopy and Microanalysis, The University of Queensland, Brisbane, QLD 4072, Australia
e-mail: j.zou@uq.edu.au

Z.-G. Chen (✉)
Centre for Future Materials, University of Southern Queensland, Springfield Central, QLD 4300, Australia
e-mail: zhigang.chen@usq.edu.au

Z. Sun and T. Liao (eds.), *Responsive Nanomaterials for Sustainable Applications*, Springer Series in Materials Science 297,
https://doi.org/10.1007/978-3-030-39994-8_9

efficiency of the entire system, and in turn suspend greenhouse gas emission and relevant pollution [1, 3–5].

Efficiency of thermoelectric materials is dominated by the dimensionless figure of merit, *zT*:

$$zT = \frac{S^2\sigma}{\kappa}T, \tag{9.1}$$

where S, σ, κ and T are the Seebeck coefficient, electrical conductivity, total thermal conductivity and absolute temperature, respectively [6–8]. $S^2\sigma$ is defined as the power factor to identify the electrical performance and κ is contributed by the electrical (κ_e) and lattice thermal conductivity (κ_l): $\kappa = \kappa_l + \kappa_e$. The $\kappa_e = L\sigma T$, where L is the Lorenz factor [9–11]. Apparently, thermal and electrical performance is interrelated with each other, so that enhancing zT lies in the optimization of electrical performance and minimizing κ_l. In terms of the electrical performance, based on the parabolic band and energy-independent scattering approximation (generally suitable for metals and degenerate semiconductors), S can be expressed as [12]:

$$S = \left(\frac{8\pi^2k_B^2}{3eh^2}\right)m^*\left(\frac{\pi}{3n_H}\right)^{\frac{2}{3}}T \tag{9.2}$$

where k_B is the Boltzmann constant, h is the Plank constant, e is the elementary charge, T is the absolute temperature and n_H is the carrier concentration. σ is also related to n_H and can be described as [13, 14]:

$$\sigma = 1/\rho = n_He\mu. \tag{9.3}$$

As can be seen, both S and σ are strongly related to n_H. Although high n_H may lead to high σ, it can also decrease S. μ is related to m^* to some degree. m^* increases when the material has the flat, narrow bands at the Fermi surface [15, 16]. The relationship between m^* and μ is complex and related to the electronic structure and the carrier scattering mechanisms [17–19]. Due to the complex relationship between these factors, optimization of electrical performance mainly lies in optimizing carrier transport properties, especially approaching an optimized n_H under certain m^*.

Efficiency of thermoelectric devices is defined on the basis of energy input and output. For example, for a thermoelectric generator, its efficiency is defined as the ratio of energy provided to the load to the heat absorbed at the hot junction. In a thermoelectric refrigerator, the efficiency is simply defined as the maximum temperature drop which can be calculated as [20]:

$$\Delta T_{max} = \frac{zT_c^2}{2}, \tag{9.4}$$

where zT_c is average zT. Efficiency of thermoelectric refrigerator may also be expressed in the form of the coefficient of performance (*COP*) [21]:

$$\mathrm{COP} = \frac{T_c}{T_h - T_c} \cdot \frac{\sqrt{1 + z\bar{T}} - \frac{T_h}{T_c}}{\sqrt{1 + z\bar{T}} + 1}, \tag{9.5}$$

where $\bar{T}$ is the average temperature between hot side (T_h) and cold side (T_c), and $z\bar{T}$ is the calculated zT under average temperature. Efficiency of thermoelectric power generator, η_e, is similar with COP of thermoelectric refrigerator [21]:

$$\eta_e = \frac{T_h - T_c}{T_h} \cdot \frac{\sqrt{1 + z\bar{T}} - 1}{\sqrt{1 + z\bar{T}} + \frac{T_c}{T_h}}. \tag{9.6}$$

Both efficiencies of thermoelectric power generator and refrigerator are limited by the Carnot efficiency, ε_c [21]:

$$\varepsilon_c = \frac{T_h - T_c}{T_h}. \tag{9.7}$$

As can be seen, η_e is mainly determined by zT, the intrinsic property of the material itself and the applied temperature, where the later part is mainly up to the application conditions and temperature stability of the material. Hence, the development of thermoelectric materials has been focused on searching for materials with naturally high zT values under certain temperature or on enhancing zT of composing thermoelectric materials. Figure 9.1a shows η_e of thermoelectric power generators as a function of the hot-side temperature under different zT of composing materials and the cold-side temperature is 300 K [22]. For the purpose of comparison, $\mathcal{E}_c$ (upper-limit efficiency of heat engine) is shown in Fig. 9.1a. Under current peak zT (~2) of most state-of-the-art thermoelectric materials and a temperature difference of ~800 K, the estimated η_e of thermoelectric power generator can approach as high as ~20%. Note that this value is still lower than $\mathcal{E}_c$ [23]. Therefore, more efforts are needed on further enhancing zT of thermoelectric materials.

9.2 Recent Advances in Thermoelectric Materials

Electrical performance of thermoelectric materials has been effectively enhanced via band convergence [24], superionic conductivity [14], modulation doping [25], electron-energy filtering effect [26], quantum confinement [27] and resonant effect [28]. Simultaneously, reducing κ_l via nanostructuring [8], phonon-glass electron-crystal (PGEC) [29], hierarchical architecture engineering [30] and nanoprecipitations [31] further increases zT values. Recently, high-performance thermoelectric materials with high zT values under various temperature range have been identified, including skutterudites [32], half-Heusler alloys [33–35], clathrates [36], semiconducting oxides [37, 38], silicides [39–42], polymers [43], chalcogenides [14, 44–47] and single-element thermoelectric materials [48], which have been summarized and

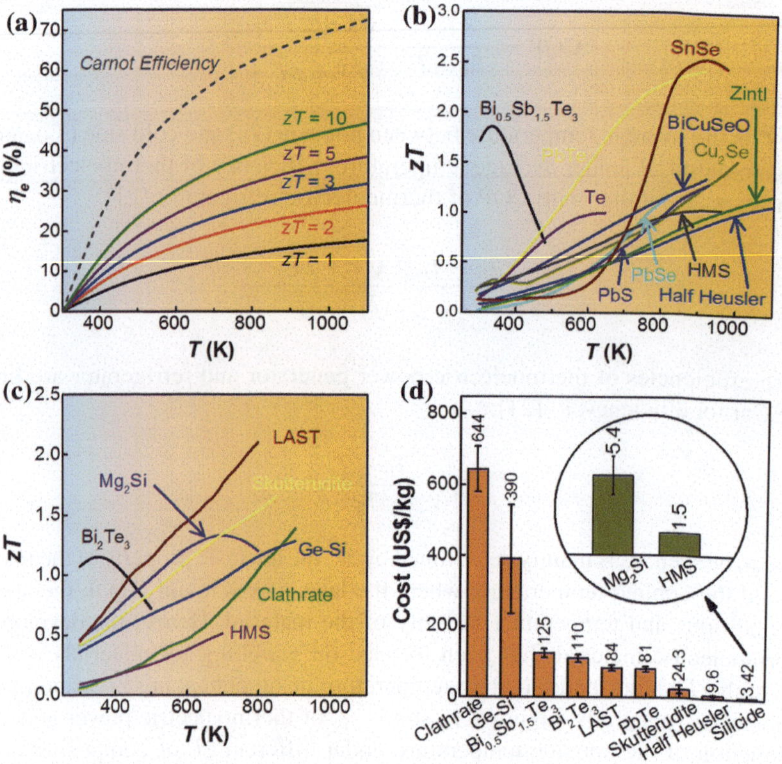

Fig. 9.1 **a** Energy conversion efficiency (η_e) of a thermoelectric power generator as a function of the hot-side temperature under different zT of the composing thermoelectric materials, **b** and **c** temperature-dependent zT of p-type and n-type state-of-art thermoelectric materials [14, 32, 33, 35, 36, 38, 41, 42, 44, 45, 48, 49, 67, 68, 70, 81–84], respectively, where HMS is higher manganese silicide and LAST stands for lead-antimony-silver-telluride. **d** Cost of thermoelectric materials [20]. Reproduced with permission [22]. Copyright 2018, John Wiley and Sons

shown in Fig. 9.1b and c. Although encouraging zT values of over 2 have been achieved in various thermoelectric materials, such as a zT value of 2.5 at ~923 K for $Pb_{0.98}Na_{0.02}Te$–8% SrTe [49], the zT values are still far below their theoretical predictions [50]. Additionally, η_e of thermoelectric devices/modules is not competitive enough when comparing with other alternative/traditional energy sources [51, 52].

Traditionally, most thermoelectric materials are highly toxic or expensive, which have limited their wide applications. In the synthesis of traditional thermoelectric materials, high-purity earth-rare elements are generally required coupled with complex synthesis process, which increases the cost. Figure 9.1d summaries the costs of thermoelectric materials, in which costs of producing 1 kg traditional $Bi_{0.52}Sb_{1.48}Te_3$ nanobulk and $AgPb_{18}SbTe_{20}$ are US$125 [53] and US$84 [53], respectively. Correspondingly, the cost of generating thermoelectric power can be estimated as ~US$5 per W [20]. Hence, regardless of the long development history of thermoelectrics, its applications have been limited in niche areas, mainly focusing on out-space missions [54] and refrigerations [55]. Furthermore, nearly all state-of-art thermoelectric materials are composed of earth-rare or toxic elements, such as expensive Bi [2, 7, 10, 38, 44, 56–65] and carcinogenic Pb [31, 49, 66–71]. In recent years, development of nontoxic and low-cost thermoelectric materials has drawn increasing attentions, especially the industrial attention. Inspired by this, earth-abundant and less-toxic thermoelectric materials, such as eco-friendly SnTe [1, 72–74], Cu_2Se [9, 75], $Cu_{12}Sb_4S_{13}$ [76], $Cu_2ZnGeSe_4$ [77], $Cu_2CdSnSe_4$ [78], and silicides [40, 79], (including Mg_2Si [41] and higher manganese silicide (HMS) [80–82]) have been widely explored in recent decades.

9.3 Electrical Performance Enhancement

Electrical performance of thermoelectric materials is generally enhanced via the carrier transport property optimization under certain m^* [16, 22]. Here, we take HMS as an example for detailed discussion. In the single parabolic band model, under the assumption of the κ_l value of ~1 W m^{-1} K^{-1} [82], the calculated zT values of HMS are shown in Fig. 9.2a, in which the reduced m^* increases the zT values and suspend the level of optimized n_H. The optimized level of n_H for HMS lies close to 1×10^{21} cm^{-3}. When m^* is ~7·m_0 (m_0 is the rest mass of an electron), under state-of-the-art lowest κ_l of HMS [82], a peak zT of as high as ~1 can be achieved. As reported, the reduced m^* can be achieved in Re-doped/alloyed HMS [85]. However, the low solubility of Re in HMS (<4%) has limited further reduction of m^* in HMS [85]. The nanoprecipitations will reduce μ_H and modify carrier scattering mechanism [85–87]. When introducing liquid quenching into HMS synthesis, the doping level of Re can be effectively enhanced in HMS, leading to further increased electrical performance and zT value [82]. A peak zT of as high as ~1 has been achieved at ~800 K with 16.5% Re in Re-doped HMS [82].

n_H optimization is the most commonly employed strategy for enhancing the electrical performance of thermoelectric materials [12]. n_H is generally tuned by introducing dopants of different valence electron counts (VECs) comparing with the matrix element. For example, VEC [88] has been employed to tune n_H of HMS via employing Al to substitute Si [89–91], and Fe [92], W [81], Cr [93–96] to substitute Mn. Kikuchi et al. [93] defined VEC of $Mn_{1-x}Cr_xSi_y$ as $\mathrm{VI}\cdot x + (1 - x)\cdot \mathrm{VII} + y\cdot \mathrm{IV}$, where VI, VII and IV are the VECs of Cr, Mn and Si, respectively, where the n_H data are not available. Referring to this definition, VEC of $Mn(Si_{1-x}Al_x)_{1.80}$ can be defined as $\mathrm{VII} + 1.80\cdot(1 - x)\cdot \mathrm{IV} + 1.80\cdot x\cdot \mathrm{III}$, where III is the VECs of Al. Figure 9.2b and c shows that decreasing VEC in p-type HMS via Al/Si substitution leads to unfilled bonding orbitals or anionic states (valence band) which would subsequently increase n_H [93, 97]. The enhancement of n_H has further increased σ and zT. By tuning VEC, n_H of HMS can be adjusted in a wide range. Based on this concept, through Fe-alloying, VEC increases subsequently [81, 92, 98], and the intrinsic p-type HMS can be even converted into n-type [99]. Limited by Fe solubility in HMS, peak zT of

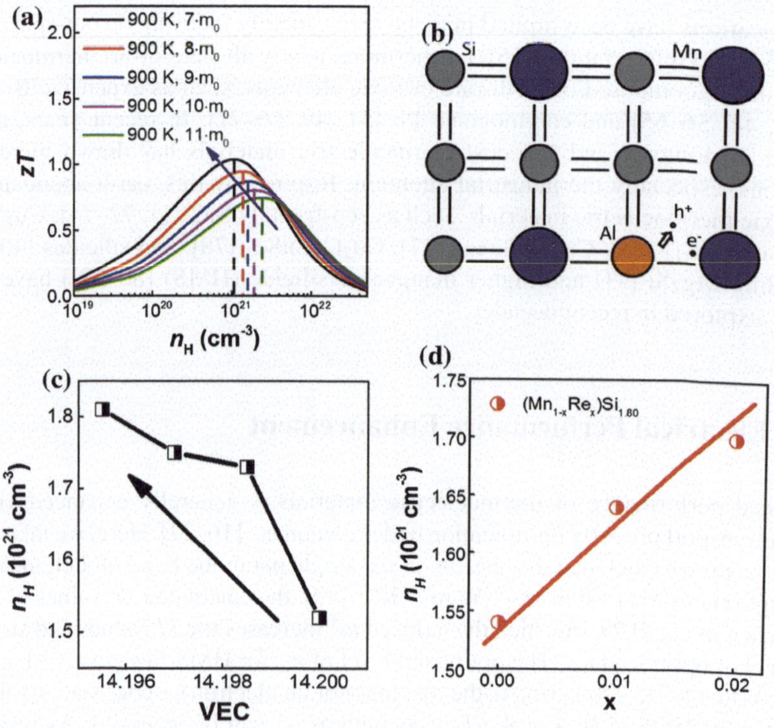

Fig. 9.2 **a** The hole carrier concentration (n_{H}) dependent dimensionless figure of merit (zT) at 900 K under different effective mass (m^*) values, **b** a schematic diagram of Al/Si substitution in higher manganese silicide (HMS) increases the n_{H} [89], **c** the n_{H} of Al doped/alloyed HMS as a function of estimated valence electron count (VEC) [89] and **d** Re doping level (x) dependent n_{H} of Re-doped HMS [85]. Reproduced with permission [22]. Copyright 2018, John Wiley and Sons

the as-synthesized n-type HMS is only ~0.07 at 650 K [92]. However, the combination with liquid quenching and W-doping has further increased the Fe solubility and maximized the zT value of n-type HMS to ~0.5 [81].

Apart from tuning VEC, n_{H} can also be adjusted via isoelectronic dopants. In HMS, isoelectronic doping can disturb the arrangement of Si/Mn sublattice, leading to the formation of stacking faults [100], which subsequently modifies the density of states (DOS) near the Fermi level [100, 101], n_{H} and σ, leading to increased $S^2\sigma$ and zT. This is possibly due to bond softening induced by either larger atomic radius or smaller electronegativity difference [85, 102]. A typical case is Re-doped/alloyed HMS, as shown in Fig. 9d. n_{H} increases with increasing the Re doping level, leading to enhancing σ, $S^2\sigma$ and zT [85]. Additionally, an ultra-high σ has been achieved via Re-supersaturation in $Re_{0.165}Mn_{0.835}Si_{1.75}$, resulting in a peak $S^2\sigma$ of ~22 W m^{-1} K^{-2} above 800 K [82]. With simultaneously reduced κ_1, a peak zT of ~1 above 800 K has been achieved in Re-supersaturated HMS. Similarly, Zhou et al. [100] and She et al. [103] have also demonstrated the increased n_H, σ, $S^2\sigma$ and zT via substituting Si/Ge substitution.

Regardless of the influence of changed m^* and optimized n_H, electrical performance of thermoelectric materials can be modified via the resonant doping. For example, in In-doped SnTe, S values of all the In-doped samples lie far above the Pisarenko line, indicating that In-doped SnTe has a modified band structure [104, 105]. According to the Mahan–Sofo theory [105]:

$$S = \frac{\pi^2}{3}\frac{k_B}{e}k_B T\left\{\frac{1}{n_H}\frac{dn_H(E)}{dE} + \frac{1}{\mu}\frac{d\mu(E)}{dE}\right\}_{E=E_F} \tag{9.8}$$

where E_F is the Fermi energy. Increasing $n(E)$, via a local increase in DOS, S increases. For example, in SnTe, the resonant doping, distortion of the DOS near the Fermi level, is effective in enhancing S without affecting n_H [1, 104, 106–109]. For a given thermoelectric material, the hybridizations of electrons between dilute impurity and the host valence or conduction band can generate the resonant energy level [107], which can locally increase DOS over a narrow energy range at the Fermi level, leading to increased S. In SnTe-based thermoelectric materials, In can produce a resonant state in the valence band and increases S due to the increased DOS near the Fermi level, as schematically illustrated in Fig. 9.3a [1]. Figure 9.3b shows the calculated DOS of In-doped and pristine SnTe [107], where high DOS is observed near the Fermi level in In-doped SnTe comparing with pristine SnTe. Figure 9.3c and d shows that In resonant doping can enhance the $S^2\sigma$ and zT of SnTe, respectively [107].

Furthermore, the band structure convergence has also enhanced S. Figure 9.4a shows a schematic diagram of doping-modified band structure of SnTe, where the significantly reduced energy offset between heavy and light valence bands ($\Delta E_{VB,\,L\text{-}\Sigma}$) by converging two valence bands has allowed the heavy hole to participate in the electron–hole transport and increased S [1, 105, 107, 110, 111]. Figure 9.4b, c and d shows temperature-dependent S, $S^2\sigma$, and zT of Mn-doped SnTe [112], Ca-doped SnTe [113], Mg-doped SnTe [111] and pristine SnTe [111]. Mg-doping can reduce the energy offset between heavy and light valence band from ~0.375 to 0.18 eV, leading to enhanced S, $S^2\sigma$ and zT [1, 111]. Similar results have also been reported in Ca- and Mn-doped SnTe with increased S, $S^2\sigma$ and zT [1].

9.4 Lattice Thermal Conductivity Suspension

To increase zT, except for carrier transport property optimization (leading to enhanced electrical performance), reducing κ is also necessary. As κ_e is directly related to σ which has been considered in the electrical performance optimization, κ can be further reduced via lowering κ_l. This is generally realized strengthening the phonons scattering with various wavelengths. This can be achieved by introducing corresponding multi-scale phonon scattering centers, including mesoscale ones

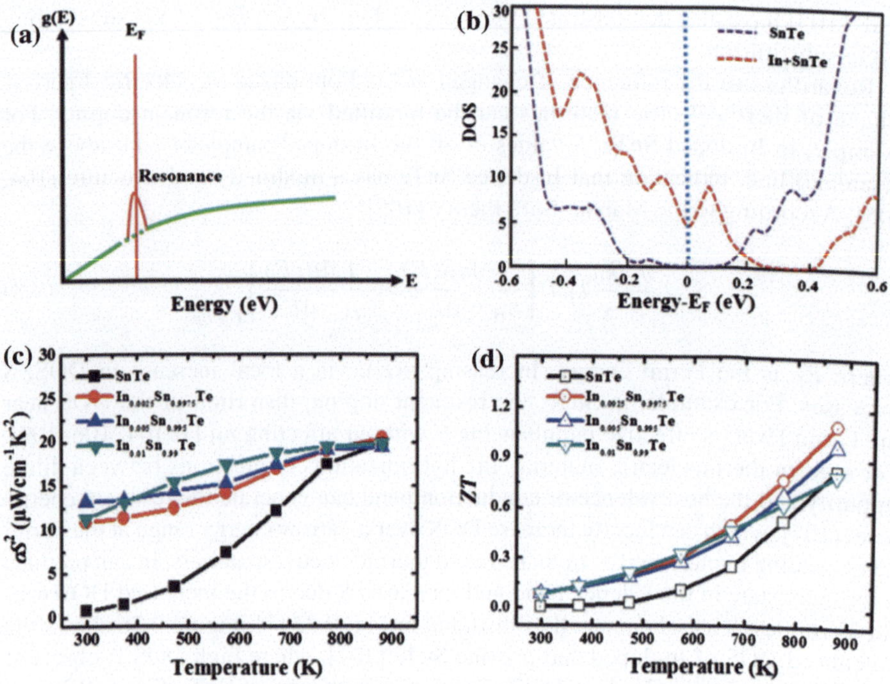

Fig. 9.3 **a** Schematic diagram of the distortion of density of states (DOS, $g(E)$). The green and red lines stand for the DOS of pristine SnTe and In-doped SnTe (with resonant energy levels from In; E_F = Fermi energy level), respectively, **b** change of DOS in SnTe due to the resonance states [107], **c** power factor ($S^2\sigma$) and **d** dimensionless figure of merit (zT) of the In-doped SnTe comparing with pristine SnTe [104]. Reproduced with permission [1]. Copyright 2017, John Wiley and Sons

(secondary phases [114] and pores [115]) to nanoscale ones (nanoprecipitates [31], nanopores [115], or nanograins [34]) and atomic-scale ones (points defects) [85].

Figure 9.5a illustrates the influence of atomic-scale point defects on phonon scattering, where the mass and radius fluctuation and subsequently rising strain field can effectively scatter short-wavelength phonons and reduce κ_l. Take an example of Cu heavily doped SnSe, where Cu-doping has induced lattice defects. Figure 9.5b and c shows transmission electron microscopy (TEM) images of the dislocations and the strain field that are possibly induced by point defects. They can effectively scatter short-wavelength phonons, leading to reduced κ_l as shown in Fig. 9.5d. The inset suggests that κ_l has nearly linear relationship with 1000/T regardless of the Cu-doping level which indicates that the phonon scattering in Cu heavily doped SnSe is still dominated by the Umklapp phonon scattering. The reduced κ_l has increased optimized zT as shown in Fig. 9.5e where proper doping results in n_H approaching to the optimized level and in turn leading to enhanced zT [116].

Figure 9.6a schematically shows the influence of meso- and nanoscale phonon scattering centers that scatter long and middle wavelength phonons to reduce κ_l

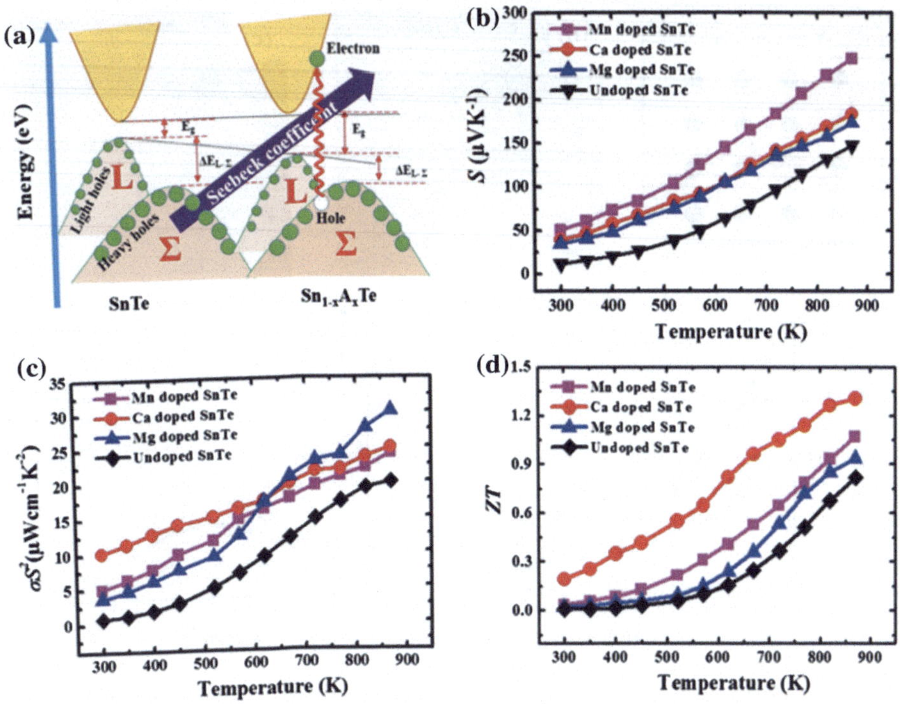

Fig. 9.4 **a** Schematic of band convergence of two valence bands, **b** Seebeck coefficient (S), **c** power factor ($S^2\sigma$) and **d** dimensionless figure of merit (zT) of Mn-doped SnTe [112], Ca-doped SnTe [113], Mg-doped SnTe [111] and pristine SnTe [111]. Reproduced with permission [1]. Copyright 2017, John Wiley and Sons

[22]. Figure 9.6b is the calculated phonon mean free path (MFP) and accumulated κ_l as a function of phonon wavelength reciprocal of HMS [117]. The accumulated κ_l sharply increased with the reducing phonon wavelength, indicating the increase of phonon energy [117]. This is attributed to higher DOS of high energy phonons [117]. Meanwhile, indicated by the slope increment of phonon MFP versus energy, the effective suspension of κ_l in HMS lies in strengthening the scattering of short-wavelength phonons (<10 nm) [117]. Additionally, as shown in Fig. 9.6b, reducing HMS grain-size smaller than 20 nm can effectively scatter phonons with high MFP, leading to reduced κ_l [117]. Chen et al. [118] also pointed out that grain-boundary scattering in the scale of ~10 nm may help to approach the glass-like κ_l in HMS.

As discussed above, nanoengineering can effectively increase the density of grain boundaries and subsequently strengthen corresponding phonon scattering, leading to reduced κ_l [34, 39, 41, 45, 49, 58, 119]. For instance, in HMS, to introduce nanoprecipitations, nanosized secondary phases, nanopores and reduce the grain size of HMS, different methods, including ball milling [120], melt spinning [87] or one-step spark plasma sintering (SPS) [121–123], have been employed. Ball milling, as a high energy and low-temperature synthesis method, has attracted extensive

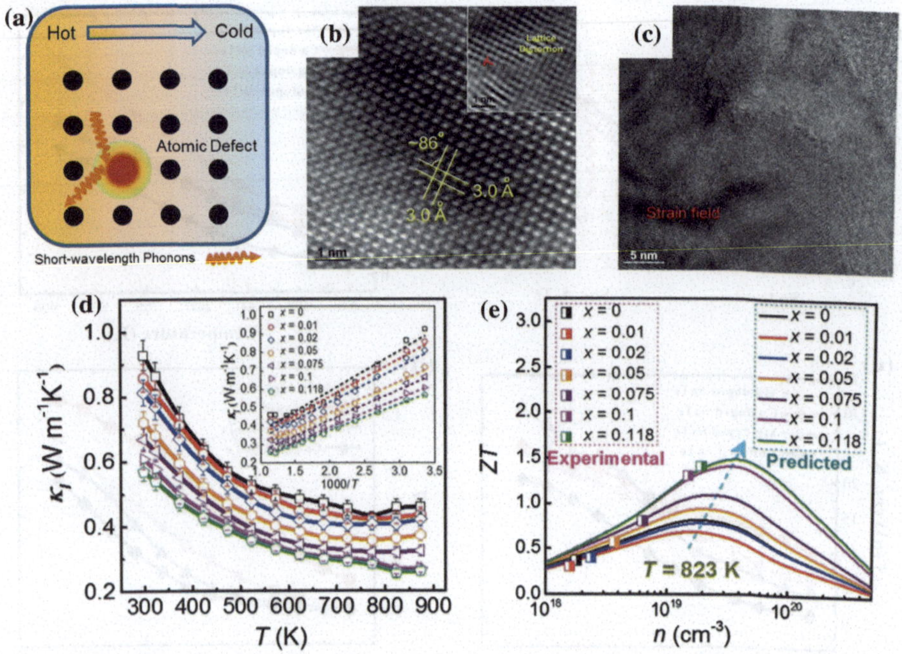

Fig. 9.5 **a** Schematic diagram of reducing lattice thermal conductivity (κ_l) by atomic-scale phonon scattering centers. Reproduced with permission [22], Copyright 2018, John Wiley and Sons. **b** High-resolution transmission electron microscope (HRTEM) image of Cu heavily doped SnSe with a dislocation shown in the inset, **c** transmission electron microscope (TEM) image of Cu heavily doped SnSe with clear strain field, **d** *T*-dependent κ_l with inset of *1000/T*-dependent κ_l of Cu heavily doped SnSe and **e** comparison of experimental zT values with predicted plots for Cu-doped SnSe pellet [116]

research interest in synthesizing HMS nanocrystals [120, 123, 124]. Chen et al. [120] demonstrated that ball milling can achieve the HMS grain size as small as ~90 nm after the milling time of ~18 h [120]. Figure 9.6c shows κ_l of ball-milled HMS with different milling time. With prolonging the milling time (from 2 to 18 h), the grain size of HMS has dramatically reduced from ~500 to ~90 nm, which has subsequently decreased κ_l from ~2.4 to ~2.0 W m^{-1} K^{-1} at 700 K [120]. Further prolonging the milling time can introduce contaminations from jars and balls as well as MnSi impurities due to decomposition of HMS [125, 126], in turn deteriorating $S^2\sigma$ and zT [120]. However, with simultaneously reduced μ_H and $S^2\sigma$, the reduction of κ_l via nanoengineering itself fails to enhance zT [120]. Additionally, as illustrated in Fig. 9.6c, nanoprecipitations are also effective in reducing κ_l of HMS. $MnSi_{1.75}$ with MnSi nanoprecipitates (~50 nm in size) can reduce κ_l from ~2 to ~1.5 W m^{-1} K^{-1} at 700 K [87].

Apart from nanograins and nanoprecipitates, nanosized pores may also effectively scatter phonons and suspend κ_l. As an example, Fig. 9.7a and b shows scanning electron microscopy (SEM) and TEM images of SnSe with a high density of nanopores,

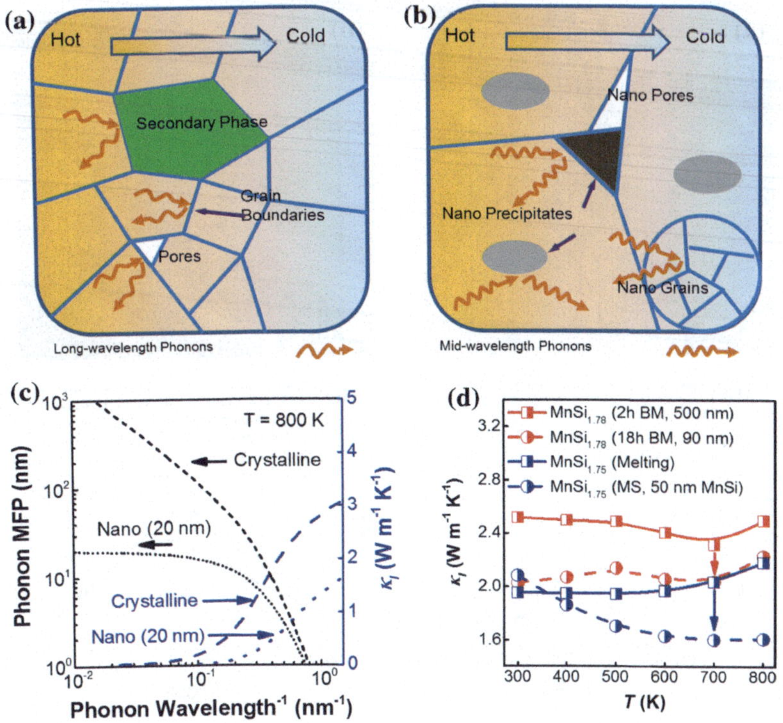

Fig. 9.6 Schematic diagram showing **a** meso- and **b** nanoscale phonon scattering centers leading to reduced lattice thermal conductivity (κ_l), **c** calculated relationship between phonon mean free path (MFP), accumulated κ_l and reciprocal of phonon wavelength of higher manganese silicide (HMS) under 800 K [117], **d** the effect of nanoengineering in reducing κ_l of HMS, where BM is ball milling and MS stands for melt spinning [85, 87, 90, 103, 120]. Reproduced with permission [22]. Copyright 2018, John Wiley and Sons

which has led to reduced κ_l (Fig. 9.7c). Such a κ_l reduction in porous SnSe is attributed to simultaneously reduced specific heat (C_V), minimum phonon mean free path (l_{min}) and sound velocity (v), as the κ_l can be expressed as:

$$\kappa_l = \frac{1}{3} C_v v l_{min}. \tag{9.9}$$

It should be noted that κ_l of porous SnSe is much lower than the calculated minimum κ_l ($\kappa_{l,min}$, 0.36, 0.33, and 0.26 W m^{-1} K^{-1} along the *b*-, *c*-, and *a*-axes of SnSe single crystals) using a classical Debye-Cahill model, respectively, due to a high density of nanopores. The inset of Fig. 9.7c shows the plot of 1000/*T*-dependent κ_l for all porous SnSe, from which all κ_l values show a roughly linear relationship, indicating that the Umklapp phonon scattering plays a critical role in SnSe. With the κ_l reduction, *zT* at the optimized n_H has been effectively enhanced (Fig. 9.7d), where n_H has not changed obviously [127].

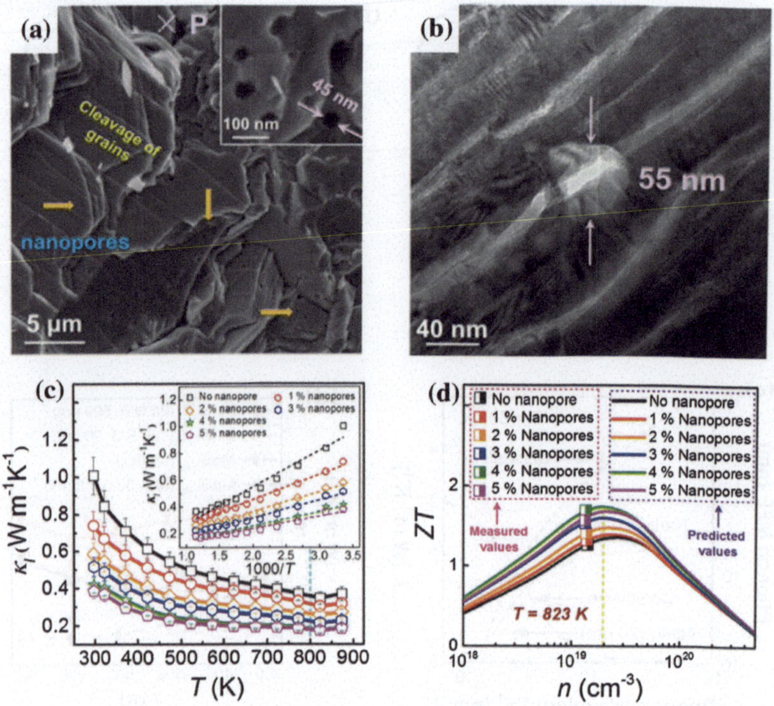

Fig. 9.7 **a** SEM image with inset of a magnified SEM image of SnSe-4% nanopores pellets; **b** magnified TEM image to see one typical nanopore of SnSe; **c** lattice thermal conductivity (κ_l) with inset of 1000/T-dependent κ_l for SnSe-x % nanopores pellets ($x = 0$, 1, 2, 3, 4 and 5); and **d** comparison between experimentally achieved dimensionless figure of merit (zT) values with predicted values via calculations. The yellow dashed line indicates the best carrier concentration (n_H) to achieve peak zT [127]. Reproduced with permission [127]. Copyright 2018, American Chemistry Society

As a combination of different scattering centers on meso-, nano- and atomic-scale, hierarchical architectural engineering can more effectively scatter phonons, leading to ultralow κ_l [30]. For example, optimizing atomic substitution in PbTe, as illustrated in Fig. 9.8a, can lead to a zT value of 1.1 at 775 K (with 2 mol% Na) [28, 30, 128, 129]. The maximum zT has increased to 1.7 at 800 K with further introducing 2–10 nm SrTe nanoprecipitations [30]. This enhanced performance stems from the fact that the nanoprecipitations may impede the heat flow in this system while maintain the carrier mobility nearly unaffected. Nanostructuring can scatter phonons with short and medium wavelengths (3–100 nm) effectively, leaving the longer wavelength phonons un-scattered. Additional κ_l reduction can be achieved by further scattering the longer wavelength phonons (0.1–1 mm, mesoscale) with grain boundaries. Experimentally, grain-boundary phonon scattering has effectively improved thermoelectric performance of PbTe [30].

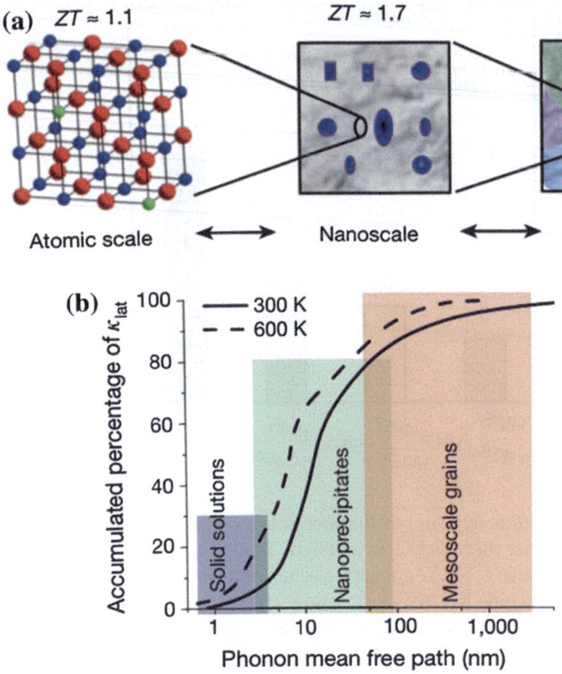

Fig. 9.8 **a** Maximum achievable dimensionless figure of merit (zT) values of PbTe-based thermoelectric materials for the respective length scales: the atomic scale (alloy scattering: red, Te; blue, Pb; green, dopant), the nanoscale (PbTe matrix, gray; SrTe nanocrystals, blue) to the mesoscale (grain-boundary scattering). By combining the effects of atomic-scale alloy doping, endotaxial nanostructuring and mesoscale grain-boundary control, maximum phonon scattering can be achieved at high temperatures and the figure of merit can be increased beyond the value possible with nanostructuring alone; **b** contributions of phonons with different mean free paths to the cumulative κ_l value for PbTe. Phonons with short, medium and long mean free paths can be scattered by atomic-scale defects, nanoscale precipitates and mesoscale grain boundaries, respectively [30]. Reproduced with permission [30]. Copyright 2012, Springer Nature

9.5 Efficiency of Prototype Thermoelectric Modules

As a prerequisite for further applications of thermoelectric materials, prototype thermoelectric modules have been constantly assembled, and their actual energy conversion efficiencies have been evaluated. Figure 9.9 summarizes η_e of thermoelectric modules made of HMS and other thermoelectric materials working from room-temperature to ~800 K. In the $FeSi_2$/HMS system under a temperature difference of 600 K, limited by low zT of $FeSi_2$ n-type leg, the η_e is only ~3% [130]. When employing $Mg_2(Si,Sn)$ as the n-type legs, simulation results illustrate that η_e of HMS and $Mg_2(Si,Sn)$ thermoelectric module can approach a value >7%, where the cold-side temperature is 298 K and temperature difference falls in the range of 400–700 K

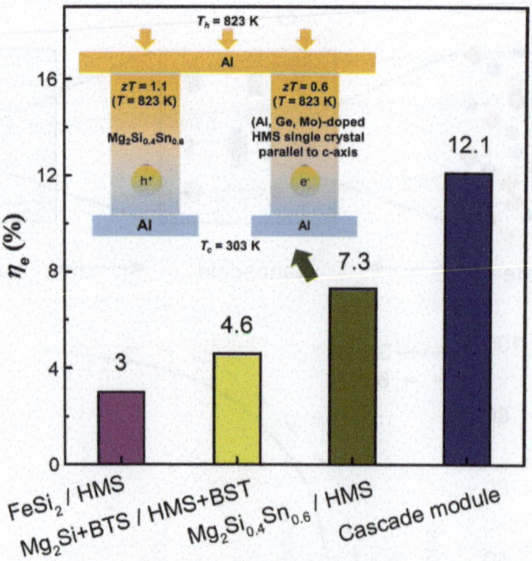

Fig. 9.9 Energy conversion efficiency (η_e) of thermoelectric modules composed of higher manganese silicide (HMS) and other thermoelectric materials [130, 132–134], where BTS stands for $Bi_{2\text{-}x}Sb_xTe_3$ and BST stands for $Bi_{2\text{-}x}Sb_xTe_3$, inset is a schematic diagram of the segmented thermoelectric module composed of n-type $Mg_2Si_{0.4}Sn_{0.6}$ and p-type Mo, Al and Ge-doped HMS single crystal parallel to *c*-axis [132]. Reproduced with permission [22]. Copyright 2018, John Wiley and Sons

[131]. Ikuto et al. [132] experimentally demonstrated that η_e achieved over 7.3%, as shown in the inset of Fig. 9.9.

Peak zT values of different thermoelectric materials lie within different temperature ranges, when taking this into consideration of maximizing η_e of thermoelectric modules, segmented design of joined thermoelectric materials with different operational temperature is a general strategy [133]. One thermoelectric module employing HMS and $Bi_{2-x}Sb_xTe_3$ as the p-type leg materials and Bi-doped Mg_2Si and $Bi_2Te_{3-x}Se_x$ as the n-type leg materials has been assembled [133]. However, the η_e value of segmented designed module is only ~4.6% under the temperature difference of 500 K [133], which has been attributed to the low zT of Bi-doped Mg_2Si (0.5) at the working temperature of ~773 K [133]. More interestingly, stacking cascade module employing Bi_2Te_3-based modules (p-type Sb_2Te_3 based alloy and halogen-doped n-type Bi_2Te_3-Bi_2Se_3 alloy) and silicide modules (Sb-doped $Mg_2Sn_{0.4}Si_{0.6}$ and Mo, Al and Ge-doped HMS) has approached an η_e as high as 12.1% [134].

9.6 Devices and Applications

Thermoelectric modules have been further assembled into devices and employed in various sectors. With wide potential applications, flexible/bendable thermoelectric devices based on both flexible thermoelectric materials and bulk inorganic thermoelectric materials have attracted extensive research interest. While employing thermoelectric generators substituting batteries (TEGs, Fig. 9.10a, b), durable power supply can be secured for devices [4, 135]. For the practical applications, TEGs need to become highly flexible and tailoring-convenient which leads to better adaptability and processability. For these goals, thermoelectric devices can be assembled via the following techniques: deposition or printing as films or on flexible bases (Fig. 9.10c), or assembling niche TEG modules on the flexible bases via advanced microfabrication techniques, as demonstrated in Fig. 9.10d and e [4, 136]. Figure 9.10e shows a flexible TEG module containing 72 thermoelectric couples, in which a high power output density of ~4.78 mW cm^{-2} is achieved when the temperature difference is only 25 K [4, 137].

Figure 9.11a shows a portable thermoelectric energy conversion unit (TECU) of a thermal envelope covering a human body, and Fig. 9.11b schematically illustrates the detailed functioning process of TECU. As can be seen, the ambient air is cooled down (under the cooling mode) or heated up (under the heating mode) within TECU before supplied to the human body through a tree-like rubber tube network that is knitted into a thermoregulatory clothing [142]. The performance of TECU has been tested on a non-sweating thermal manikin under 298 K. Figure 9.11c and d shows the

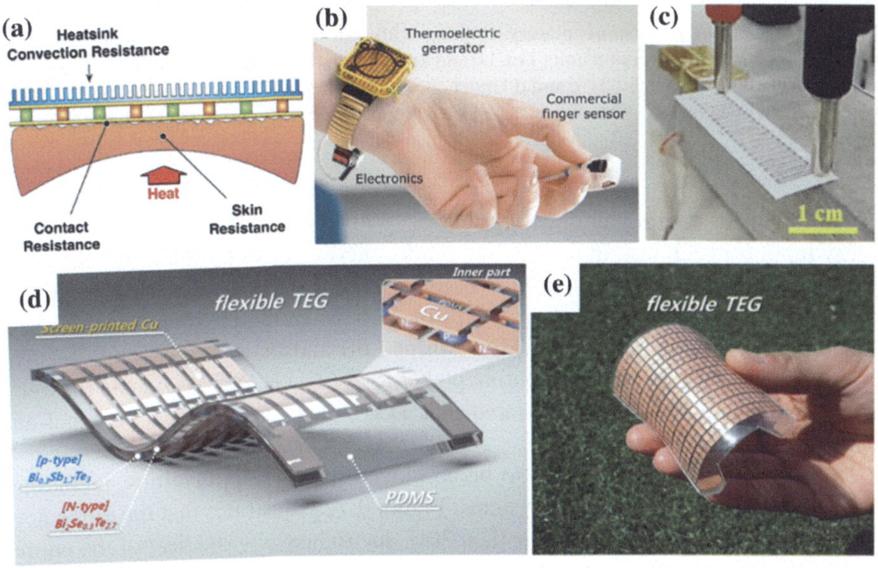

Fig. 9.10 **a** Schematic diagram of wearable thermoelectric generator (TEG) working mechanism. Reproduced with permission [138]. Copyright 2016, Royal Society of Chemistry. **b** An example of TEG powered wearable device. Reproduced with permission [139]. Copyright 2009, AIP Publishing. **c** An example of paper-based TEG device. Reproduced with permission [140]. Copyright 2017, Elsevier. **d** and **e** Schematic and photo of a flexible TEG assembled with inorganic modules [4]. Reproduced with permission [141]. Copyright 2016, American Chemical Society

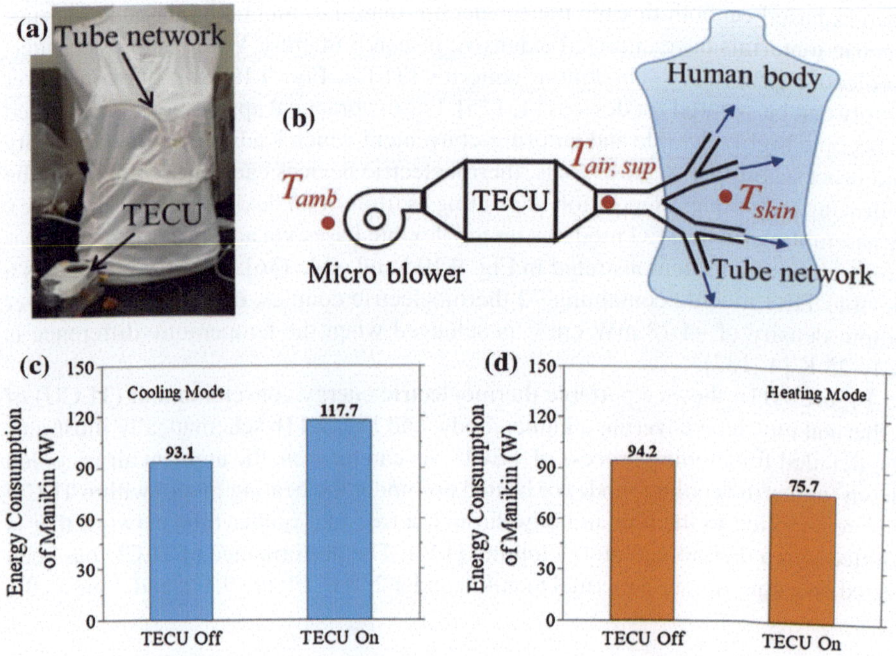

Fig. 9.11 **a** Photo of a thermoregulatory clothing with tree-like tube network connected to the thermoelectric energy conversion unit (TECU) and knitted into the clothes. **b** Schematic diagram of working mechanism of the TECU. **c** and **d** Under cooling/heating mode, energy consumption of the manikin with TECU turned off and turned on [142]. Reproduced with permission [142]. Copyright 2018, Elsevier

test results under cooling and heating modes. While employing the same electrical energy input for each test, in the cooling mode, energy consumption of the manikin with TECU turned off and on is 93.1 W and 117.7 W, respectively. Hence, TECU has approached a personal cooling power of 24.6 W. In the heating mode, energy consumption of the manikin with TECU turned off and on is 94.2 W and 75.7 W, respectively. Thus, the achieved heating power for TECU is 18.5 W. As can be seen, the thermal manikin test demonstrated that TECU can function as a personal air conditioner effectively [142].

Thermoelectric modules can also be employed in other applications, including vehicles and water condensation systems. Figure 9.12a illustrates that the efficiency of the combustion engine is smaller than 30% due to heavy waste heat of the entire system. With the thermoelectric devices, the waste heat can be collected and re-used effectively, leading to improve the system efficiency and reducing the CO_2 emission. The re-collected waste heat could be used as electrical energy and powering the electrical systems in vehicles, such as air conditioner [4, 143].

As discussed above, exhaust TEGs are capable of converting waste heat into electricity from the exhaust gas, where ~40% of the energy are from fuel combustion (Fig. 9.12a). The temperature of exhaust gas can approach as high as 1073 K. Even

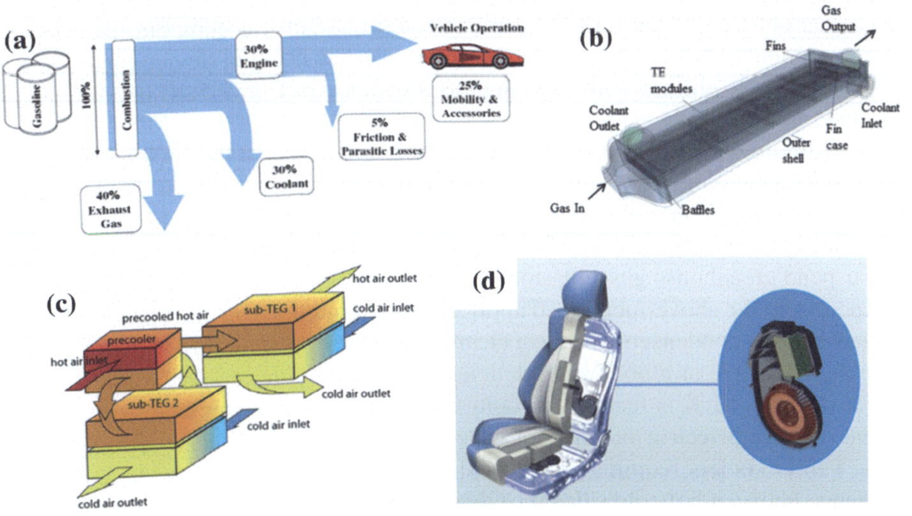

Fig. 9.12 **a** Typical energy path for combustion engines. Reproduced with permission [148]. Copyright 2009, Springer USA. **b** Schematic diagram of a heat exchanger for exhaust gas energy recollection employing thermoelectric generators. Reproduced with permission [149]. Copyright 2017, Elsevier. **c** A schematic concept of regenerative thermoelectric generators. Reproduced with permission [150]. Copyright 2017, Elsevier. **d** Automotive air-conditioning seat in vehicles. Reproduced with permission [151]. Copyright 2013, Nature Publishing Group

in the downstream, it is still higher than 500 K. Such a high temperature is sufficient to secure a stable hot side for thermoelectric module [4, 143]. As illustrated in Fig. 9.12b, in an exhaust TEG, the thermoelectric modules are highly integrated with the heat exchangers whose function is identifying the thermal flux of the exhaust gas, heating the hot side of the thermoelectric modules, cooling down the cold side and enabling the optimized working condition of the thermoelectric modules to achieve a high-density power output. To maximize the heat gain, heat exchangers should be composed of high κ materials. To secure effective heat transfer and sufficient exhaust flow, rational geometry design of the heat exchangers is also necessary. Furthermore, as illustrated in Fig. 9.12c, the exhaust gas can be recirculated to suit thermoelectric modules working under low temperature and further enhancing the energy conversion efficiency of the entire system [4].

The dominating challenge of TEG-commercialization lies in their low efficiency (<5%). In the 1990s, by employing commercial Bi_2Te_3-based thermoelectric modules with an efficiency of ~5%, 1 kW exhaust TEGs were designed [144]. By the 2000s, the maximum power point tracking was applied to TEG systems, which can maximize the electrical power stored in the battery and improve the power output [4]. Since then, extensive efforts have been devoted to optimizing the design of exhaust TEGs for higher efficiency. Furthermore, the energy efficiency of vehicle-applicable TEGs has been improved by two ways: (1) employing coolant water (363–383 K) as well as charge air cooler outlet gases (423 K) as the heat source for low-temperature thermoelectric modules and (2) introducing the thermoelectric cooling device (also known as the automotive climate control system as illustrated in Fig. 9.12d) in automotive air-conditioning [145–147].

In the future, with the development of high *zT* and low-cost thermoelectric materials, novel technology and rational design, efficiency of the exhaust TEGs can be further increased. When efficiency of TEGs reached ~20%, it should reach to the point that can be widely employed and significantly resolve environmental pollution and suspend greenhouse gas emission [4].

Apart from the above-mentioned applications, thermoelectric devices can also be applied in water condensers to collect clean water from moist air directly, and in turn solve the challenge of global water deficiency. For the Peltier cooler, a temperature drop as large as 81 K is feasible at room-temperature, which can enable the applications of thermoelectric modules for water condensers [4]. Efficiency optimization of the Peltier coolers requires proper control on the current input and temperature difference between hot/cold sides. For this reason, a temperature controller assembled to the system is necessary. In addition, employing solar cells as power source can achieve self-sufficient solar-driven water collection. Such a design will make the system more attractive and competitive, especially under extreme environment conditions without extrinsic power supply. Figure 9.13a illustrates such integrated water condenser systems that can supply as much as 1 L of water per hour in high humidity areas [4, 152].

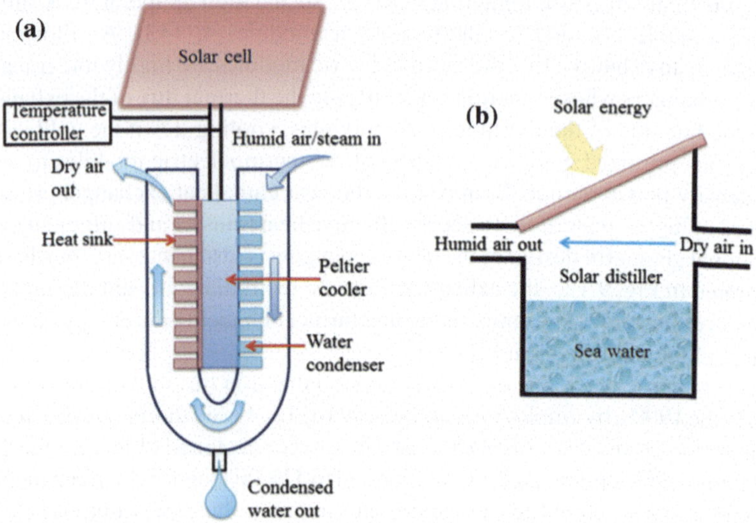

Fig. 9.13 Schematic diagrams of **a** a thermoelectric water condensation system powered by solar cells, **b** a solar water distiller to generate humid air. Reproduced with permission [4]. Copyright 2018, John Wiley and Sons

Additionally, thermoelectric modules can also convert undrinkable water into drinkable ones, such as seawater. Figure 9.13b shows the schematic diagram of solar distiller system, in which the seawater can be effectively evaporated by the sunlight. Subsequently, the moisture can be transported into thermoelectric condenser systems, which will condense as drinkable water. This system can be applied the situation where the climate is dry with undrinkable water sources. It should be noted that further purification may be also necessary when the undrinkable water is heavily contaminated. In the future, with the ever-growing demand of drinkable water, the market of thermoelectric water condensers and relative assemble systems must attract increasing attention with the advantage of utilizing undrinkable water sources or moist air.

Figure 9.14a shows that thermoelectric module powdered devices can be also employed in niche areas, such as implantable medical devices. Since the electrical energy level required by implantable/wearable devices is only several microwatts to milliwatts, TEGs may easily fulfill this requirement and power relevant devices. Figure 9.14b illustrates that the power source of implantable TEGs lies in the temperature difference between human body cores and the skin surfaces. In comparison, the power source of wearable TEGs is mainly temperature difference between the skin and surrounding air [4]. Limited by the low zT values of thermoelectric

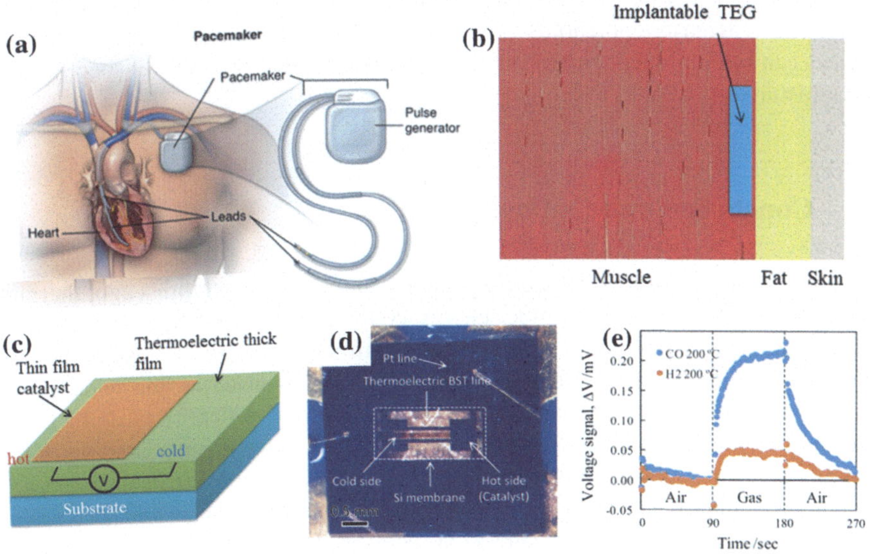

Fig. 9.14 **a** Schematic diagram of implantable pacemaker, **b** schematic diagram of an implantable thermoelectric generator in muscle and human body, **c** schematic diagram of the thermoelectric gas sensor. Reproduced with permission [4]. Copyright 2018, John Wiley and Sons. **d** Optical image of a micro gas sensor assembled by Au/SnO_2–Co_3O_4 catalyst on the Si membrane and **e** its voltage signal against varied CO and H_2 gas concentration. Reproduced with permission [154]. Copyright 2016, Elsevier

materials while working under human body temperatures and the small temperature difference across the device, the output power of implantable/wearable TEGs is not competitive. In recent years, with the development of high-performance thermoelectric materials as well as advanced microfabrication technologies, application of implantable/wearable thermoelectric devices is becoming more promising.

In general, implantable/wearable thermoelectric devices have also been applied in other niche areas, such as signal detectors. Capable of heat signal detecting from catalytic reactions, thermoelectric devices can be applied in gas sensors. Figure 9.14c illustrated a thick film NiO-based H_2 sensor [153]. In this H_2 sensor, half of the surface is coated with Pt thin film acting as catalyst, which will react with H_2 and release heat, create a hot region, subsequently generate a thermoelectric voltage across the hot/cold side and alert the leakage of H_2. Such a film gas sensor has shown sensitivity for promising applications in gas control and alarm systems. Besides, SiGe film, monodispersed SnO_2 nanoparticles, Pt/alumina catalyst and chalcogenide nanowire arrays [4] have also been developed for H_2 sensing. Note that sensors designed based on monodispersed SnO_2 nanoparticles are also response to ethanol and NO_x. Additionally, as shown in Fig. 9.14d and e, double AuPtPd/SnO_2 and Pt/Al_2O_3 catalyst and Au/SnO_2–Co_3O_4 on a micro-thermoelectric sensor are employed for CO gas sensors. For the purpose of detecting hydrocarbon, porous alumina-based sensors were developed. Dual-catalyzed Pd/θ-Al_2O_3 and Pt/α-Al_2O_3 system, capable of H_2 and CH_4 detecting, can be applied in monitoring fuel gas combustion. These gas sensors have extensive potentials for commercialization. Moreover, thermoelectric devices can be combined with a heat-pulse method to measure the soil specific heat and moisture.

9.7 Conclusion and Outlook

Since the discovery of thermoelectric phenomena more than centuries ago, thermoelectric materials, capable of direct energy conversion between heat and electricity without direct environment pollution, have attracted extensive research interest. However, limited by the performance and high cost of material, the wide applications of thermoelectric materials still require further in-depth research. Low-cost thermoelectric materials, such as silicides, are future promising candidates for commercialization, although their performance is not competitive currently.

The performance enhancement of thermoelectric materials lies in two parts: electrical performance optimization and κ_l suspension. Decreasing the effective mass can lead to enhanced optimized zT value. With proper n_H optimization via various doping, optimized zT values could be approached. In recent years, strategies, including band convergence and resonant doping, have been proposed which could further enhance electrical performance of thermoelectric materials. Suspending κ_l of traditional thermoelectric materials mainly lies heavy element doping induced mass/radius fluctuations and strain field. Nanoengineering including nanograins, nanoprecipitates and nanopores could effectively reduce κ_l. The combination of point defects, nanograins and macro-grains, named hierarchical architectural engineering, κ_l of thermoelectric materials could be further reduced. With simultaneously reduced κ_l and optimized electrical performance, the zT value of thermoelectric materials could be effectively enhanced. Composing thermoelectric materials into modules properly could effectively utilize employed thermoelectric materials and approach the peak energy conversion efficiency.

Under appropriate cost and performance consideration, thermoelectric materials could be applied in various areas. With bendable matrix material, thermoelectric modules can be assembled into bendable/flexible device and applied as power generator for niche applications, such as watches. Traditional thermoelectric devices could also be applied in other areas, such as individual air-conditioning clothes, vehicle waste-energy recovery, water condensing system, implantable pacemakers and gas sensors.

References

1. R. Moshwan, L. Yang, J. Zou, Z. Chen, Adv. Funct. Mater. **27**, 1703278 (2017)
2. S. Cao, Z.Y. Huang, F.Q. Zu, J. Xu, L. Yang, Z.G. Chen, ACS Appl. Mater. Interf. **9**, 36478–36482 (2017)
3. Z. Chen, X. Shi, L. Zhao, J. Zou, Prog. Mater Sci. **97**, 283–346 (2018)
4. L. Yang, Z. Chen, M.S. Dargusch, J. Zou, Adv. Energy Mater. **8**, 1701797 (2017)
5. G. Han, Z.G. Chen, J. Drennan, J. Zou, Small **10**, 2747–2765 (2014)
6. M. Hong, Z.G. Chen, L. Yang, J. Zou, Nanoscale **8**, 8681–8686 (2016)
7. L.N. Cheng, Z.G. Chen, L. Yang, G. Han, H.Y. Xu, G.J. Snyder, G.Q. Lu, J. Zou, J. Phys. Chem. C **117**, 12458–12464 (2013)
8. Z.-G. Chen, G. Han, L. Yang, L. Cheng, J. Zou, Prog. Nat. Sci. **22**, 535–549 (2012)
9. L. Yang, Z.G. Chen, G. Han, M. Hong, Y.C. Zou, J. Zou, Nano Energy **16**, 367–374 (2015)
10. G. Han, Z.G. Chen, L. Yang, M. Hong, J. Drennan, J. Zou, ACS Appl. Mater. Interf. **7**, 989–995 (2015)
11. G. Han, Z.G. Chen, L. Yang, L. Cheng, K. Jack, J. Drennan, J. Zou, Appl. Phys. Lett. **103**, 5 (2013)
12. G.J. Snyder, E.S. Toberer, Nat. Mater. **7**, 105–114 (2008)
13. X. Hanhui, W. Heng, P. Yanzhong, F. Chenguang, L. Xiaohua, S.G. Jeffrey, Z. Xinbing, Z. Tiejun, Adv. Funct. Mater. **23**, 5123–5130 (2013)
14. H. Liu, X. Shi, F. Xu, L. Zhang, W. Zhang, L. Chen, Q. Li, C. Uher, T. Day, G.J. Snyder, Nat. Mater. **11**, 422–425 (2012)
15. J. Xin, Y. Tang, Y. Liu, X. Zhao, H. Pan, T. Zhu, NPJ Quant. Mater. **3**, 9 (2018)
16. T. Zhu, Y. Liu, C. Fu, J.P. Heremans, J.G. Snyder, X. Zhao, Adv. Mater. **29**, 1605884 (2017)
17. Y. Xiao, L.-D. Zhao, NPJ Quant. Mater. **3**, 55 (2018)
18. G. Tan, L. Zhao, M.G. Kanatzidis, Chem. Rev. **116**, 12123–12149 (2016)
19. L.D. Zhao, J.Q. He, D. Berardan, Y.H. Lin, J.F. Li, C.W. Nan, N. Dragoe, Energy Environ. Sci. **7**, 2900–2924 (2014)
20. S. LeBlanc, S.K. Yee, M.L. Scullin, C. Dames, K.E. Goodson, Renew. Sust. Energy Rev. **32**, 313–327 (2014)
21. I. Sarbu, A. Dorca, Int. J. Energ. Res. **42**, 395–415 (2018)
22. W.D. Liu, Z.G. Chen, J. Zou, Adv. Energy Mater. **8**, 1800056 (2018)
23. F.L. Curzon, B. Ahlborn, Am. J. Phys. **43**, 22–24 (1975)
24. W. Liu, X.J. Tan, K. Yin, H.J. Liu, X.F. Tang, J. Shi, Q.J. Zhang, C. Uher, Phys. Rev. Lett. **108**, 5 (2012)

25. M. Zebarjadi, G. Joshi, G. Zhu, B. Yu, A. Minnich, Y. Lan, X. Wang, M. Dresselhaus, Z. Ren, G. Chen, Nano Lett. **11**, 2225–2230 (2011)
26. D.-K. Ko, Y. Kang, C.B. Murray, Nano Lett. **11**, 2841–2844 (2011)
27. T.C. Harman, P.J. Taylor, M.P. Walsh, B.E. LaForge, Science **297**, 2229–2232 (2002)
28. J.P. Heremans, V. Jovovic, E.S. Toberer, A. Saramat, K. Kurosaki, A. Charoenphakdee, S. Yamanaka, G.J. Snyder, Science **321**, 554–557 (2008)
29. B.C. Sales, D. Mandrus, B.C. Chakoumakos, V. Keppens, J.R. Thompson, Phys. Rev. B **56**, 15081–15089 (1997)
30. K. Biswas, J.Q. He, I.D. Blum, C.I. Wu, T.P. Hogan, D.N. Seidman, V.P. Dravid, M.G. Kanatzidis, Nature **489**, 414–418 (2012)
31. L.-D. Zhao, J. He, C.-I. Wu, T.P. Hogan, X. Zhou, C. Uher, V.P. Dravid, M.G. Kanatzidis, J. Am. Chem. Soc. **134**, 7902–7912 (2012)
32. X. Shi, J. Yang, J.R. Salvador, M. Chi, J.Y. Cho, H. Wang, S. Bai, J. Yang, W. Zhang, L. Chen, J. Am. Chem. Soc. **133**, 7837–7846 (2011)
33. C.G. Fu, T.J. Zhu, Y.T. Liu, H.H. Xie, X.B. Zhao, Energy Environ. Sci. **8**, 216–220 (2015)
34. G. Joshi, X. Yan, H. Wang, W. Liu, G. Chen, Z. Ren, Adv. Energy Mater. **1**, 643–647 (2011)
35. H. Li, X.F. Tang, Q.J. Zhang, C. Uher, Appl. Phys. Lett. **94**, 3 (2009)
36. A. Saramat, G. Svensson, A.E.C. Palmqvist, C. Stiewe, E. Mueller, D. Platzek, S.G.K. Williams, D.M. Rowe, J.D. Bryan, G.D. Stucky, J. Appl. Phys. **99**, 1–5 (2006)
37. G. Kieslich, G. Cerretti, I. Veremchuk, R.P. Hermann, M. Panthofer, J. Grin, W. Tremel, Phys. Status Solidi a Appl. Mater. Sci. **213**, 808–823 (2016)
38. J. Sui, J. Li, J. He, Y.-L. Pei, D. Berardan, H. Wu, N. Dragoe, W. Cai, L.-D. Zhao, Energ. Environ. Sci. **6**, 2916–2920 (2013)
39. X.W. Wang, H. Lee, Y.C. Lan, G.H. Zhu, G. Joshi, D.Z. Wang, J. Yang, A.J. Muto, M.Y. Tang, J. Klatsky, S. Song, M.S. Dresselhaus, G. Chen, Z.F. Ren, Appl. Phys. Lett. **93**, 193121 (2008)
40. D. Narducci, S. Frabboni, X. Zianni, J. Mater. Chem. C **3**, 12176–12185 (2015)
41. H.P. Ning, G.D. Mastrorillo, S. Grasso, B.L. Du, T. Mori, C.F. Hu, Y. Xu, K. Simpson, G. Maizza, M.J. Reece, J. Mater. Chem. A **3**, 17426–17432 (2015)
42. B. Yu, M. Zebarjadi, H. Wang, K. Lukas, H.Z. Wang, D.Z. Wang, C. Opeil, M. Dresselhaus, G. Chen, Z.F. Ren, Nano Lett. **12**, 2077–2082 (2012)
43. O. Bubnova, Z.U. Khan, A. Malti, S. Braun, M. Fahlman, M. Berggren, X. Crispin, Nat. Mater. **10**, 429–433 (2011)
44. W. Shanyu, L. Han, L. Ruiming, Z. Gang, T. Xinfeng, Nanotechnology **24**, 1–12 (2013)
45. S.I. Kim, K.H. Lee, H.A. Mun, H.S. Kim, S.W. Hwang, J.W. Roh, D.J. Yang, W.H. Shin, X.S. Li, Y.H. Lee, G.J. Snyder, S.W. Kim, Science **348**, 109–114 (2015)
46. X.L. Shi, Z.G. Chen, W.D. Liu, L. Yang, M. Hong, R. Moshwan, L.Q. Huang, J. Zou, Energy Storage Mater. **10**, 130–138 (2018)
47. M. Hong, Z.-G. Chen, L. Yang, Z.-M. Liao, Y.-C. Zou, Y.-H. Chen, S. Matsumura, J. Zou, Adv. Energy Mater. 1702333-n/a
48. S. Lin, W. Li, Z. Chen, J. Shen, B. Ge, Y. Pei, Nat Commun **7**, 1–6 (2016)
49. G. Tan, F. Shi, S. Hao, L.-D. Zhao, H. Chi, X. Zhang, C. Uher, C. Wolverton, V.P. Dravid, M.G. Kanatzidis, Nat. Commun. **7**, 1–9 (2016)
50. Y.M. Lin, M.S. Dresselhaus, Phys. Rev. B **68**, 14 (2003)
51. T.Y. Kim, A.A. Negash, G. Cho, Energy Convers. Manage. **124**, 280–286 (2016)
52. D.K. Aswal, R. Basu, A. Singh, Energy Convers. Manage. **114**, 50–67 (2016)
53. C. Dames, Scripta Mater. **111**, 16–22 (2016)
54. W. He, G. Zhang, X. Zhang, J. Ji, G. Li, X. Zhao, Appl. Energy **143**, 1–25 (2015)
55. B.C. Sales, Science **295**, 1248–1249 (2002)
56. W. Liu, K.C. Lukas, K. McEnaney, S. Lee, Q. Zhang, C.P. Opeil, G. Chen, Z. Ren, Energy Environ. Sci. **6**, 552–560 (2013)
57. W.-S. Liu, Q. Zhang, Y. Lan, S. Chen, X. Yan, Q. Zhang, H. Wang, D. Wang, G. Chen, Z. Ren, Adv. Energy Mater. **1**, 577–587 (2011)

58. B. Poudel, Q. Hao, Y. Ma, Y. Lan, A. Minnich, B. Yu, X. Yan, D. Wang, A. Muto, D. Vashaee, X. Chen, J. Liu, M.S. Dresselhaus, G. Chen, Z. Ren, Science **320**, 634–638 (2008)
59. D.-Y. Chung, T. Hogan, P. Brazis, M. Rocci-Lane, C. Kannewurf, M. Bastea, C. Uher, M.G. Kanatzidis, Science **287**, 1024–1027 (2000)
60. Y.C. Zou, Z.G. Chen, E.Z. Zhang, F.T. Kong, Y. Lu, L.H. Wang, J. Drennan, Z.C. Wang, F.X. Xiu, K. Cho, J. Zou, Nano Res. **11**, 696–706 (2018)
61. B. Zhu, Z.Y. Huang, X.Y. Wang, Y. Yu, L. Yang, N. Gao, Z.G. Chen, F.Q. Zu, Nano Energy **42**, 8–16 (2017)
62. M. Hong, Z.G. Chen, L. Yang, J. Zou, Nano Energy **20**, 144–155 (2016)
63. M. Hong, T.C. Chasapis, Z.G. Chen, L. Yang, M.G. Kanatzidis, G.J. Snyder, J. Zou, ACS Nano **10**, 4719–4727 (2016)
64. L. Yang, Z.G. Chen, M. Hong, G. Han, J. Zou, A.C.S. Appl, Mater. Interf. **7**, 23694–23699 (2015)
65. G. Han, Z.G. Chen, D.L. Ye, L. Yang, L.Z. Wang, J. Drennan, J. Zou, J. Mater. Chem. A **2**, 7109–7116 (2014)
66. H.J. Wu, L.D. Zhao, F.S. Zheng, D. Wu, Y.L. Pei, X. Tong, M.G. Kanatzidis, J.Q. He, Nat. Commun. **5**, 4515 (2014)
67. L.-D. Zhao, J. He, S. Hao, C.-I. Wu, T.P. Hogan, C. Wolverton, V.P. Dravid, M.G. Kanatzidis, J. Am. Chem. Soc. **134**, 16327–16336 (2012)
68. H. Wang, Y.Z. Pei, A.D. LaLonde, G.J. Snyder, Adv. Mater. **23**, 1366–1370 (2011)
69. Y. Pei, J. Lensch-Falk, E.S. Toberer, D.L. Medlin, G.J. Snyder, Adv. Funct. Mater. **21**, 241–249 (2011)
70. K.F. Hsu, S. Loo, F. Guo, W. Chen, J.S. Dyck, C. Uher, T. Hogan, E.K. Polychroniadis, M.G. Kanatzidis, Science **303**, 818–821 (2004)
71. L. Yang, Z. Chen, M. Hong, L.H. Wang, D.L. Kong, L.Q. Huang, G. Han, Y.C. Zou, M. Dargusch, J. Zou, Nano Energy **31**, 105–112 (2017)
72. L. Wang, S. Chang, S. Zheng, T. Fang, W. Cui, P.P. Bai, L. Yue, Z. Chen, ACS Appl. Mater. Interfaces **9**, 22612–22619 (2017)
73. Y.C. Zou, Z.G. Chen, F.T. Kong, J. Lin, J. Drennan, K. Cho, Z.C. Wang, J. Zou, ACS Nano **10**, 5507–5515 (2016)
74. Y.C. Zou, Z.G. Chen, F. Kong, E. Zhang, J. Drennan, K. Cho, F. Xiu, J. Zou, Nanoscale **8**, 19383–19389 (2016)
75. L. Yang, Z.G. Chen, G. Han, M. Hong, L.Q. Huang, J. Zou, J. Mater. Chem. A **4**, 9213–9219 (2016)
76. T. Barbier, S. Rollin-Martinet, P. Lemoine, F. Gascoin, A. Kaltzoglou, P. Vaqueiro, A.V. Powell, E. Guilmeau, J. Am. Ceram. Soc. **99**, 51–56 (2016)
77. C.P. Heinrich, T.W. Day, W.G. Zeier, G.J. Snyder, W. Tremel, J. Am. Chem. Soc. **136**, 442–448 (2014)
78. M. Ibanez, D. Cadavid, R. Zamani, N. Garcia-Castello, V. Izquierdo-Roca, W.H. Li, A. Fairbrother, J.D. Prades, A. Shavel, J. Arbiol, A. Perez-Rodriguez, J.R. Morante, A. Cabot, Chem. Mater. **24**, 562–570 (2012)
79. W. Liu, K. Yin, Q. Zhang, C. Uher, X. Tang, Natl. Sci. Rev. **4**, 611–626 (2017)
80. Z.L. Li, J.F. Dong, F.H. Sun, S. Hirono, J.F. Li, Chem. Mater. **29**, 7378–7389 (2017)
81. S. Ghodke, N. Hiroishi, A. Yamamoto, H. Ikuta, M. Matsunami, T. Takeuchi, J. Electron. Mater. **45**, 5279–5284 (2016)
82. Y. Akio, G. Swapnil, M. Hidetoshi, I. Manabu, N. Yoichi, M. Masaharu, T. Tsunehiro, Jpn. J. Appl. Phys. **55**, 020301 (2016)
83. L.D. Zhao, S.H. Lo, Y.S. Zhang, H. Sun, G.J. Tan, C. Uher, C. Wolverton, V.P. Dravid, M.G. Kanatzidis, Nature **508**, 373–377 (2014)
84. S.R. Brown, S.M. Kauzlarich, F. Gascoin, G.J. Snyder, Chem. Mater. **18**, 1873–1877 (2006)
85. X. Chen, S.N. Girard, F. Meng, E. Lara-Curzio, S. Jin, J.B. Goodenough, J. Zhou, L. Shi, Adv. Energy Mater. **4**, 1–10 (2014)
86. X. Chen, J. Zhou, J.B. Goodenough, L. Shi, J. Mater. Chem. C **3**, 10500–10508 (2015)
87. W. Luo, H. Li, Y. Yan, Z. Lin, X. Tang, Q. Zhang, C. Uher, Intermetallics **19**, 404–408 (2011)

88. Y. Kikuchi, T. Nakajo, K. Hayashi, Y. Miyazaki, J. Alloy. Compd. **616**, 263–267 (2014)
89. X. Chen, A. Weathers, D. Salta, L. Zhang, J. Zhou, J.B. Goodenough, L. Shi, J. Appl. Phys. **114**, 1–10 (2013)
90. W. Luo, H. Li, F. Fu, W. Hao, X. Tang, J. Electron. Mater. **40**, 1233–1237 (2011)
91. D.-K. Shin, S.-W. You, I.-H. Kim, J. Korean Phys. Soc. **64**, 1412–1415 (2014)
92. Y. Miyazaki, Y. Saito, K. Hayashi, K. Yubuta, T. Kajitani, Jpn. J. Appl. Phys. **50**, 5 (2011)
93. Y. Kikuchi, Y. Miyazaki, Y. Saito, K. Hayashi, K. Yubuta, T. Kajitani, Jpn. J. Appl. Phys. **51**, 1–5 (2012)
94. J.M. Higgins, A.L. Schmitt, I.A. Guzei, S. Jin, J. Am. Chem. Soc. **130**, 16086–16094 (2008)
95. D.-K. Shin, S.-C. Ur, K.-W. Jang, I.-H. Kim, J. Electron. Mater. **43**, 2104–2108 (2014)
96. V. Ponnambalam, D.T. Morelli, S. Bhattacharya, T.M. Tritt, J. Alloy. Compd. **580**, 598–603 (2013)
97. E.S. Toberer, A.F. May, G.J. Snyder, Chem. Mater. **22**, 624–634 (2010)
98. V. Ponnambalam, D.T. Morelli, J. Electron. Mater. **41**, 1389–1394 (2012)
99. A. Allam, P. Boulet, M.C. Record, J. Electron. Mater. **43**, 761–773 (2014)
100. A.J. Zhou, T.J. Zhu, X.B. Zhao, S.H. Yang, T. Dasgupta, C. Stiewe, R. Hassdorf, E. Mueller, J. Electron. Mater. **39**, 2002–2007 (2010)
101. D.B. Migas, V.L. Shaposhnikov, A.B. Filonov, V.E. Borisenko, N.N. Dorozhkin, Phys. Rev. B **77**, 1–9 (2008)
102. X. Su, P. Wei, H. Li, W. Liu, Y. Yan, P. Li, C. Su, C. Xie, W. Zhao, P. Zhai, Q. Zhang, X. Tang, C. Uher, Adv. Mater. 1602013-n/a (2017)
103. X. She, X. Su, H. Du, T. Liang, G. Zheng, Y. Yan, R. Akram, C. Uher, X. Tang, J. Mater. Chem. C **3**, 12116–12122 (2015)
104. Q. Zhang, B. Liao, Y. Lan, K. Lukas, W. Liu, K. Esfarjani, C. Opeil, D. Broido, G. Chen, Z. Ren, Proc. Natl. Acad. Sci. **110**, 13261–13266 (2013)
105. W. Li, Y. Wu, S. Lin, Z. Chen, J. Li, X. Zhang, L. Zheng, Y. Pei, ACS Energy Lett., 2349–2355 (2017)
106. X. Tan, X. Tan, G. Liu, J. Xu, H. Shao, H. Hu, M. Jin, H. Jiang, J. Jiang, J. Mater. Chem. C (2017)
107. G. Tan, F. Shi, S. Hao, H. Chi, L.-D. Zhao, C. Uher, C. Wolverton, V.P. Dravid, M.G. Kanatzidis, J. Am. Chem. Soc. **137**, 5100–5112 (2015)
108. G. Tan, F. Shi, J.W. Doak, H. Sun, L.-D. Zhao, P. Wang, C. Uher, C. Wolverton, V.P. Dravid, M.G. Kanatzidis, Energy Environ. Sci. **8**, 267–277 (2015)
109. L. Hu, Y. Zhang, H. Wu, J. Li, Y. Li, M. Mckenna, J. He, F. Liu, S. J. Pennycook, X. Zeng, Adv. Energy Mater. 1802116
110. W. Li, L. Zheng, B. Ge, S. Lin, X. Zhang, Z. Chen, Y. Chang, Y. Pei, Adv. Mater. **29**, 1605887-n/a (2017)
111. A. Banik, U.S. Shenoy, S. Anand, U.V. Waghmare, K. Biswas, Chem. Mater. **27**, 581–587 (2015)
112. J. He, X. Tan, J. Xu, G.-Q. Liu, H. Shao, Y. Fu, X. Wang, Z. Liu, J. Xu, H. Jiang, J. Jiang, J. Mater. Chem. A **3**, 19974–19979 (2015)
113. R. Al Rahal Al Orabi, N. A. Mecholsky, J. Hwang, W. Kim, J.-S. Rhyee, D. Wee, M. Fornari, Chem. Mater. **28,** 376–384 (2016)
114. Z.H. Ge, L.D. Zhao, D. Wu, X.Y. Liu, B.P. Zhang, J.F. Li, J.Q. He, Mater. Today **19**, 227–239 (2016)
115. H. Mun, S.M. Choi, K.H. Lee, S.W. Kim, Chemsuschem **8**, 2312–2326 (2015)
116. X.L. Shi, K. Zheng, M. Hong, W.D. Liu, R. Moshwan, Y. Wang, X.L. Qu, Z.G. Chen, J. Zou, Chem. Sci. **9**, 7376–7389 (2018)
117. P. Norouzzadeh, Z. Zamanipour, J.S. Krasinski, D. Vashaee, J. Appl. Phys. **112**, 1–7 (2012)
118. X. Chen, A. Weathers, J. Carrete, S. Mukhopadhyay, O. Delaire, D.A. Stewart, N. Mingo, S.N. Girard, J. Ma, D.L. Abernathy, J.Q. Yan, R. Sheshka, D.P. Sellan, F. Meng, S. Jin, J.S. Zhou, L. Shi, Nat. Commun. **6**, 1–9 (2015)
119. W. Kim, J. Mater. Chem. C **3**, 10336–10348 (2015)
120. X. Chen, L. Shi, J. Zhou, J.B. Goodenough, J. Alloy. Compd. **641**, 30–36 (2015)

121. S. Muthiah, R.C. Singh, S.D. Pathak, A. Dhar, Scripta Mater. **119**, 60–64 (2016)
122. T.H. An, S.M. Choi, W.S. Seo, C. Park, I.H. Kim, S.U. Kim, J. Electron. Mater. **42**, 2269–2273 (2013)
123. D.K. Shin, K.W. Jang, S.C. Ur, I.H. Kim, J. Electron. Mater. **42**, 1756–1761 (2013)
124. Y. Sadia, L. Dinnerman, Y. Gelbstein, J. Electron. Mater. **42**, 1926–1931 (2013)
125. A.J. Zhou, T.J. Zhu, X.B. Zhao, E. Mueller, J. Mater. Res. **26**, 1900–1906 (2011)
126. A.J. Zhou, X.B. Zhao, T.J. Zhu, T. Dasgupta, C. Stiewe, R. Hassdorf, E. Mueller, Intermetallics **18**, 2051–2056 (2010)
127. X. Shi, A. Wu, W. Liu, R. Moshwan, Y. Wang, Z.G. Chen, J. Zou, ACS Nano **12**, 11417–11425 (2018)
128. Z.-Z. Luo, X. Zhang, X. Hua, G. Tan, T.P. Bailey, J. Xu, C. Uher, C. Wolverton, V.P. Dravid, Q. Yan, M.G. Kanatzidis, Adv. Funct. Mater. **28**, 1801617 (2018)
129. J. He, L.-D. Zhao, J.-C. Zheng, J.W. Doak, H. Wu, H.-Q. Wang, Y. Lee, C. Wolverton, M.G. Kanatzidis, V.P. Dravid, J. Am. Chem. Soc. **135**, 4624–4627 (2013)
130. E. Groβ, M. Riffel, U. Stöhrer, J. Mater. Res. **10**, 34–40 (2011)
131. A. Zhou, H. Cui, J. Li, T. Zhu, X. Zhao, Procedia Eng. **27**, 94–102 (2012)
132. A. Ikuto, K. Hiromasa, R. Lutz, K. Toshio, M. Mika, S. Seijiro, T. Toshihide, Jpn. J. Appl. Phys. **44**, 4275 (2005)
133. H.S. Kim, K. Kikuchi, T. Itoh, T. Iida, M. Taya, Mater. Sci. Eng., B **185**, 45–52 (2014)
134. H. Kaibe, I. Aoyama, M. Mukoujima, T. Kanda, S. Fujimoto, T. Kurosawa, H. Ishimabushi, K. Ishida, L. Rauscher, Y. Hata, S. Sano, In Development of thermoelectric generating stacked modules aiming for 15% of conversion efficiency, in *ICT 2005. Conference. 24th International Conference on Thermoelectrics, Clemson, SC, USA, June, 2005; Clemson, SC, USA*, pp. 242–247
135. C.S. Kim, H.M. Yang, J. Lee, G.S. Lee, H. Choi, Y.J. Kim, S.H. Lim, S.H. Cho, B.J. Cho, ACS Energy Lett. **3**, 501–507 (2018)
136. J.-H. Bahk, H. Fang, K. Yazawa, A. Shakouri, J. Mater. Chem. C **3**, 10362–10374 (2015)
137. S. Yang, K. Cho, Y. Park, S. Kim, Nano Energy **49**, 333–337 (2018)
138. F. Suarez, A. Nozariasbmarz, D. Vashaee, M.C. Öztürk, Energy Environ. Sci. **9**, 2099–2113 (2016)
139. V. Leonov, R.J.M. Vullers, J. Renew. Sustain. Energy **1**, 062701 (2009)
140. J.P. Rojas, D. Conchouso, A. Arevalo, D. Singh, I.G. Foulds, M.M. Hussain, Nano Energy **31**, 296–301 (2017)
141. S.J. Kim, H.E. Lee, H. Choi, Y. Kim, J.H. We, J.S. Shin, K.J. Lee, B.J. Cho, ACS Nano **10**, 10851–10857 (2016)
142. D. Zhao, X. Lu, T. Fan, Y.S. Wu, L. Lou, Q. Wang, J. Fan, R. Yang, Appl. Energy **218**, 282–291 (2018)
143. J. Yang, T. Caillat, MRS Bull. **31**, 224–229 (2011)
144. J.C. Bass, N.B. Elsner, F.A. Leavitt, AIP Conf. Proc. **316**, 295–298 (1994)
145. S. Wang, T. Xie, H. Xie, Appl. Therm. Eng. **130**, 847–853 (2018)
146. R.A. Kishore, P. Kumar, S. Priya, Sustain. Energy Fuels **2**, 175–190 (2018)
147. L.G. Chen, F.K. Meng, F.R. Sun, Sci. China-Technol. Sci. **59**, 442–455 (2016)
148. J. Yang, F.R. Stabler, J. Electron. Mater. **38**, 1245–1251 (2009)
149. R. Stobart, M.A. Wijewardane, Z. Yang, Appl. Therm. Eng. **112**, 1433–1444 (2017)
150. S. Huang, X. Xu, Appl. Energ. **185**, 119–125 (2017)
151. J.P. Heremans, M.S. Dresselhaus, L.E. Bell, D.T. Morelli, Nat. Nano. **8**, 471 (2013)
152. H. Yang, P. Reichert, K.C. Abbaspour, A.J.B. Zehnder, Environ. Sci. Technol. **37**, 3048–3054 (2003)
153. W. Shin, K. Imai, N. Izu, N. Murayama, Jpn. J. Appl. Phys. **40**, L1232–L1234 (2001)
154. T. Goto, T. Itoh, T. Akamatsu, N. Izu, W. Shin, Sens. Actuators B: Chem. **223**, 774–783 (2016)

Index

Z. Sun and T. Liao (eds.), *Responsive Nanomaterials for Sustainable Applications*, Springer Series in Materials Science 297,
https://doi.org/10.1007/978-3-030-39994-8

The manufacturer's authorised representative in the EU is Springer Nature Customer Service Centre GmbH, Europaplatz 3, 69115 Heidelberg, Germany. If you have any concerns regarding our products, please contact ProductSafety@springernature.com

Printed and bound by CPI Group (UK) Ltd, Croydon, CR0 4YY
30/09/2026
02225646-0013